T0383332

Social Participation in Water Governance and Management

Social Participation in Water Governance and Management

Critical and Global Perspectives

Edited by
Kate A. Berry and Eric Mollard

London • Sterling, VA

First published by Earthscan in the UK and USA in 2010

ISBN: 978-1-84407-885-1

Typeset by FiSH Books, Enfield
Cover design by Dan Bramall

For a full list of publications please contact:

Earthscan
Dunstan House
14a St Cross St
London, EC1N 8XA, UK
Tel: +44 (0)20 7841 1930
Fax: +44 (0)20 7242 1474
Email: earthinfo@earthscan.co.uk
Web: **www.earthscan.co.uk**

22883 Quicksilver Drive, Sterling, VA 20166-2012, USA

Earthscan publishes in association with the International Institute for Environment and Development

A catalogue record for this book is available from the British Library

Library of Congress Cataloging-in-Publication Data has been applied for

At Earthscan we strive to minimize our environmental impacts and carbon footprint through reducing waste, recycling and offsetting our CO_2 emissions, including those created through publication of this book. For more details of our environmental policy, see www.earthscan.co.uk

This book was printed in the UK by TJ International, an ISO 14001 accredited company. The paper used is FSC certified and the inks are vegetable based.

This book is dedicated to our children:
Miles, Brice, Audrey and Arnaud

Contents

List of Figures and Tables

Figures

Tables

List of Contributors

Sophie Allain is a Senior Researcher in Sociology at the French National Institute for Agricultural Research. She conducts research on environmental planning and policy analysis according to negotiation theory. Along these lines, Dr Allain works on environmental conflicts, participatory planning, public debates and mediation practices. She is especially concerned with the field of water management.
Email: allain.sophie@gmail.com

Eneida Assis is currently the Vice Coordinator of the College of Social Sciences at the *University Federal do Pará*, Pará State, Brazil. She received her Master's degree in anthropology at the University of Brasilia and her PhD in political science from the *Instituto Universitário de Pesquisas do Rio de Janeiro*. Dr Assis has worked with indigenous peoples of the Brazilian Amazon for over 30 years and presently coordinates a programme in ethnodevelopment among the Tembé Indians in conjunction with the non-governmental organization (NGO) *Pobreza and Meio Ambiente* (Poverty and Environment).
Email: kiavnu2@yahoo.com.br

Kate Berry is an Associate Professor of Geography at the University of Nevada, Reno. She is interested in the cultural and political dimensions of water issues; intergovernmental relations and water policy; indigenous geographies; and identity studies. Dr Berry has served as president of the Association of Pacific Coast Geographers and is currently an editorial board member of *Water History*. She serves as chairperson of the Department of Geography at the University of Nevada, Reno.
Email: kberry@unr.edu

Pranita Bhushan Udas is an agriculture graduate with a Master's degree in water management from Wageningen University, The Netherlands. She is currently involved in PhD research on 'Gendered Participation in Water Management in Nepal: Discourses, Policies and Practices' at Wageningen University funded by The Netherlands Organization for Scientific Research (Science for Global Development) (NWO-WOTRO). Her work focuses on mainstreaming gender in the water sector. She has worked in various development and research projects with organizations such as Eco-Himal, WATCH, SUTRA Centre in Nepal and SaciWATERs, GLOCAL and GIDR in India. Most recently, she has done a study on women water professionals for the South Asia Consortium for Interdisciplinary Water Resources Studies (SaciWATERs) and a study on the use of

gender guidelines and manuals by water professionals in Nepal for the Gender and Water Alliance.
Email: pranitabhushan@yahoo.com

Rutgerd Boelens is an Associate Professor and Senior Researcher with the Irrigation and Water Engineering Group at Wageningen University, the Netherlands. He coordinates the programme WALIR (Water Law and Indigenous Rights) and the InterAndean programmes Concertación and Struggling for Water Security and the global comparative research programme Justicia Hidrica. Dr Boelens has published books and articles on the links between water rights, cultures, politics, law and reforms, and power, property relations, interventions, collective action and mediation processes.
Email: Rutgerd.Boelens@wur.nl

Yue Chaoyun is an Assistant Researcher of Economics at the Yunnan Academy of Social Science.
Email: chaoyun.yue@sina.com

Liang Chuan is an Associate Researcher at the Yunnan Academy of Social Science with a Master's degree in sociology. She has engaged in research and action on sustainable development from the perspective of social science for more than a decade.
Email: liangchuan66@163.com

Barbara Durham is the Timbisha Shoshone historic preservation officer. Ms Durham was formerly the Timbisha Tribal Administrator and a member of the Timbisha Land Restoration Committee. She was involved in all phases of the Homeland Act negotiations.
Email: dvdurbarbara@netscape.com

Madeline Esteves is a current and former tribal council member of the Timbisha Shoshone. Ms Esteves currently is a member of the Timbisha Historic Preservation Committee, and is the Timbisha Tribal Council secretary. She was a member of the Timbisha Land Restoration Committee.

Pauline Esteves is a former chair and tribal council member of the Timbisha Shoshone. Ms Esteves was a member of the Timbisha Land Restoration Committee and the lead negotiator for the Timbisha Shoshone during the Homeland Act negotiations. She is currently an advisor to the Tribal Historic Preservation Office, a member of the Timbisha Historic Preservation Committee, and an executive member of the Timbisha Tribal Council.

Terry Fisk was the hydrologist for Death Valley National Park from 2001 to 2008. He is currently attending graduate school at the University of Nevada, Reno with a focus on hydrogeology. While at Death Valley he managed the park's

water resources programme, including water rights protection, research into local and regional water resources and coordinating with other government agencies and a variety of other organizations. Before working in Death Valley, he had almost 14 years' experience in the consulting engineering field in Washington and Oregon, focused primarily on water resource and environmental contaminant investigation and remediation. Terry has a BSc in geology from the University of Washington and is a registered professional geologist in Oregon.
Email: terry.fisk@dri.edu

Louis Forline is an Assistant Professor of Anthropology at the University of Nevada, Reno, where he has taught and conducted research since 2004. He is Brazilian-American and has been working in the Brazilian Amazon among peasant and indigenous groups since 1987. Dr Forline's main areas of interests are indigenous ethnology, traditional resource management, sustainable development, racial and ethnic identity, health and foodways. He is currently engaged in a longitudinal study of the Guaja Indians of the eastern Amazon.
Email: forline@unr.edu

Palitha Jayaweera attended the Water Engineering Development Centre (WEDC) of the University of Loughborough, UK. He is currently a consultant and researcher in the water and environmental sanitation sector and manager of Community Self Improvement (COSI), a Sri Lankan NGO. His main interest is in assisting communities to solve their challenges through training and empowerment.
Email: palitha.jayaweera@cosi.org.lk

Stéphane La Branche is a political scientist who works as a research and teaching Fellow at the Energy and Environmental Policy Department, Université Pierre-Mendès France in Grenoble. His research centres on the acceptability of environmental policies, measures and technologies. He also publishes on pedagogical innovations in political science and in 2007 won the European Political Science Network Innovation in Teaching Award.
Email: asosan95@hotmail.com

Richard Laster is a Professor of Environmental Law at the Hebrew University faculties of Law, Geography, and Environmental Sciences. He is also a partner in the law firm of Laster and Gouldman in Jerusalem. The firm's clients include several drainage and river authorities, including the Yarqon River Authority.
Email: richard@laster.co.il

Laurent Lepage is a Professor at the Institute of Environmental Sciences at the Université du Québec à Montréal (UQAM) and has held the Chaire d'études sur les écosystèmes urbains since 2000. He is a member of the editorial committee of the interdisciplinary journal *Natures, Sciences et Sociétés*. His research deals with the implementation of sustainable development and the analysis of environmental policies. Dr Lepage's recent published work examines the link

between science, public decision-making and local communities in the new area of integrated management and climate change adaptation.
Email: lepage.laurent@uqam.ca

Dan Livney is an attorney at the law firm of Laster and Gouldman in Jerusalem. His work focuses mainly on environmental law research on both the national and international level. Dan is also an olive farmer on Kibbutz Gezer.
Email: dani@laster.co.il

Carmen Maganda is a Research Associate in the Laboratoire de Sciences Politiques, Identités, Politiques, Sociétés, Espaces (IPSE) research unit, at the University of Luxembourg. Previously, she was the coordinator of the Bi-national Border Water Project for the Center for US–Mexican Studies at the University of California, San Diego; a HERMES Fellowship Programme of the Foundation Maison des sciences de l'homme fellow at Centre lillois d'études et de recherches sociologiques et économiques (Clersé), Université de Lille 1; a research fellow at the Center for US–Mexican Studies at the University of California, San Diego; and a visiting scholar at the Institute for Mexico and the US. Dr Maganda's present research focuses on water management in cross-border basins in Europe and the Americas, social participation in water politics and elite behaviour and decision-making processes in water management.
Email: carmen.maganda@uni.lu

Nicolas Milot received his PhD at the Institute of Environmental Sciences at the Université du Québec à Montréal (UQAM), where his thesis focused on social dynamics related to Québec watershed governance. He has been doing postdoctoral research on ethical issues related to the implementation of collaborative approaches in environmental governance. Dr Milot is also a lecturer in the Environmental Sciences Master Program of the UQAM where he teaches about environment and natural resources governance.
Email: milot.nicolas@uqam.ca

Eric Mollard is a Senior Researcher in environmental sociology and political ecology at the French Research Institute for Development (IRD) based in Montpellier. Trained in agroeconomy, he first studied rural development in Latin America and West Africa. Dr Mollard then focused his research on associative water management, conflicts from local to transboundary level, and negotiations in Thailand and Mexico. His current research is on democracy as the way to make negotiations and environmental protection work.
Email: Eric.Mollard@ird.fr

David Pargament has been the Director General of the Yarqon River Authority since 1993. He is an agronomist, and holds a doctorate from the University of Haifa on the subject of watershed management.
Email: david@yarqon.org.il

Sylvain Roger Perret is an Agronomist and Water Economist in the G-Eau research unit at the *Centre de coopération internationale en recherche agronomique pour le développement (*CIRAD), a French international research and co-operation organization, based in Montpellier, France. His research focus is on water management, economics and governance. Dr Perret has been working on these issues in South Africa for the last ten years. He has been seconded to the Asian Institute of Technology in Bangkok, Thailand since 2007.
Email: sylvain.peret@cirad.fr

Margaret Shanafield is currently completing doctoral research in hydrology/hydrogeology at the University of Nevada, Reno. Her dissertation work involves interactions between surface and groundwater sources with an application to seepage from water delivery channels. Her involvement in international water issues began with a project to survey handpumps in the Leogane region of Haiti. Recently, she has been involved with rural water development in Sri Lanka and water quality research in the Aral Sea Basin of Uzbekistan.
Email: mshanafield@yahoo.com

Sergio Vargas Velázquez is a researcher in the anthropology of water at the Social Participation Department in the Mexican Institute for Water Technology, Cuernavaca, Mexico. He completed his PhD in Social Anthropology in 2008. His current research is on social participation in water issues, irrigation management transfer policies and participatory development applied to user organizations for ground and surface water.
Email: erontskuri@gmail.com

Juana Vera Delgado is an agricultural engineer from Peru with an MSc degree in Anthropology and Social Sciences from Wageningen University in The Netherlands. She is currently doing a PhD study at the same university, supported by The Netherlands Organisation for Scientific Research (NWO). She has vast experience as a practitioner and researcher in thematic areas, including irrigation management and rural development, water and gender, and water and ethnicity in the Andes of Latin America. She is member of different gender networks, including Gender and Water Alliance (GWA) and the Gender and Water Network of the Andean Countries (RAGPA) of Latin America.
Email: juana.vera@wur.nl

Philippus (Flip) Wester is an Assistant Professor of Water Reforms at Wageningen University, Wageningen, The Netherlands. Trained as an interdisciplinary water management researcher, he has studied water governance processes in Senegal, Pakistan, The Netherlands, Bangladesh and Mexico. His current research focuses on water reforms, river basin governance and environmental and institutional change processes.
Email: Flip.Wester@wur.nl

Zoë Wilson is External Associate to the Pollution Research Group at the University of KwaZulu Natal, South Africa, and is based in Nairobi, Kenya, part-time. She is an active practitioner in the water and sanitation sector in Africa, working frequently with local government and private sector in multidisciplinary collaboration across the sciences and social sciences. Zoë's expertise relates to the social feasibility issues associated with the implementation of appropriate and eco-technologies. Most recently, she was appointed external evaluator for the World Health Organization's Household Water Treatment Safe Storage Network.
Email: wilsonz@ukzn.ac.za

Margreet Zwarteveen is a Lecturer and Researcher at the Irrigation and Water Management Group, at the Centre for Water and Climate and is a member of the Gender Analysis Group (all at Wageningen University, The Netherlands). She has worked with the International Water Management Institute and is currently involved in various capacity building and research projects on gender and integrated water management in South Asia and the Andes. Dr Zwarteveen's interests include:

- the linkages between gender and water;
- equity and justice aspects of water policies and management in rural areas and agriculture;
- critical (feminist) analyses of water policies and professional water cultures;
- masculinities in the water sector and water-related injustices linked to processes of appropriation of water.

She was a co-editor of the book *Liquid Relations*.
Email: Margreet.Zwarteveen@wur.nl

Acknowledgements

We appreciate the help and inspiration of everyone who assisted and contributed to this project. They are too numerous to list but special thanks are due to Cassandra Hansen, Ashley York and Kathryn Mann of the University of Nevada, Reno, for making the maps, figures and index for the book. Support from the University of Nevada, Reno, French Research Institute for Development (IRD) in Montpellier and the US Fulbright Foundation made it possible to work on this project.

List of Abbreviations

ABRINORD	Agence de bassin de la rivière du nord
ADR	Alternative dispute resolution
ANC	African National Congress
ATDR	Administración Técnica de Riego (Technical Administration of Irrigation Districts)
BAPE	Bureau d'audiences publiques sur l'environnement
BECC	Border Environment Cooperation Commission
CAP	Ceyhan Aslantas Project
CAWT	Central American Water Tribunal
CBO	community-based organizations
CECOP	Consejo de Ejidos y Comunidades Opositores a la Presa la Parota (Council of Cooperatives and Communities Opposing the Parota Dam)
CELPA	Centrais Elétricas do Pará
CFE	Federal Electricity Commission
CIMI	Conselho Indigenista Missionário (Brazil's Indigenous Missionary Council)
CIRAD	Centre de coopération internationale en recherche agronomique pour le développement
CLE	Commission locale de l'eau
CLERSÉ	Centre lillois d'études et de recherches sociologiques et économiques
CMA	Catchment management agencies
CNA	National Water Commission (Comisión Nacional del Agua)
CNDP	Commission Nationale du Débat Public
CNRH	National Water Resources Board
CODOCAL	Corporation of Rural Organizations of Licto
COSATU	Congress of South Africa Trade Unions
COSI	Community Self Improvement
CSBCT	Committee for the Evaluation of Historical Impacts of Dams
Eletronorte	Centrais Elétricas do Norte do Brasil
Espacio DESC	Espacio de Derechos Economicos, Sociales y Culturales
EU	European Union
FADESP	Fundação de Amparo ao Desenvolvimento da Pesquisa (the University's Grant Administration Office)
FEDURIC	Federation of Irrigation User Groups of Cotopaxi
FUNAI	Fundação Nacional do Índio
GAP	Greater Anatolian Project

GEAR	Growth Employment And Redistribution strategy
GMD	Group for Management and Distribution
GWA	Gender and Water Alliance
IBAMA	Instituto Brasileiro do Meio Ambiente e dos Recursos Naturais Renováveis
IBGE	Instituto Brasileiro de Geografia e Estatística (Brazil's Institute for Geography and Statistics)
IDASA	Institute for Democracy in South Africa
IFP	Inkatha Freedom Party
IMT	Irrigation Management Transfer
INCRA	Instituto Nacional de Colonização e Reforma Agrária (Brazilian Land Settlement and Agrarian Reform Institute)
INE	Instituto Nacional de Ecología
INERHI	Ecuadorian Institute of Water Resources
IPSE	Identités, politiques, société, espaces
IR	International Relations
IRC	International Water and Sanitation Centre
IRD	Research Institute for Development (France)
IWM	Integrated watershed management
IWRM	Integrated water resources management
JBF	Jal Bhagirathi Foundation
LAWT	Latin American Water Tribunal (Tribunal Latinoamericano del Agua)
LSI	Laws of the State of Israel
LTTE	Liberation Tigers of Tamil Eelam
MAB	Movimento de Atingidos por Barragens (Movement of People Impacted by Dams)
MDG	Millennium Development Goal
MEG	Monitoring and Evaluation Group
MIG	Municipal Infrastructure Grant
MPF	Ministério Público Federal (Federal Prosecution Service)
MW	Megawatts
NGO	Non-governmental organization
NIMBY	Not in my backyard
NPS	National Park Service
NRA	Nature Reserves and National Parks Authority
NWA	National Water Act
NWO	The Netherlands Organisation for Scientific Research
OECD	Organisation for Economic Co-operation and Development
PGC	Carajás Mining Project
PGE	Plan de gestion des Étiages
PIN	Projeto de Integração Nacional (Programme for National Integration)
QWP	Québec Water Policy
RAGPA	Gender and Water Network of the Andean Countries

RDP	Reconstruction and Development Programme
SaciWATERs	South Asia Consortium for Interdisciplinary Water Resources Studies
SACP	South African Communist Party
SAGE	Schéma d'Aménagement et de Gestion des Eaux
SAMWU	South African Municipal Workers Union
SDAGE	Schéma Directeur d'Aménagement et de Gestion des Eaux
SDC	Swiss Agency for Development and Cooperation
SECTAM	Secretaria de Ciência e Tecnologia da Amazônia (Environmental Secretariat)
SECTUR	Statistic Information Center DATATUR
SEMARNAT	Ministry of Environment and Natural Resources
SENAGUA	National Water Secretary
SPD	Sustainable and Participatory Development
SPNI	Society for the Protection of Nature
STF	Supremo Tribunal Federal (Brazil's Supreme Court)
TBS	Tarun Bharat Sangh
TOR	Terms of Reference
UFPA	Universidade Federal do Pará
UKZN	University of KwaZulu Natal
UNDP	United Nations Development Programme
UNESCO	United Nations Educational, Scientific and Cultural Organization
UNICEF	United Nations Children's Fund
UPA	Union des Producteurs Agricoles
UQAM	Université du Québec à Montréal
WALIR	Water Law and Indigenous Rights
WAMB	Water Affairs Management Bureau
WAMS	Water Affairs Management Station
WCD	World Commission on Dams
WEDC	Water Engineering Development Centre
WES	Water and Environmental Sanitation
WHO	World Health Organization
WSA	Water Services Act
WSSCC	Water Supply and Sanitation Collaborative Council
WUA	Water User Association
WWC	World Water Council
WWF	World Water Forum
WWF	World Wide Fund for Nature (formerly World Wildlife Fund)
YRA	Yarqon River Authority

Introduction

Social Participation in Water Governance and Management

Kate A. Berry and Eric Mollard

There is no longer room for doubt – social participation in water management and governance is a reality today. Many envision an era of enhanced citizenship and dream of putting participation into practice to facilitate this. Enthusiasm for social participation extends beyond speeches as solutions are being crafted to water scarcity, bankruptcy of municipal operations, inequality, health and new distributions among sectors in ways that explicitly engage citizens, water users or anyone who may cause or prevent water problems. At the same time, questions have begun to emerge about whether this planetary dream might actually be a nightmare in which democracy, social justice or the environment suffers because of the conditions of governance involved.

A World Bank publication defined participation as 'a process through which stakeholders influence and share control over development initiatives and the decisions and resources which affect them' (Bhatnagar et al, 1996, pxvi). Yet, this is rather incomplete as the ramifications of social participation in many instances extend beyond the interests of those directly involved and beyond the sphere of development issues. Characterized by the direct involvement of an array of people in decision-making and implementation of water policy or management, at a minimum, social participation involves individuals and/or collectives having an opportunity to express their voices and articulate their arguments in public forums (Arnstein, 1969). As such, the notion of social participation is captured in varying degrees in the rubric of community participation, decentralized management and participatory development.

Social participation occurs at a variety of scales from grassroots through international levels. Participation may arise from the bottom-up (rather than the top-down) as people struggle to be heard or increase the visibility of a particular issue to make it public. At such times people involved in water matters may take social participation farther, demanding to be involved in final decisions or implementation. Approaches that focus on these participatory mechanisms tend to emphasize issues of empowerment, stress the needs of the marginalized, suggest a distrust of the state and celebrate local knowledge (Henkel and Stirrat, 2001). Yet, social participation encompasses more than activism; it cannot be restricted to referendums or social movements although it may be associated with both. In many instances, extensive involvement in informal associations or official committees is required of participants. Within the complex realm of water governance this is particularly true because of longstanding dominance of professionals (scientists, engineers, economists, attorneys and politicians) in water decision-making as well as the need for sustained work on water management.

While the potential virtues of social participation may be numerous, there has been a dramatic disjunct between some of the magnified hopes and initial outcomes of increased participation. Moreover, little has emerged in the way of analysis that critically evaluates social participation in water governance or attempts to theorize the conditions and objectives necessary for it to be realized (D'Aquino, 2007). While there have been many advocates for increasing social participation and involving individuals and communities directly in water management, few over-arching research projects on the specifics in water management have yet emerged and little has been offered in the way of the critical approaches to understanding participation in the context of water management around the world. Many research questions remain to be addressed for water management and governance, such as:

- how to effectively balance administrative control and reforms with social participation;
- why some parties are involved and others are excluded;
- what sorts of historical cycles and geographic patterns are associated with social participation;
- how power differentials affect participation;
- how rhetorical appeal meshes with actual experiences;
- how to encourage effective, open public decision-making.

These issues underscore that social participation is inherently political as well as economic, embedded with stresses that arise among competing values, rights and interests. Tensions between consensus-seeking and co-optation are frequently at play and inevitably balances must be struck between participation and authority, as the knowledge of lay citizens intersects with technical and managerial expertise (López Cerezo and Garcia, 1996). Issues of representation also arise as individuals, groups or coalitions seek to speak on behalf of others or frame their position as representing the public's interests. Moreover, in engaging the public and water users in participation, political changes, such as the need for administrative reform of the state, are often coupled with economic matters, such as altering the capacity of the private sector.

Given that the dimensions in which power, social equity and democracy-building are engaged within social participation have not been well vetted for water matters, this book aims to satisfy two expectations: to reveal the extent and challenges of social participation within water management and governance and to take stock of this initial period. This collective work opens up debate about social participation, presenting a variety of water cases from around the world and analysing these cases conceptually. Many different water management topics are addressed including:

- water rights definition;
- hydropower dam construction;
- urban river renewal;
- irrigation organizations;
- water supply development;
- river basin management;
- water policy implementation;
- judicial decision-making in water conflicts.

Some authors in this volume are more optimistic than others, yet these assessments do not unquestioningly embrace the virtues of social participation, because it is not clear that the conditions needed for effective participation are being achieved in many situations. The wide diversity of approaches and interpretations also reflects the plural nature of the object of study, variations between disciplines and the absence of common frameworks.

In organizing this book, we have framed understandings about participation around social dimensions that influence the complexity of water management today. Accordingly, the book is structured into five sections with chapters written by authors with expertise in different parts of the world. These sections address different issues within contemporary water management that influence and, in turn, are influenced by social participation, including:

- indigenous water governance;
- dynamics of gender in water management;
- river basin governance;
- implementation of water management;
- the politics of water governance.

Part I

This section probes dimensions of social participation and indigenous water governance. Contemporary struggles for indigenous self-determination are often explicitly connected with water matters because the spiritual value, social meaning, customary access and political significance of water still resonates (Berry, 1998; Nakashima and Chiba, 2006). While carving out space for meaningful native participation remains challenging, knowledge about the specificities of indigenous identity and demands for native governance of water has the potential to enrich understanding of social participation (Corpuz, 2006). For example, indigenous demands for participation on their own terms may reveal how social identity and mobilization can be significant in redefining participatory processes. The two chapters in Part I assess how participation in water decision-making resulted from structures of indigenous governance.

Chapter 1 focuses on the participatory dynamics that arose during water rights negotiations between an American Indian Tribe, the Timbisha Shoshone of the western United States, and the federal US government. Waiting for the right time to achieve results, the Tribe benefited through collaboration with legal counsel, non-governmental organizations (NGOs) and the media, as well as by tapping into a political environment conducive to settlement negotiations. The Tribe's insistence that longstanding inequities be acknowledged and redressed also shifted the balance of power in their favour.

Chapter 2 examines the case of nine indigenous groups in the Brazilian Amazon who opposed the construction of a hydropower dam on the Xingu River. An indigenous coalition was mobilized against the proposed dam and they had assistance from allies in the media, non-indigenous professionals and international NGOs who brought public scrutiny to bear on the project. While indigenous collective action shaped the terms of social participation and resulted in successful litigation, the project proponent, a national power agency, recently co-opted participation for its own ends as it tries to push forward with water project development.

Part II

The contributions in Part II delve into social participation and the dynamics of gender in water management. As participatory development has evolved to become a catchphrase in water management, much of the impetus has been the goal of addressing gender inequities. It is not uncommon for practitioners, policy makers and academics to assert the significance of gender in water management, drawing particular attention to the plight of poor women because, on the one hand, women develop expertise from their work on water matters but, on the other hand, are often not given the opportunity for active participation in decision-making about water matters (Bhatia 2004). In this section of the book, two chapters address the implications of participatory water management for gender matters.

In Chapter 3 the dynamics of gender in participatory water initiatives are considered for a locally based NGO in western Rajasthan. Through initiatives to enhance women's participation in water projects, NGOs assume pivotal roles in mediating traditional gender roles that are influenced by three factors:

- the NGO's relationship with governmental and international donors;
- the context of communities in which customary practices of men and women arise;
- through the NGO's own orientation and approaches.

The chapter suggests that gender norms and practices may be actively and passively restructured through participatory water initiatives as a result of these factors that influence a NGO.

Chapter 4 critically examines the conventional wisdom that women's exclusion from irrigation water user organizations is a reliable indicator of gender inequity in water matters. This assessment of communities in Nepal and Peru concludes that while gender inequities still exist, women's invisibility within formal irrigation organizations may actually be an asset, providing greater freedom from institutional responsibility but not precluding women's inclusion in significant on-the-ground, operational decisions, such as the timing of irrigation rotations or mobilization of labour for maintenance.

Part III

These chapters explore social participation in river basin governance. River basin organizations of all sorts began to develop in the 1980s and now they contain many of the ingredients of current water governance doctrines throughout much of the world. Not only are river basins frequently seen by many water experts as the relevant scale for governance but they also are actively used to facilitate integrated water resource management, decentralization and participation. The three chapters in this section evaluate social participation within river basin and watershed organizations.

Chapter 5 looks at stratified water governance in France with responsibilities spread across various geographic levels. Social participation has been engaged in a number of ways because there has been a great deal of leeway and few standards to follow in developing water management plans for river basins and watersheds. The conclusion is that social participation in French water management is best viewed as a tool designed to improve negotiation processes, rather than an aim in itself.

In Chapter 6 a view into Mexican river basin organizations suggests that effective democracy is a precondition for effective participation. This reverses the argument of those who identify with the standard model of participation, which idealizes participation but views much of it as a façade due to virtually unchecked federal administrative power. It is concluded that Mexican river basin organizations could make more progress if there was cross regulation, a combination of interdependence between independent actors with collective dependence and legitimacy, in which trust is built in institutions over time.

Chapter 7 focuses on recent experiences with social participation in watershed and river basin management in Quebec, Canada. In these cases a variety of stakeholder interests were represented within the local watershed management committees, whose size often became unwieldy. Some participants were not well invested in the committee's work, while others preferred to address water management issues through alternative means that had proved successful in the past. Even seemingly benign processes involved in building consensus risked alienating those who raised different issues or who wanted immediate action. Despite this, a climate of trust seemed to evolve among participants in the three organizations studied.

Part IV

The authors in this section consider the implications of social participation in the implementation of water management plans and programmes. Social participation has gained widespread acclaim within water regulatory circles as it has become a mantra, not only for river basin governance, but for a variety of management objectives and scales. Participation has been engaged with varying measures of success in curbing illegal agricultural diversions, pricing municipal water supplies, enhancing water-based recreational activities, protecting against flood hazards and other initiatives designed to implement new or modify existing water management programmes. Yet in cities, as well as in rural areas, social participation raises challenges as such water management initiatives are being implemented. These issues and their implications are the object of four chapters in this section.

Chapter 8 focuses on participation amongst rural irrigators in Yunnan Province, China after a shift from an exclusively centralized system to a bureaucracy framed around participatory water management. Even when they were not elected, local representatives and leaders interacted with government administrators as they attempted to strike a balance in water governance between actors' decisions, rules of the state and customary social practices. Consequently, information circulated, timely irrigation was secured and certain abuses were stopped. This enabled participation within the formal structure to co-exist alongside relatively benign civil resistance of irrigators refusing to pay their water fees.

Chapter 9 takes stock of the difficulties and examines misunderstandings between expected effects of participation and social reality as water management was implemented in South Africa during the post-apartheid period. Genuine participation and representation seemed to be scarce, but even with these in place, equitable and sustainable outcomes are not guaranteed. Differentials in power, traditional social practices and the characteristics of conflicts may hamper fair and inclusive participatory processes. Facilitation and capacity building are needed, particularly with regard to the flow of information and the co-production of knowledge.

Chapter 10 reveals a local NGO dedicated to participatory approaches in developing rural water supply and wastewater management in Sri Lankan villages. Working under the auspices of the United Nations Millennium Development Goals, this Sri Lankan NGO established an approach structured on community mobilization and village participation. Characterized by both their cultural and technical expertise, a new professional class of facilitators within the NGO have become brokers for development to stimulate local interest in water and wastewater management.

The theme of Chapter 11 is participation within the Yarqon River Basin Authority of Israel. Despite early cycles of environmental degradation along the Yarqon River, the state-designated Authority was able to leverage participation to work on water quality improvement, which in turn made it possible to extend discussions to more stakeholders and to increase the sophistication of information collected and disseminated. The successes in improving water quality and environmental conditions along the Yarqon River are seen as being the result of active social participation within the framework of adaptive management practices.

Part V

This final section examines the social participation and the politics of water governance. While the preceding sections address political alongside other factors in social participation, this section focuses on the political realities that structure speech and practice surrounding participation. Spanning all manner of partisan politics in many countries, parties from right, centre and left support participation and have employed it in their own ways. Understanding the role of social participation may reveal democratic bases in societies by exposing sources of power, hidden processes in previously opaque negotiations and the relative significance of law and formal structures. Three chapters in this section bring to light different governance processes, assess social processes and identify varied powers that influence decisions separate from more formal institutions and legal procedures.

In Chapter 12 the rationality behind norms and local forms of irrigation management for peasants and indigenous rural communities in the central Ecuadorian Andes is explored. Battles over material control of water systems were simultaneously struggles over the right to culturally define, politically organize and discursively shape the existence of people in these rural communities. It is argued that attempts by the state to rein in and control social participation cannot simply be viewed as management decentralization nor as benign inclusion of local beneficiaries in national water development programmes. The initiatives of indigenous and peasant communities to reshape participation have the potential to define terms that will sustain their livelihoods through controlling water usage.

Chapter 13 introduces the Latin American Water Tribunal, recently formed by a group of Latin American NGOs as an alternate means to achieve justice over water conflicts. Similar in many respects to a court, the Tribunal presides over water conflicts brought to them by small groups (often minority or impoverished) against government agencies or multinational corporations. While the Tribunal has been able to provide moral resolution and may facilitate conflict resolution, it has been unable to effect legal resolution for lack of a mandate and has not attracted many of the defendants to engage in this alternate litigation process. Nevertheless, the appeal is clear as the Tribunal incorporates a mix of governance and resistance, serving as a means to integrate justice as a conceptual goal while regaining control over the mechanisms of participation.

Chapter 14 addresses another way by which social participation has been configured through a review of the opposition to two proposed dam projects in Turkey. Recourse to the courts and engagement of independent media facilitated the success of the intellectuals who fought this proposed development. These intellectual elites gave voice to the concerned populations, leveraging their legitimacy, knowledge and motivation to explain objections to the projects in terms the government and courts understood. They argued about appropriate international protocols and deployed their own research studies to examine the potential for loss of historical and environmental resources. Instead of collective mobilization, actual participation in these water conflicts was rather limited. Somewhat ironically, the media found the participatory dimensions of the project's opposition to be compelling in contrast to the government's approach to decision-making which was seen as undemocratic.

The book's conclusions draw attention to the political and critical implications of social participation, relating some of the challenges and looking into some solutions to balance powers. Calls for inclusion and participation have not disappeared and are unlikely to, particularly for those engaged with democratization. Social participation continues to be connected with the goals of rectifying social inequities, responding appropriately to environmental disturbances and transforming structures of power. Attempts to level the terrain of social equity through participatory water governance remain appealing largely because *genuine* participation of the disenfranchised in water management may build bases of power and change networks of social equity (Kurnia et al, 2000). Yet the constraints to genuine broad-based social participation are undeniable (Cooke and Kothari, 2001; Hickey and Mohan, 2004). As translating social demands and coping with public interests within water management have become a reality in many parts of the globe during recent decades, many challenges have cropped up. It is our hope that the discussions in this book will build bridges between water experts in various disciplines studying participation, point to ways that study methods could be standardized, contribute to interpreting generalizable patterns and suggest avenues to advance social participation beyond the constraints that have evolved.

References

Arnstein, S. R. (1969) 'A ladder of citizen participation', *Journal of the American Planning Association*, vol 35, no 4, pp216–224

Berry, K. A. (1998) 'Race for Water? American Indians, Eurocentrism, and Western Water', in D. Camacho (ed) *Environmental Injustices, Political Struggles: Race, Class, and the Environment*, Duke University Press, Durham, NC

Bhatia, R. (2004) 'Role of women in sustainable development – a statistical perspective', in *Proceedings of National Seminar on Gender Statistics and Data Gaps,* New Delhi, India: Central Statistical Organisation, Government of India, February, pp233–241

Bhatnagar, B., Kearns, J. and Sequeira, D. (1996) *World Bank Participation Sourcebook,* The World Bank, Washington, DC

Cooke, B. and Kothari, U. (eds) (2001) *Participation: The New Tyranny?* Zed Books, London

Corpuz, V. T. (2006) 'Indigenous peoples and international debates on water: reflections and challenges', in R. Boelens, M. Chiba and D. Nakashima (eds) *Water and Indigenous Peoples*, UNESCO, Paris, pp24–36

D'Aquino, P. (2007) 'Empowerment et participation. comment mieux cadrer les effets possible des démarches participatives?', http://hal.archives-ouvertes.fr/hal-00157747

Henkel, H. and Stirrat, R. (2001) 'Participation as spiritual duty: empowerment as secular subjectation', in B. Cooke and U. Kothari (eds) *Participation: The New Tyranny?*, Zed Books, London, pp168–184

Hickey, S. and Mohan, G. (eds) (2004) *Participation: From Tyranny to Transformation?* Zed Books, London

Kurnia, G., Avianto, T. W. and Bruns, B. R. (2000) 'Farmers, factories and the dynamics of water allocation in West Java', in: B. R. Bruns and R. S. Meinzen-Dick (eds) *Negotiating Water Rights*, International Food Policy Research Institute, London, pp292–314

López Cerezo, J. A. and Garcia, M. G. (1996) 'Lay knowledge and public participation in technological and environmental policy', *Philosophy and Technology*, vol 2, no 1, pp53–72

Nakashima, D. and Chiba, M. (2006) 'Introduction', in R. Boelens, M. Chiba and D. Nakashima (eds) *Water and Indigenous Peoples*, UNESCO, Paris, pp12–16

Part I

Participation and Indigenous Water Governance

1

Participation of the Timbisha Shoshone Tribe in Land and Water Resource Management Decisions in Death Valley National Park, California and Nevada, US

Terry T. Fisk, Pauline Esteves, Barbara Durham and Madeline Esteves

Introduction

The contrast between a participatory culture and a highly judicial dominant culture is illustrated by examining the relationship between the Timbisha Shoshone Tribe and the US government during negotiations relating to land and water. The negotiations culminated in the signing of the Timbisha Shoshone Homeland Act by President Clinton in November 2000. The Homeland Act provides the Timbisha Shoshone with a permanent land base and water rights in California and Nevada, in and around Death Valley National Park. The location of Death Valley National Park is shown in Figure 1.1. Effective social participation, particularly by the Timbisha Shoshone and, during the latter stages of negotiation, by government representatives, enabled the negotiations to conclude successfully. Looking back at the history of the negotiations, the social participation practised by the Timbisha Shoshone was superior to that of the more dominant judicial culture. It was not until the US government adopted a more participatory stance that negotiations proceeded to a positive conclusion. The processes of effective participation still influence interaction between the Timbisha Shoshone and the dominant culture. Although other government agencies were involved in the negotiations, this study focuses on the National Park Service (NPS), a bureau within the US Department of the Interior, because the NPS played the most critical role with respect to allocation of land and water between the US government and the Timbisha Shoshone.

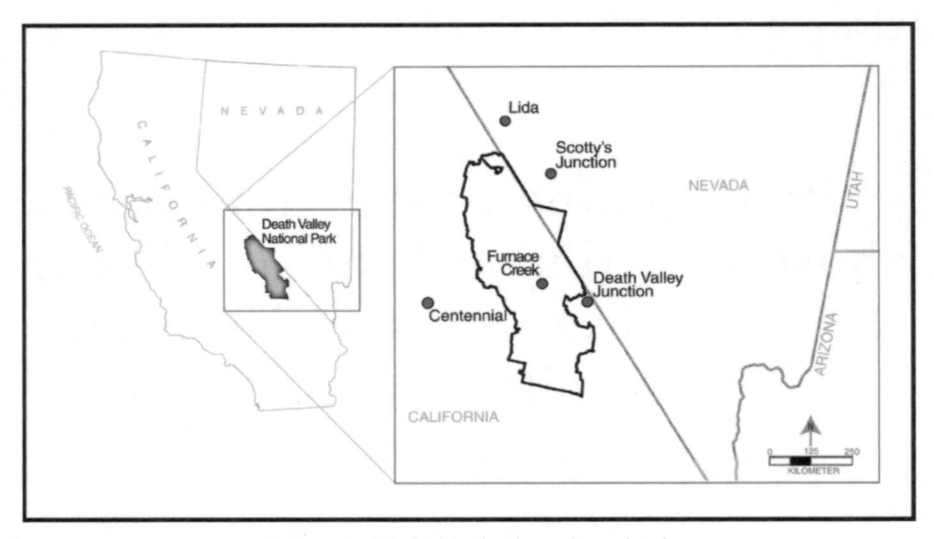

Figure 1.1 Timbisha Shoshone homelands

In the western US, management of water resources by the dominant culture is an iron triangle controlled by government, water supply agencies and private enterprise. Water is a powerful economic and political force. Historically, water agencies ranging from enormous public suppliers to local irrigation districts have been the only arenas in which social participation occurred. More recently, non-governmental organizations (NGOs) and an engaged citizenry have increased social participation within the dominant culture. Water management by the Timbisha Shoshone provides a marked contrast to the dominant culture. Water is a sacred resource, not an avenue to power, and is a central part of the Timbisha Shoshone culture and relationship with the land.

The settlement between the US government and the Timbisha Shoshone is indicative of water settlements in recent years involving indigenous peoples in the western United States. The negotiations also illustrate the evolving relationship between the US government and Indian tribes, particularly in relation to management and governance of water resources. The characteristics that enabled the Timbisha Shoshone to successfully conclude land and water negotiations with the US government are consistent with characteristics demonstrated by other tribes participating in similar events.

Timbisha Shoshone Society and Water

Death Valley, located at the intersection of the southern Great Basin and the eastern Mojave Desert, has a fearsome reputation for heat and aridity – and rightfully so. Annual temperatures at Furnace Creek, the location of the

Timbisha Shoshone village and Death Valley National Park headquarters, range from approximately -4°C to 55°C. Average annual rainfall in the park ranges from approximately 40mm at Furnace Creek to 300mm at elevations greater than 3000m. Terrain in Death Valley National Park ranges between elevations of -85m at Badwater to greater than 3350m in the high peaks. Death Valley National Park claims hundreds of spring-fed water resources and riparian and wetland habitats. Without this water, life in Death Valley would be virtually impossible.

The Timbisha Shoshone's elders say the elders of their elders taught them that the Timbisha Shoshone originated in their homeland and they have always been here. The Timbisha Shoshone do not name their homeland Death Valley, which is the name applied by the dominant culture. The Timbisha Shoshone call their homeland: Tümppisha (phonetic) or Tümppisa (linguistic). It is the centre of their world, not a place inhabited only temporarily and for material gain. The Timbisha Shoshone elders and their elder's elders talked about the different features throughout the Tümppisha region and they know the landscape and the names of the places. Elders say the Tribe does not have its history in black and white like this 'new' society (the dominant culture) does. Instead, their history is in the mountains, in the water and in all the belongings of Mother Earth.

Water is a sacred resource and is a central part of the Timbisha Shoshone cultural being. Timbisha Shoshone ceremonies recognize that the land and all earth's belongings need water. Water is not a commodity. Instead, the relationship with water engages the whole culture and life of the Timbisha Shoshone. The Timbisha Shoshone give back to the land to show respect, to acknowledge to the Creator that they are asking for a blessing and to guide them in the right way. Before contact with European culture, the Timbisha Shoshone settled around areas where water was available. They lived some distance away from water sources and hauled water for their drinking, cooking, laundry and bathing needs. Usually, the Timbisha Shoshone only entered water for healing purposes, although children played in the water. Waste was never discharged into springs or streams. Springs were cleaned to prevent them being choked with vegetation and to keep them flowing. The Timbisha Shoshone believe that water is intended to have a surface flow for a multitude of animals and plants, not just those plants that crowd out other life if a spring is untended. Periodically, the Timbisha Shoshone burned vegetation around springs, particularly those at lower elevations with relatively large ponds. Fire reduced overall vegetation quantity, produced more open water for waterfowl and encouraged growth of vegetation preferred by the Timbisha Shoshone for basketry and food.

There was no individual ownership of a spring or other water source within the tribal culture. After contact with western society, individual members of the Timbisha Shoshone filed for water rights to springs in accordance with western water law. The purpose of filing for rights as individuals, against their custom, was to ensure the Timbisha Shoshone retained legal rights to ancestral water resources that were increasingly being claimed by miners and ranchers moving into the Death Valley region (P. White, pers. comm., 2007).

The Timbisha Shoshone traditional style of government was for the heads of family groups, the elders, to get together, discuss issues and make decisions. Today's tribal government is based on a structure imposed on many tribes by the US government in the 1930s to mirror the federal government. The Tribe had written a constitution before they were formally recognized as a tribe by the US government. The Timbisha Shoshone government consists of a general council and a tribal council. The general council consists of the entire population that has reached the voting age of 16 years. There are approximately 330 members of the Tribe and approximately 200 voting members. The tribal council is elected by the general council in yearly elections. The tribal council consists of the chair, vice-chair, secretary-treasurer and two other members (Timbisha Shoshone and US Department of Interior, 1999). Positions on the tribal council last two years and the terms of the council members are staggered so that in any given year, two or three positions are on the ballot. An election committee organizes and implements the election. A neutral third party provides assistance to ensure valid elections. If elections are contested, the Bureau of Indian Affairs has the final authority to determine the legality of the tribal council. The Tribe has excellent participation in elections, generally 65–75 per cent.

Tribal members are dispersed throughout southern California and Nevada, although a majority live within a day's drive of Death Valley. Occupations range from tribal government positions to nursing; however, many tribal members have positions categorized as trades people or unskilled labour (Timbisha Shoshone and US Department of Interior, 1999). Several individuals in the tribe attend a university or are university graduates. Recent data are not available; however, in the early 1990s, approximately 40 per cent of tribal members were unemployed, and of those employed, many worked only part-time. In 1993, earnings of most families were less than the official US poverty level.

Dominant Culture and Water

In the western US, water is power and control of water is absolute power. It is said that 'whiskey is for drinking and water is for fighting'. Another saying is that 'water ignores the laws of gravity and flows uphill toward political power and money' (Reisner, 1986).

Western water has invariably been controlled for decades by the iron triangle of committees and subcommittees of the US Congress, federal and state agencies and private interest groups (McCool, 1994; Wilkinson, 1992). The US Congress funds water projects and western members of Congress are adept at bringing water projects to their home districts. Federal agencies such as the Bureau of Reclamation, and state bureaucracies created to support and administer water law, are the second point of the triangle. Private sector interests include developers and commercial entities, often closely associated with municipal interests, who require assured quantities of water for their proposed actions (Wilkinson, 1992). Iron triangles can be extremely stable because, while the benefits of the triangle's efforts are concentrated, the costs are dispersed widely. As a result, the iron triangles often operate in low-conflict environments that offer stable long-term relationships (McCool, 1994).

Western water law and administration

Western water was, and is, free for the taking, for those who get to it first (Wilkinson, 1992). The foundation of western water law is the prior appropriation doctrine and is most 'out of kilter of all the 19th century laws that rule our lives' (Wilkinson, 1992). Prior appropriation depends entirely on the concept of 'first in time, first in right' for the beneficial use of water, and is based on the need of the individual, not the needs of society. Prior appropriation also depends on the concept of 'use it or lose it'. Thus, water not diverted from a stream or pumped from the ground is often considered wasted because it is not used for a 'beneficial purpose' (Wilkinson, 1992).

The 'use it or lose it' concept is still very much alive and well in the western US. For example, in 2007, the city of Las Vegas, Nevada was granted preliminary rights to 74,008,860m^3 of groundwater per year from Spring Valley, Nevada, 400km northeast of Las Vegas (State of Nevada, Office of the State Engineer, 2007a). Las Vegas' rights were granted because the city got to the water first. The local population (admittedly minimal) and the water-dependent natural resources in Spring Valley lost the water because, legally, they were not using it. From a water law perspective the groundwater was not used beneficially because it had not been legally claimed or appropriated even though it supported (and supports) social needs of the local rural communities and also ecosystem needs.

Even though usually administered by states, water rights are property rights of individuals. Government, whether federal, state or local, cannot take away vested water rights (property) without full compensation. Senior water right holders are always superior to junior water right holders no matter what the beneficial use and manner of using the water (Wilkinson, 1992).

Water development

Throughout the western US, the goal of the dominant society has been to conquer nature and extract natural resources. Nearly unfettered private resources development was allowed, encouraged and often subsidized. However, 'Indians and Hispanics were resolutely dealt out' of water projects large and small (Wilkinson, 1992). There was no commonly perceived need for any environmental policy until recent decades. Federal environmental laws passed during and since the 1970s, such as the National Environmental Policy Act,

Clean Water Act and Endangered Species Act have opened the decision-making process to new constituents outside of the traditional iron triangle.

The public trust doctrine has been added recently to the vocabulary of westerners. The public trust is the concept that the public has overriding interests on major water resources that cannot be granted away to any single segment of the population (Wilkinson, 1992). NGOs and individuals have become more involved in decisions regarding water resources in recent years. They represent the public trust, and agencies and courts typically recognize their involvement (Western Water Policy Review Advisory Commission, 1998). More states are beginning to incorporate the public trust doctrine into their legal codes and water rights rulings. For example, in its 2007 ruling on groundwater right applications in the Amargosa Desert, east of Death Valley National Park, the Nevada State Engineer invoked the public trust doctrine as part of its rationale for denial of the applications (State of Nevada, Office of the State Engineer, 2007b).

Death Valley National Park water rights

Water rights of the park are owned by the US government and consist of federal reserved rights, state appropriative rights and riparian rights. When the government reserves land for a specific purpose, it also reserves, by implication, sufficient unappropriated water necessary to accomplish the purposes for which the land was reserved: these are federal reserved water rights (National Park Service, 1998). State appropriative rights are based on the prior appropriation doctrine, as discussed above. Riparian rights apply only to land adjacent to the source of water and the water must be used on riparian lands. There is no prior appropriation doctrine with riparian rights – all users share alike in times of shortage or surplus.

In the case of Death Valley National Park, Congress reserved sufficient water for the purposes of the park when Congress created the park. The park also has a specific federal reserved water right, upheld by the US Supreme Court, at Devils Hole, Nevada. The park participates in state administrative water right procedures in California and Nevada when necessary to protect its federal reserved rights or state appropriative rights.

Timbisha Shoshone Homeland Act and Water Agreement

President Clinton signed the Timbisha Shoshone Homeland Act into law on November 1, 2000. Although the Timbisha Shoshone have lived in Death Valley since time immemorial, the Tribe received federal recognition only in 1983. Recognition, however, did not include designation of a tribal homeland. The Timbisha Shoshone homelands had been appropriated by the dominant culture since the late 1800s: informally by mining, ranching, agriculture and resorts and formally by the 1933 creation of Death Valley National Monument. Specific reasons why homelands were not included when the Tribe received formal recognition in 1983 were not identified in this study. However, research conducted for this study suggests that the NPS would have objected strongly, and perhaps did object, to any suggestion that a homeland for the Timbisha Shoshone be created within what was, at the time, Death Valley National Monument.

The Timbisha Shoshone Homeland Act provides the Timbisha Shoshone with a land base on which they can live in perpetuity and from which they can engage in traditional cultural, governmental, economic and social practices. Moreover, in this parched landscape, the Act guarantees water rights for the Tribe. Without water, the land is essentially useless for many purposes. Under the Water Agreement negotiated between the Timbisha Shoshone and the NPS, the government recognizes the Tribe's water rights on trust lands and the Tribe's need for adequate water to support the Tribe's interests (Timbisha Shoshone Tribe and National Park Service, 2000).

The genesis of the Timbisha Shoshone Homeland Act was the California Desert Protection Act (1994). Section 705(b) of the Desert Protection Act specifies that the Department of Interior must conduct a study to identify lands within and without Death Valley National Park suitable for a reservation for the Timbisha Shoshone (Timbisha Shoshone and US Department of Interior, 1999). Section 705(b) of the Desert Protection Act is the foundation of the Tribe's push for land and water resources, without which no homeland would yet exist.

The passion of the Timbisha Shoshone leaders was to obtain a homeland and water. They determined they would take control of their destiny and establish their sovereignty as a Tribe. Formulating the language of the Desert Protection Act during the early 1990s spurred the Timbisha Shoshone into action. The Desert Protection Act was chosen as the vehicle because it appeared to be the best, and perhaps final, opportunity for the US Congress to consider the Tribe's land and water claims. Paradoxically, the Timbisha Shoshone were greatly concerned that, without their intervention, the Desert Protection Act would forever remove land important to the Tribe from consideration as homelands. The concern lay in the fact that by adding approximately 607,030ha to Death Valley National Monument, changing the designation from monument to park (indicating a greater range of natural and cultural features under protection of the NPS) and designating much of the land as wilderness, the Tribe would be forever locked out of homelands in their ancestral territory. In fact, the Tribe opposed early versions of the Desert Protection Act because these versions did not require the Department of Interior to study the suitability of a homeland for the Tribe. Only Congress has the authority to make substantial boundary changes to national parks. Similarly, only Congress can establish trust lands for Indian tribes. From the Timbisha Shoshone perspective, it seemed a perfect fit: a major piece of land-use legislation coming before Congress with a high probability of passing. There could not be a better opportunity to establish homelands in their ancestral territory of Death Valley. No one knew when, or if, Congress would ever again make such sweeping land use changes and contemplate establishing trust lands in the Tribe's traditional homeland areas.

The Timbisha Shoshone Homeland Act requires the US to hold in trust for the Timbisha Shoshone five parcels of land totalling 3138ha with total reserved water rights of 625,745m^3 per year, as shown in Table 1.1 and Figure 1.1. The water may be used for domestic, commercial or irrigation purposes. The total water right of the Timbisha Shoshone is put in perspective by considering that the private corporation owning land and resort facilities at Furnace Creek, Death Valley National Park has a water right of almost 1.85 million m^3 per year.

Table 1.1 Timbisha Shoshone homelands and water rights

Location	Land Area (hectares)	Water Rights (m^3/year)
Furnace Creek, Death Valley National Park, California	127	113,480
Centennial, California	259	12,335
Death Valley Junction, California	405	18,626
Lida, Nevada	1214	18,132
Scotty's Junction, Nevada	1133	463,172

Source: Timbisha Shoshone Homeland Act, 2000

In addition, the NPS and Tribe agreed that over 100,000ha of land in the park would be designated as the Timbisha Shoshone Natural and Cultural Preservation Area. The exact amount of land included in the preservation area remains unspecified; however, it is depicted generally on maps included in the Timbisha Shoshone Homeland Act supporting documents, and includes a substantial portion of park lands. The preservation area also is shown on park maps provided to park visitors. The preservation area provides the opportunity for the Tribe and NPS to cooperatively manage land and resources of significant cultural value to the Tribe and provides tribal members access to these lands for traditional cultural practices.

Fostering good relationships with the California Congressional delegation and other powerful politicians in Washington DC was critical to writing language into the Desert Protection Act that required the US Department of the Interior to evaluate homelands for the Tribe. Representative Jerry Lewis of California worked closely with the Tribe's leaders to include acceptable language into the Desert Protection Act. Representative Lewis and the California Desert Protection Act gave the Tribe its foothold. Senators Daniel Inouye of Hawaii and Ben Nighthorse Campbell of Colorado as well as Senators Barbara Boxer and Diane Feinstein of California helped lead Congress to pass the Desert Protection Act in 1994 and the Timbisha Shoshone Homeland Act in 2000.

Homeland Act and Water Agreement Negotiations

Formal negotiations between the Timbisha Shoshone and US government regarding land and water occurred in two primary episodes between 1995 and 2000. The first episode in 1995 and 1996 involved primarily local and regional NPS staff representing the government. Episode two, between 1998 and 2000 included high-level NPS and Department of Interior decision-makers in California and Washington DC.

Members of the Timbisha Shoshone negotiating team included the tribal chair, two members of the tribal council and the tribal administrator. The chief spokesperson of the Tribe was involved in the 1995 and 1996 negotiations. With the exception of one member of the tribal council, all of the participants lived in the village at Furnace Creek. Other members of the tribe were not prohibited from involvement; however, there was very little involvement by others. The tribal council had not been voted into office on a platform of 'We're going to Washington DC and we're going to get land.' Even today, members of the negotiating team are unsure why there was limited involvement by others in the Tribe. Possible reasons include a lack of interest on the part of those living outside Death Valley, a lack of understanding of the potential benefits of gaining homelands and water resources and uncertainty that the negotiations would actually result in a positive outcome for the Tribe.

Social participation methods adopted by the Timbisha Shoshone included political support in the US Congress, a public relations campaign, support from American Indian organizations, direct communication to the President and one cabinet member and fund raising. The most important participatory action of tribal leaders was telling their story in a compelling way. Social participation by the US government consisted primarily of communication between US Department of the Interior bureaux. For the NPS, communication was between the park, Pacific West Region, and Washington DC offices and the Bureau of Land Management.

1995–1996 negotiations

Negotiations between the Tribe and NPS began in May 1995. The first meeting included representatives from the Timbisha Shoshone, Indian organizations supporting the Tribe, the NPS and a host of other federal and state agencies. Working groups were established to examine the issues of:

- lands that would be suitable for the Timbisha Shoshone;
- potential uses of the suitable lands and prospects for economic development;
- legal issues regarding 'conversion' of lands from the NPS or Bureau of Land Management to the status of reservation lands held in trust for the Timbisha Shoshone by the US government;
- cooperative management agreements allowing access of the Timbisha Shoshone to traditional use areas, and implementing traditional methods of land, water and vegetation management.

The meeting had a very positive atmosphere and stressed collaborative efforts using staff expertise (L. Greene, pers. comm., 2007).

The initial tribal request was that 344,000ha be designated as reservation trust lands, including 2025ha at Furnace Creek. The Timbisha Shoshone also proposed that an additional 809,370ha be co-managed by the Timbisha Shoshone and federal agencies (Timbisha Shoshone Tribe, 1995). However, converting this amount of public land to tribal land was not acceptable to the US government and the Tribe eventually accepted much less.

Additional meetings led by a high-level planner from the NPS Pacific West Region were held frequently in 1995 and early 1996. As time progressed, they became more confrontational and ineffective: a planning-style meeting was not the proper forum for discussion of tribal homelands (P. Parker, pers. comm., 2007). During this period, a primary obstacle to effective social participation by the Timbisha Shoshone was negotiating with NPS staff who did not have authority to make decisions (L. Greene, pers. comm., 2007). A complicating factor was that the NPS planner who led the negotiating sessions was not trained in Indian affairs and relations. The ability to communicate objectively and unemotionally was very difficult and resulted in another barrier to effective participation. The Timbisha Shoshone, in particular, expressed a great deal of what is justifiably called historical outrage (P. Parker, pers. comm., 2007) at their treatment by the US government for over 60 years. The NPS did not fully understand, at that time, that they were dealing with deep historical, moral, social and legal issues that were of tremendous importance to the Tribe (P. Parker, pers. comm., 2007). Unfortunately, in many respects the 1995–1996 negotiations did not follow the Clinton Administration's Indian policy stating that relationships between the US government and tribes would be as government-to-government.

The Timbisha Shoshone disseminated information to tribes across the country and received support from many Indian organizations. Along with the Miccosukee Seminole Nation in Florida, the Navajo Nation (Arizona, New Mexico and Utah) and several tribes among the New Mexico pueblos, the Timbisha Shoshone formed the Alliance to Protect Native Rights in National Parks. The Alliance to Protect Native Rights in National Parks was very effective from the Tribe's point of view because it brought the homeland issue onto a larger stage and demonstrated to the NPS and other tribes the commitment of the Timbisha Shoshone. The National Congress of American Indians passed resolutions in support of the Timbisha Shoshone, invited the Tribe to speak at various gatherings and encouraged its members to write letters of support.

The Tribe conducted an effective public relations campaign as the negotiations proceeded. In 1995 and 1996 local and national newspapers printed many articles sympathetic to the Timbisha Shoshone's desire for homelands and their struggle to maintain cultural practices. Newspapers that covered the negotiations ranged from local and state (*Inyo County Register*, *Modesto Bee*) to national (*Associated Press*, *USA Today*, *Christian Science Monitor*).

Correspondence and meetings between the Timbisha Shoshone and NPS regarding water rights and water resources occurred throughout 1995 and 1996. After homelands, water rights and sufficient water for economic development were the most critical issues for the Tribe. Water was also a critical issue for the NPS, which was already concerned about upgradient water withdrawals in Nevada. The NPS was very concerned that future Timbisha Shoshone water use from proposed homelands would exacerbate what the NPS considered to be the ongoing problem of too much groundwater pumping upgradient from the park (National Park Service, 1998).

Finally, in a death blow to negotiations, the NPS informed the Timbisha Shoshone in March 1996 that no lands within Death Valley National Park would ever be considered for a tribal homeland. In response the Timbisha Shoshone ceased negotiations and stated forcefully that until the question of the Tribe's ancestral homelands within the park was resolved satisfactorily, no serious consideration would be given to lands identified outside the Park (Timbisha Shoshone to G. Thompson, 1996a; Timbisha Shoshone to R. Martin, 1996c). By July 1996, after internal discussions of which the Tribe was unaware, the NPS and Bureau of Land Management had agreed that they could potentially convert approximately 4650ha to tribal homelands. However, none of this land was inside Death Valley National Park and the Tribe was unwilling to resume negotiations.

Not surprisingly, a substantial portion of the NPS social participation consisted of internal dialogue regarding lands that could conceivably be converted to Timbisha Shoshone homelands. The NPS did not regard any lands within the Park as suitable and steadfastly refused to consider that option. Of all the obstacles to effective social participation, the issue of homelands within Death Valley National Park was the largest and most forbidding. Without access to park land the Tribe was locked out of its most important area, and the NPS considered giving up land in the park as non-negotiable. Other long-term solutions were considered that would allow the Timbisha Shoshone to continue residence at Furnace Creek without actually converting any land to tribal homelands. These options, however, were unacceptable to the Tribe. They required land of their own.

Based on the March 1996 NPS rejection of a homeland within the park, the Timbisha Shoshone reinvigorated its public relations campaign and concentrated on building support with the public. A letter writing campaign from indigenous cultures and citizens around the world was fostered and provided much support to the Timbisha Shoshone. Letters went to the US Department of the Interior, the NPS and Bureau of Land Management and members of Congress. The Timbisha Shoshone created brochures in Japanese, German and French and handed them out whenever and wherever they had the opportunity. These presented a brief discussion of the Timbisha Shoshone's history and described the struggle to gain land and water. The materials accused the US government of trying to evict the Timbisha Shoshone from their ancestral homelands and prevent them from practising their cultural heritage.

Death Valley National Park has an annual visitation between 800,000 and 1,000,000 people, a substantial percentage of whom are foreign. Therefore, the Tribe had reasonable access to individuals from the US and other nations who might otherwise have never known of the Timbisha Shoshone.

There was almost no public outreach by the US government except the minimum required by law. The NPS and Bureau of Land Management completed Environmental Impact Statements and were required to seek public comment regarding the alternatives for conversion of land from ownership by the US government to tribal land. The state of Nevada supported a reservation for the Tribe (State of Nevada, 2000). However, the politics of western water meant that Nevada strongly criticized the NPS and Congress for not quantifying the Tribe's water rights (quantification of the Tribe's water rights was eventually completed in August 2000).

The Timbisha Shoshone made 27 Freedom of Information Act requests to the Department of the Interior to learn the rationale for excluding land within Death Valley National Park from consideration as homelands. They wrote to President Clinton in September 1996 reminding him of his previous support for Indian cultures (Timbisha Shoshone Tribe, 1996b). The letter accused the Secretary of the Interior of cultural genocide by attempting to destroy the Timbisha Shoshone as a sovereign people and requested the President intervene on their behalf. The Tribe followed this by writing to the Secretary of the Interior reminding the Secretary that the NPS and Bureau of Land Management were failing in their legal obligations to implement federal law (Section 705(b) of the Desert Protection Act) with respect to the Timbisha Shoshone homeland (Timbisha Shoshone Tribe, 1996d).

National Park Service internal and external communications expressed substantial disappointment with the Tribe's accusations of cultural genocide. In essence, the NPS claimed that by distorting the intentions of the US government the Tribe was not playing fair. Eventually, in direct response to the Timbisha Shoshone, the Secretary of the Interior pledged to continue to work with the Timbisha Shoshone and assured the Tribe that the Department of the Interior 'remained committed to accommodating the continued residency of the Timbisha in Death Valley' (US Department of the Interior, 1996). Even in this response though, the chasm between homeland and residency remained. There was no substantive movement by the NPS during late 1996 and 1997 (National Park Service, 1997). This was a period of kitchen table informal communications between a few representatives of the Timbisha Shoshone and one or two NPS staff in Washington DC (P. Parker, pers. comm., 2007).

1998–2000 negotiations

The second round of negotiations began in January 1998, sponsored by the Department of the Interior Assistant Secretary's Office. Participants included the Timbisha Shoshone, the NPS, the Bureau of Land Management, the Bureau of Indian Affairs and the US Department of the Interior Office of Trust

Responsibility. To minimize obstacles that bedevilled effective participation in the 1995–1996 negotiations three key decisions were made:

- only decision-makers would take part in the negotiations;
- a moderator would lead the negotiations;
- the Tribe and US government would negotiate as sovereign nations.

The Timbisha Shoshone also stipulated that NPS personnel who had taken part in the 1995–1996 negotiations could not be involved directly because of the tension and disputes that marred the first negotiations.

A mediator, Charles Wilkinson of the University of Colorado School of Law, was brought into the proceedings and stayed through to the end. Professor Wilkinson understood the Timbisha Shoshone point of view and also the US government's needs and legal restrictions (L. Greene, pers. comm., 2007). He was a powerful advocate that the Timbisha Shoshone should be treated as a nation during the negotiations. Professor Wilkinson also stipulated that senior people involved on both the government and tribal sides had to remain personally committed to the negotiations.

Involvement by the Director of the NPS Pacific West Region, John Reynolds, was fundamental to a successful outcome because he was an advocate for negotiating in good faith:

> At the very first meeting we went around the table and John [Reynolds] made a powerful statement about the Park Service's land ethic and that he wouldn't for a minute say that the Park Service had a deeper land ethic than tribes, but that the Park Service does have a deep land ethic. It was a very powerful statement. Very respectful, but very powerful. And it got around to Pauline [Esteves, tribal Chair], and you can see this scene, it gets around to Pauline and she looks right at John and she says 'I have a question for you. Are you going to respect our rights? Are you going to be our trustee? Are you going to respect our sovereignty? Are you going to respect the horrible history of when you moved us around this Park? Are you going to appreciate how hard it has been for us, what it is like for a Tribe to be landless?' And she went on for, it felt like an hour, asking this one question that had 80 parts to it. She stopped. John looked at her and said 'Yes.' (C. Wilkinson, pers. comm., 2007)

New legal counsel for the Tribe, California Indian Legal Services, brought a degree of competence and professionalism to the Tribe's legal representation that had been lacking previously. The new Timbisha Shoshone legal counsel stressed interest-based negotiations. This was a valuable concept, because once the interests of each side were stated, it became apparent how much both sides had in common regarding respect for natural and cultural resources (L. Greene, pers. comm., 2007). Interest-based negotiations, often labelled alternative dispute resolution (ADR) emphasizes direct participation, consensus building, collaborative problem-solving and a focus on interests rather than positions (McCool, 2002). Representative shared interests of the Timbisha Shoshone and the US government are shown in Table 1.2.

Table 1.2 Shared interests: Timbisha Shoshone and US government

Tribal sovereignty and self-determination
• Recognition of the sovereignty of the Timbisha Shoshone. • Establishment of a permanent land base for the Timbisha Shoshone in traditional ancestral homelands. • Preservation and development of the Timbisha Shoshone's dynamic indigenous culture. • Involvement of the Tribe in economic and employment development activities. • Establishment of the Tribe's government and administrative headquarters at Furnace Creek.
Natural and cultural resource protection
• Recognition of the common interests of the Timbisha Shoshone and the United States in the conservation and preservation of the plants, animals, land, air and water and their natural relationships in the Death Valley area. • Recognition of the Tribe's historic responsibility for protection and preservation of the environmental and cultural resources of the Death Valley area.
Tribal presence in the park
• Co-existence and a dynamic relationship between the Tribe and the NPS. • Recognition of the interests of the Timbisha Shoshone in presenting and interpreting the Timbisha Shoshone's own history and culture to visitors. • Use by the Timbisha Shoshone of traditional lands for continuing cultural practices. • Recognition that any future development of land in Furnace Creek will be conditioned by availability of water and by jointly established design standards.

Source: Timbisha Shoshone and US Department of the Interior, 1999

As during 1995–1996, the most difficult obstacle was the issue of Timbisha Shoshone land within Death Valley National Park. The decision early in 1998 by Regional Director Reynolds that all lands, including those within Death Valley National Park, were open for discussion removed the major roadblock to achieving a satisfactory agreement (C. Wilkinson, pers. comm., 2007).

Water resources were also an extremely critical issue and tied closely to any discussion regarding suitable lands for the Tribe. Major water-related discussions revolved around the amount of water that might be available and needed by the Tribe on each parcel of land considered for homeland status. Because of the overriding need for water as part of any settlement, the Timbisha Shoshone requested that the expertise of the Bureau of Reclamation be available for consultation during the negotiations, and that the Bureau evaluate the potential water resources on land of interest to the Tribe.

Eventually, the two sides worked on multiple drafts of a Secretarial Report to Congress and settled on a total of 3138ha to become tribal homelands,

127ha inside Death Valley National Park. To this day, it is not fully clear how, when and why the Tribe's initial demands of 344,000ha, including 2025ha within Death Valley National Park were whittled down so drastically. Much of the land requested by the Tribe is designated wilderness, and while that land would be acceptable for many traditional cultural practices, with a few exceptions it is not viable for economic enterprises. It is possible that the NPS could have been convinced to create a small homeland within the park before 1999; however, 2025ha was simply too much. The NPS culture is to accept and retain land, never to give it up. The Secretarial Report was submitted to Congress in late 1999 as supporting documentation for the Timbisha Shoshone Homeland Act (Timbisha Shoshone and US Department of Interior, 1999). The Act was passed by Congress and signed in November 2000 by President Clinton.

Effects and Influences of Social Participation

Indian tribes have a universal belief that land is the salvation of Indian life and that without it their people will scatter and their culture will wither. 'For Indian tribes, virtually nothing could be more threatening to these place-based peoples than the expropriation of their land' (Wilkinson, 2005). The Timbisha Shoshone regained lost homelands and water resources because they conducted the classic campaign that other successful tribes and tribal leaders have carried out in recent years. They explained their rights and their history in a way that was both intellectually and emotionally powerful (C. Wilkinson, pers. comm., 2007). In doing so, the Timbisha Shoshone leaders, consciously or unconsciously, emulated the processes that recently allowed other tribes to survive and even thrive. Tribal sovereignty forms the bedrock of modern court decisions regarding Indian rights (Wilkinson, 2005). The leaders of the Timbisha Shoshone established the Tribe as a sovereign government, one capable of influencing the Desert Protection Act and negotiating passage of the Timbisha Shoshone Homeland Act.

The Timbisha Shoshone effectively used social participation with their legal counsel, political friends, public and media relations and non-governmental organizations. These are essentially the same tactics used by other tribes, such as the Menominee Tribe of Wisconsin and the Taos Pueblo of New Mexico, in achieving the return of land and water resources that had been appropriated by the dominant culture (Wilkinson, 2005). Social participation as practised by the US government and Timbisha Shoshone during the Timbisha Shoshone Homeland Act negotiations is far different from the former standard of practice. In the past, social participation was typically a closed affair, occurring between the economic development interests of the dominant culture and the Department of the Interior to negotiate contracts for extraction of resources from Indian land. The government then presented the contracts to tribes for their approval (Wilkinson, 2005). Tribes were not able to participate as partners in the negotiations, but were expected to swiftly agree with the results. For example, in the 1960s, the Hopi Tribe of Arizona could not stop an agreement negotiated between Peabody Coal and the US government for a mine on Hopi land (Wilkinson, 2005).

In the wider context of western water, social participation, and relationships between tribes and the US government, the Timbisha Shoshone Homeland Act may be viewed as part of a continuum of recent water settlements. Daniel McCool (2002) argues that we are in the second treaty era between Indians and the US government. The first era, the 19th century, was primarily about land. The second era is focused on water and can be just as difficult as the first. Quantification of water rights is troublesome for tribes because they must settle for whatever they can get as part of the legal process of the dominant culture. This is contrary to tribal cultural perceptions of water, their lifeblood, an indivisible community resource (McCool, 2002). Nevertheless, in the past 20 years, water settlements have been negotiated with the Jicarilla Apache, Northern Cheyenne, Northern Ute, San Carlos Apache, Tohono O'odham, Assiniboine and Sioux Tribes of the Fort Peck Indian Reservation and others (McCool, 2002). The process is not over either: the Navajo Nation recently concluded a water settlement with the state of New Mexico regarding Navajo rights to Colorado River water in New Mexico. The Navajo are now preparing for negotiations with Arizona and Utah, and possibly the US government, about their rights to Colorado River water in those states (Jenkins, 2008).

Because the Timbisha Shoshone were formally recognized only recently, their settlement involved land and water at the same time. Similar to other recent water settlements, the Homeland Act and Water Agreement offer legal and quantified water rights and a level of tribal control and administration over water resources that did not previously exist. For the first time in over a century, the Tribe now has a legal voice in the management of water resources (and other natural and cultural resources) within and outside Death Valley National Park. The Homeland Act requires that the NPS and Timbisha Shoshone jointly develop and agree upon a groundwater monitoring programme to evaluate whether tribal use of water east of the park will have a detrimental impact on groundwater flow into Death Valley. Both the Timbisha Shoshone and NPS have completed installation of monitoring wells that were agreed upon under the terms of the Water Agreement. A written water monitoring plan that will be implemented jointly is being developed and exchanges of groundwater monitoring reports between the Timbisha Shoshone and NPS are occurring. The Timbisha Shoshone and NPS are also developing an agreement for cooperative management of springs in several areas of the Park. The initial group of springs chosen for cooperative management was selected by the Tribe as demonstration projects where traditional management practices will be undertaken. A comprehensive cooperative management programme for a larger and more diverse group of springs will be developed based on the findings at the demonstration springs.

The use of a negotiated settlement to arrive at the Homeland Act and Water Agreement follows a recently developed pattern in the western US.

Frequently, negotiations are the first substantive opportunity for the non-Indian community to begin understanding and appreciating the needs and viewpoints of the tribes with whom they are interacting (McCool, 2002). Negotiated settlements, based on common interests, create the opportunity to establish constructive relationships and deal with practical problems of managing common water resources and they can be tailored to specific circumstances (McCool, 2002). Such settlements are considered superior to the results of litigation because of the establishment of economic and governmental relationships that remain after the negotiations are concluded. Courts are adversarial in nature, producing winners and losers, and generating animosity. Negotiations are never a complete win–win proposition, but are more accurately thought of as a 'give-and-take' process in which each party must give something up (McCool, 2002).

The Timbisha Shoshone Homeland Act has had a positive effect on tribal members, but so far not as much as the elders would like. The Timbisha Shoshone have a greater sense of optimism regarding their future and their ability to control their own destiny. Participation in tribal elections is greater now than before the Homeland Act. Tribal members are realizing that with land and water resources, the tribal government will be able to accomplish more on their behalf because many governmental and economic assistance vehicles available to Indian tribes are based on tribes having a homeland. The elders hope more Timbisha Shoshone will come back to live at Furnace Creek or build on the new trust lands outside the Park; however, that has not yet happened.

The Timbisha Shoshone optimism is based primarily on economic prospects. Economic actions contemplated by the Timbisha Shoshone are focused on enhancing the economic, cultural and social well-being of the tribal members (Watt, 1996). The centrepiece of the economic development plan is a dramatic enhancement of the tribal village at Furnace Creek. In 2008, the Tribe completed building a new government and administrative centre in the tribal village. Plans for the future include a health clinic, library, cultural centre, retail outlet for Timbisha Shoshone art and a desert-style lodge for visitor accommodations. Economic development concepts on tribal lands outside Death Valley National Park range from a fueling station and convenience stores to campgrounds, motels and restaurants, and to enhancing natural resources such as springs and big game animal populations. Economic growth has been slow to come. Nevertheless, the Tribe is continuing with programmes to foster more economic development.

Since passage of the Timbisha Shoshone Homeland Act, the Tribe has joined other bands of the Western Shoshone peoples and testified several times before the United Nations Committee for the Elimination of Racial Discrimination. Testimony by tribal members has been on behalf of land and water claims by the Western Shoshone and other American Indian tribes (J. Kennedy, pers. comm., 2007, 2008). The actions before the United Nations indicate an increasing sense of power and opportunity: the Tribe is taking advantage of participation on a regional and even global scale.

Passage of the Timbisha Shoshone Homeland Act has led Death Valley National Park to be much more aware that the Timbisha Shoshone should be consulted on almost all issues related to NPS actions regarding cultural and natural resources in the park. Death Valley National Park and the NPS learned they could not rely on judicial means and dominant decision-making in order to continue with the old ways of participating with the Tribe (L. Greene, pers. comm., 2007). That realization is a change; for many years, the Park dealt with the Timbisha Shoshone almost exclusively on administrative matters and the relationship was often very tense (L. Greene, pers. comm., 2007).

Conclusion

The Timbisha Shoshone's ability to develop a complex and effective social participation network made their dream of homelands and water resources a reality. In addition to having justice and historical right on their side, the Tribe's leaders told a compelling and powerful story. As a result, for the first time in more than a century the Timbisha Shoshone have the power and authority to manage their own land and their own water resources. In addition, they have the power and authority to influence management actions by Death Valley National Park. The successful negotiation of the Homeland Act is most likely a contributing factor in the Timbisha Shoshone willingness to expand their influence, as embodied by their testimony before the United Nations.

It is impossible to visualize similar results if the Tribe had not been able to marshal effective participation from others on their behalf. The support and active involvement of other Indian and non-Indian organizations was very beneficial for the Tribe. Indeed, the Tribe would not have been successful without support from these organizations. No matter how passionate its leaders and how determined they were, they did not have the required expertise to engage the US government and regain rights to at least some of their historic land and water.

For the NPS, significant aspects of the Timbisha Shoshone Homeland Act include the recognition that:

- it is a negotiated settlement between two sovereign nations;
- it explicitly recognizes Timbisha Shoshone contributions to the history and ecology of the Death Valley region;
- it directs the NPS to manage the park in ways that accommodate Timbisha Shoshone cultural practices;
- most importantly, it provides a measure of justice to a people who have been culturally, economically and socially marginalized for over 150 years. (L. Green, pers. comm., 2007; P. Parker, pers. comm., 2007)

It is also impossible to visualize successful negotiation and passage of the Timbisha Shoshone Homeland Act without the leadership demonstrated by the Timbisha Shoshone and the NPS during the 1998–2000 negotiations. Strong leadership enabled the Tribe to give up claims to huge tracts of land in and around Death Valley National Park. It also enabled the NPS to give up the concept that no land within the Park would ever become a homeland for the Timbisha Shoshone. Freely paraphrasing Maria Kaika (2004), effective leadership was responsible for bringing land and water to the Timbisha Shoshone, while keeping content all parties involved in the Timbisha Shoshone Homeland Act and Water Agreement negotiations.

References

California Desert Protection Act (1994) Public Law 103-433, 103rd Congress, 2nd session, Washington, DC

Jenkins, M. (2008) 'Seeking the water jackpot', *High Country News*, 16 March, p10, Paonia, CO

Kaika, M. (2004) 'Water for Europe: the creation of the European Water Framework Directive' in J. Trottier and P. Slack (eds) *Managing Water Resources, Past and Present,* Oxford University Press, Oxford

McCool, D. (1994) *Command of the Waters, Iron Triangles, Federal Water Development, and Indian Water*, University of Arizona Press, Tucson, AZ

McCool, D. (2002) *Native Waters, Contemporary Indian Water Settlements and the Second Treaty Era*, University of Arizona Press, Tucson, AZ

National Park Service (1997) 'Recommended next steps', *Timbisha Shoshone Reservation Suitability Study*, Pre-decisional Document, Death Valley, CA

National Park Service (1998) *Ground Water Resource Issues of Death Valley National Park Related to Timbisha Shoshone Proposed Reservations,* W. L. Werrell (ed), Death Valley, CA

Reisner, M. (1986) *Cadillac Desert, The American West and its Disappearing Water'*, Viking Press, New York, NY

State of Nevada, Division of Water Resources (2000) Letter to Superintendent, Death Valley National Park, Death Valley, CA

State of Nevada, Office of the State Engineer (2007a) 'In the matter of Applications 54003 through 54021, inclusive, filed to appropriate the underground water of the Spring Valley Hydrographic Basin (184) White Pine County, Nevada', Ruling 5726, Carson City, NV

State of Nevada, Office of the State Engineer (2007b) 'In the matter of Applications 59352, 62529, 66072, 66077, 66078, 66079, and 66071 filed to appropriate the public waters of an underground source within the Amargosa Desert Hydrographic Basin (230) Nye County, Nevada', Ruling 5750, Carson City, NV

Timbisha Shoshone Homeland Act (2000) Public Law 106-423, 106th Congress, 2nd session, Washington DC

Timbisha Shoshone Tribe (1995) *Outline of a Proposed Study to be Conducted by the United States Department of Interior in Consultation with the Timbisha Shoshone Land Restoration Committee of Lands Suitable To Be Reserved for the Timbisha Shoshone Tribe as provided for in The California Desert Protection Act Of 1994*, Public Law 103-433 Tribal Headquarters, Death Valley, CA

Timbisha Shoshone Tribe (1996a) Letter to Greg Thompson, Resources Staff Chief, Bureau of Land Management, Ridgecrest, CA, Tribal Headquarters, Death Valley, CA

Timbisha Shoshone Tribe (1996b) Letter to President Clinton, Tribal Headquarters, Death Valley, CA

Timbisha Shoshone Tribe (1996c) Letter to Richard Martin, Superintendent, Death Valley National Park, Tribal Headquarters, Death Valley, CA

Timbisha Shoshone Tribe (1996d) Letter to Secretary of Interior Babbitt, Tribal Headquarters, Death Valley, CA

Timbisha Shoshone Tribe and National Park Service (2000) *Water Agreement Regarding Federal Reserved Water Rights for the Designated Tribal Trust Lands*, Death Valley National Park, Death Valley, CA

Timbisha Shoshone Tribe and US Department of the Interior (1999) *The Timbisha Shoshone Tribal Homeland, a Draft Secretarial Report to Congress to Establish a Permanent Tribal Land Base and Related Cooperative Activities*, Washington, DC

US Department of the Interior, Office of the Secretary (1996) Letter to Timbisha Shoshone Tribe, Washington, DC

Watt, D. E. (1996) *Water Supply and Water Development Potential for the Proposed Timbisha Shoshone Trust Lands*, US Bureau of Reclamation, Lower Colorado River Region, Boulder City, NV

Western Water Policy Review Advisory Commission (1998) *Water in the West: Challenge for the Next Century*, National Technical Information Service, Springfield, VA

Wilkinson, C. (1992) *Crossing the Next Meridian, Land, Water and the Future of the West*, Island Press, Washington, DC

Wilkinson, C. (2005) *Blood Struggle, The Rise of Modern Indian Nations*, W. W. Norton & Company, New York, NY

2

For Whom the Turbines Turn: Indigenous Citizens as Legitimate Stakeholders in the Brazilian Amazon

Louis Forline and Eneida Assis

Introduction

Hydroelectric dams have frequently been slated as a be-all and cure-all for developing countries. Proponents for dams argue that there would be an almost limitless and enduring supply of energy to build infrastructure and provide the necessary goods and services that would accrue from power generation. In developing countries, such as Brazil, the growing demand for energy would make the use of its large supply of hydrologic resources seem as only a natural and logical step in converting this potential into a much needed power supply, responding to the demands of its industry and growing population. Yet this position has been challenged by many, as the impacts of dam building have often been downplayed and the benefits to be gained from hydroelectric power can be limited and selectively earmarked for determined segments of the population. As the voices of local stakeholders have become more prominent in the wake of heavy development, the unintended and unforeseen circumstances of dam building and power generation have been revealed.

In Brazil's Amazon region social movements have emerged in the wake of dam building and their impacts are steadily coming to the fore. Additionally, a number of non-governmental organizations (NGOs) have been created in response to the large-scale development currently under way in this region. Both peasant and indigenous peoples of the Amazon have found a voice through a series of mechanisms that have developed since Brazil's reopening to democracy in the 1980s. As the environmental impacts of development have put many of these communities at risk, these projects have invariably placed such peoples on the map. In this chapter, the mechanisms involved in the engagement of these peoples will be discussed and we will consider how they are currently manifesting their concerns vis-à-vis hydroelectric projects. As a prime case study we will examine the issues surrounding the proposed hydroelectric projects programmed for the Xingu river near the city of Altamira, Pará state, Brazil (Figure 2.1). This region encompasses the mid to lower courses of the Xingu basin, an area which is home to nine different indigenous ethnic groups. These groups depart from the general category of generic Indian and represent a range of people, from those who have little contact with mainstream society to those who have mingled with Brazilian national society and reside in urban areas. While all of these groups became the key stakeholders in the controversy surrounding the Kararaô complex of dams for the Altamira area, this latter group was perhaps the most decisive in stalling recent efforts to build the Belo Monte dam in 2001. Formerly not recognized as legitimate indigenous peoples by the Brazilian state, they have re-emerged as key players in deciding some of the development regimes in the region of Altamira. Thus, while still considered legitimate stakeholders vis-à-vis development issues, their categorization as indigenous peoples adds yet another dimension to the process of informed prior consent.

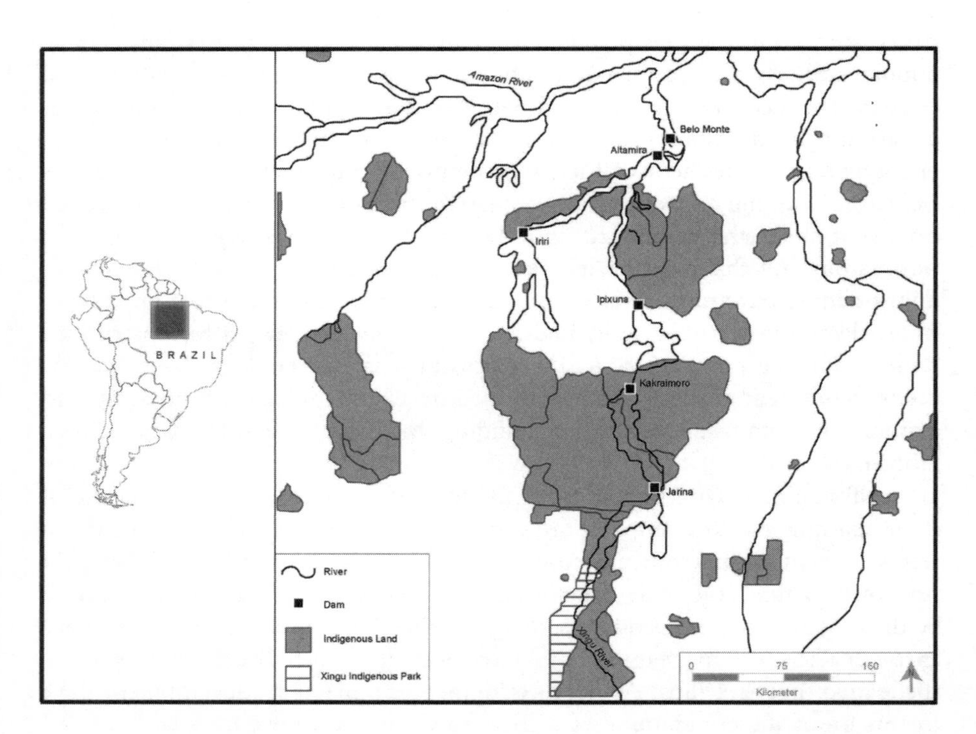

Figure 2.1 Brazil with inset showing site of proposed dam and affected areas

The Brazilian Amazon: Development and Potential for Hydroelectric Power

The Amazon region represents the world's largest freshwater reserve. The water volume flow of the Amazon River is 20 per cent of the joint volume of all rivers on earth. The Amazon River alone discharges 60 times more water per day than the Nile and 11 times more than the Mississippi. Similarly, many of its tributaries are major rivers in their own right and dwarf European rivers such as the Danube and Seine. This river network is rich in biodiversity and home to many peasant and indigenous communities, supplying a large gamut of resources for these people. In addition to its abundant supply of fish, this hydrological basin supplies nutrients for the traditional farming methods utilized along its banks. Seasonal silt deposits provide the nutrient base for the soils upon which slash and burn horticulturalists clear land and plant and harvest crops. Similarly, riverine areas represent a very rich ecological zone of the game animals that feed and breed there, and these areas draw together many resources that are managed and utilized by the inhabitants of the region. Even the interfluvial zones of the Amazon basin are cut by a series of seasonal streams that provide a multitude of resources for indigenous and peasant populations. Similarly, the black water ecosystems of this region are teaming with biodiversity when compared to temperate climate rivers, although they are considered 'rivers of hunger' in the Amazon (Moran, 1991). Additionally, rivers of the Amazon region are the main conduit for transportation for local populations. Despite the massive road building projects of the 1970s, river courses in this area provide its peoples with a natural transportation infrastructure to sustain their livelihoods and travel between communities to visit relatives and peddle their produce in local towns and cities.

Rivers of the region are also imbued with many myths regarding the origins of the earth's people and its creatures. Many indigenous legends persist till this day in various forms and some observers consider that a number of myths help safeguard natural resources from overpredation and exploitation. Thus, these myths serve as a 'primitive Environmental Protection Agency' for these peoples (McDonald, 1977, p734). For example, the concept of *panema* holds that greedy hunters would incur supernatural sanctions for overkilling animals and, in turn, would be cursed with bad luck. As such, observance of taboos and ritual protocol would keep hunters in favour with animal spirits and other animate forces of nature.

According to some accounts, most of the indigenous people in the Amazon region resided along its watercourses prior to the arrival of Europeans. As previously noted, it is not too startling that greater numbers of people would be drawn to rivers, streams and lacustrine areas, given the bounty of resources that converge upon these ecological zones. Thus, even in times past it is speculated that these areas were disputed between pre-Columbian inhabitants. In this scenario, populations residing in interfluvial zones would have been residing there not necessarily as a matter of choice but because they were forced out of riverine areas as a result of competitive exclusion from stronger and more established groups (Sponsel, 1989). Later, during the time of conquest, European powers mapped over previously established indigenous communities settled along river courses and, in fact, a number of present-day cities of the Amazon region were formerly inhabited by large indigenous communities, such as Belém, Manaus and Altamira. As Europeans settlers and their descendants quickly discovered, riverine areas are choice spots not only for their high degree of biodiversity, resources and potential for transportation, but are also advantageous for military positioning. Provided that one has the military clout to secure such areas, they can serve as a sentinel zone for the dominating powers. While in previous times weaker groups found themselves vulnerable if they were exposed in such open areas, since colonial times disease and military might enabled Europeans and their descendants to seize these locations and maintain them under their control. (Coupled with forced overthrows, disease decimated substantial portions of the indigenous Amazonian population, opening the way for European occupation and settlement of fertile lands along river courses.)

In recent times the Amazon region has been coveted for reasons of national security, geopolitical concerns and natural resources. Since the 1960s the Brazilian government has engineered projects aimed at settling the region to ensure its sovereignty over a resource rich area that is still sparsely populated and larger than Western Europe. During the heyday of Brazil's economic miracle of the 1970s the government energetically jockeyed for position to establish land settlement projects for people from other areas of Brazil. The hallmark project of this era, Projeto de Integração Nacional (PIN, or Programme for National Integration) was touted as 'land for men for men without land'. With this project and other ventures Brazil's then military government hoped to accomplish two things:

1 It would help diffuse civil unrest in other regions by siphoning off people to the Amazon and deeding out parcels of land.
2 As development planners pointed out, settlement in the region would occupy an area thought to be targeted by foreign interests. Official slogans such as 'integrar para não entregar' (integrate the country lest we give it away) became popular mottos at this time.

While the short-lived economic miracle fizzled out, settlement in the Amazon region proceeded apace and large-scale projects were begun. In addition to government and private colonization ventures, road building and massive mining projects were established. Along with these ventures a number of illegal activities surged in the wake of officially sponsored development, particularly gold mining, land grabbing and logging. Given the large volume of water present year-round in this region, developers saw in the Amazon a large potential for generating hydroelectric power.

All of these projects invariably affected Brazil's first inhabitants, its indigenous peoples. The PIN project impacted approximately 96 different indigenous groups, or about 56 per cent of the known Indian communities of the Brazilian Amazon (Ramos, 1984). For its part, the Carajás Mining Project (PGC) impacts, directly and indirectly, 40 indigenous communities along its 900km railway (Treece, 1987). Likewise, the Balbina dam forced contact with a number of indigenous groups who had few or no dealings with Brazilian mainstream society. As Sheldon Davis (1978) pointed out many of these groups turned out to be victims of the Brazilian economic miracle of the 1970s, but as development steadily progressed one project would finally mark a turn-around for Brazil's indigenous peoples in the face of unbridled and unquestioned development.

The Kararaô Project

Since the 1960s Brazilian power companies had examined the possibilities of dam building in the Amazon. Despite its distance from Brazil's main urban centres the Amazon could eventually be linked up to the national energy grid once dams were built and power transmission lines were in place. Two small-scale dams were built in the 1960s and 1970s, Curuauna and Balbina, respectively. Curuauna currently powers the small city of Santarém located near the mid-lower courses of the Amazon River and generates approximately 30MW for this community of about 250,000 people and its vicinity. For its part Balbina supplies power to the city of Manaus situated on the confluence of the Rio Negro and Amazon. This dam generates 250MW, supplying electricity to a city of nearly 1.4 million people and neighbouring communities. Dam building continued and in the 1980s the Brazilian government finished construction of the Tucuruí dam generating roughly 4000MW for the city of Belem and local industries and communities located in the eastern Amazon. In addition it is also connected to the national energy grid, and provides electricity to other regions of the country when they experience shortages. All of these projects were instituted during the period of Brazil's military regime, which governed the country from 1964 to 1985. They were established with the no-questions-asked policy instituted during this period of martial law.

Participatory decision-making processes were not contemplated in this phase nor were any impact assessments genuinely conducted to apprise the populace of the consequences of dam building. While the Curuauna dam's impacts were minimal in comparison to the other two dams, the only impact assessment assigned to it was conducted nearly 30 years after it went into operation (Forline, 2004). For its part the Balbina dam flooded nearly 2400km^2, including Indian and peasant areas, and emitted a significant amount of greenhouse gases during its early years of generation. Similarly, the Tucuruí dam formed a reservoir of over 2850km^2 even though it was projected to inundate an area less than half of that size. Moreover, it displaced almost 40,000 people and is one of the main reasons mounting resistance was levelled at hydroelectric projects. In response to these impacts, a social movement was created a few years later culminating in the creation of an NGO called *Movimento de Atingidos por Barragens* (MAB or Movement of People Impacted by Dams).[1] As Robert Goodland aptly pointed out, it was evident at that time that the Brazilian government adhered to the heavy-handed policy that 'you can't make an omelette without breaking eggs' (Goodland, 2005, p183).

In the wake of these dams another hydroelectric project was scheduled for the Xingu River near the city of Altamira. This occurred at a time when Brazil's military government had recently relinquished its mandate and power was gradually being turned over to civilian rule. Social movements were on the rise and many formerly disenfranchised groups were steadily finding a voice in matters pertaining to their well-being. However, the transitional government was still planning a number of projects without due consent from the public. By the 1980s the Brazilian government began conducting impact studies to assess the feasibility of two dams (Babaquara and Kararaô) planned for the middle and lower courses of the Xingu River, near the city of Altamira in the state of Pará.

The state-owned power company, Eletronorte (or Centrais Elétricas do Norte do Brasil), was responsible for plans to meet Brazil's energy needs by building hydroelectric dams in the Amazon. The earlier dams it built in this region drew much criticism for mismanagement of funds and poor planning that led to unforeseen consequences, not to mention the little concern it had for the people impacted by its projects. As noted above, for example, the Tucuruí dam built on the Tocantins River in the eastern Amazon formed a huge, 2850km^2 reservoir, more than twice the size anticipated by Eletronorte. The forced relocation and dispersal of many riverine dwellers and indigenous peoples to the peripheries of regional urban centres led to an increase in poverty and regional violence. Other environmental impacts caused by the dam include diminished fish production, a proliferation of mosquitoes and the diseases they carry, and acidification of the reservoir water (Fearnside, 1999). Over half of the electricity generated in Pará State (primarily from the Tucuruí dam) is sold to other Brazilian states and multinational corporations (Anonymous, 2001). Although ordinary citizens pay (relatively) higher prices for their electricity, Eletronorte continues to finish each fiscal year in the red, a loss absorbed by the Brazilian state. Again, this dam was planned and built during the days of Brazil's military regime without public support or the benefit of impact studies to assess its social and environmental feasibility.

Against this backdrop nobody was surprised that Eletronorte's project for building the Babaquara and Kararaô dams sparked such a large protest in the late 1980s after people learned of their plans. The protests were partly based on the power company's poor track record and partly because of the heavy, top-down authoritarian approach adopted by the Brazilian government, even if it was presumably transitioning towards civilian rule. Members of local communities that would be impacted by the dams were determined to have their voices heard.

The Brazilian government was depending on the World Bank for a US$500 million loan to build the dams. In the meantime, anthropologist Darrell Posey was hired by Eletronorte to help with the impact studies. When Posey attended a conference at the University of Florida to discuss the impact of the dams on indigenous people, he brought two Kayapó Indians with him, Payakan and Kube-í. Later, they flew to Washington, DC to discuss the project further with World Bank officials, informing them that the indigenous communities that would be impacted were not consulted. After meeting Posey and the two Kayapó, the World Bank decided to suspend full disbursement of its loan to the Brazilian government. The Brazilian government responded with a lawsuit against Posey, Payakan and Kube-í charging them with meddling in its internal affairs. It was ironic, indeed, that members of Brazil's first peoples were being charged with interfering in Brazilian internal matters. When the press picked up the story a worldwide protest ensued against the Brazilian government for pressing charges against Payakan and Kube-í and the building of the dams. Members of Brazil's civil society along with international human rights and environmental organizations mounted a campaign against the hydroelectric project, culminating in a gathering at Altamira in February, 1989.

Representatives for Brazil's indigenous groups also gathered and staged events that included a face-to-face encounter between a Kayapó woman, Tuíra, and Eletronorte's then director, José Antônio Muniz Lopes. The picture of Tuíra whisking a machete across the director's face circled the globe and became symbolic of the victory scored by Brazilian Indians. Although the Brazilian government dropped charges against Posey, Payakan and Kube-í before the protest, the gathering in Altamira succeeded in bringing about the cancellation of the Babaquara and Kararaô dam projects.

Political Gains and Fallout of the Kararaô Protest

One positive outcome of the Kararaô outcry (Grito Kararaô) was to put Brazil's indigenous peoples on the map. Additionally, Indians better understood the power of the media to shape public opinion, and gained confidence in their ability to stand up for their rights. After Altamira, greater global attention was given to the disregard for environmental and human rights by the Brazilian military government during the decade of destruction in the 1980s. Brazil's new constitution, ratified in 1988, was another major event that helped reshape many policies regulating the environment, traditional knowledge and the democratization of public decision-making processes.

However, there was a downside. The Brazilian government fine-tuned its policies and laws regarding research authorizations so that foreign researchers encountered more difficulties in carrying out research, especially in the Amazon region. The government justified its regulations citing concerns over biopiracy, intellectual property rights and unequal exchanges of scientific information. This stalled efforts to initiate new research projects and discouraged many ongoing research partnerships between Brazilian institutions and their foreign counterparts. The government also formed an investigative committee in the Brazilian Congress to watch over the activities of NGOs. Members of Brazilian Congress who instituted this investigative committee claimed that many of these NGOs were nothing more than fronts for foreign interests with designs on the Amazon and its abundant resources. Regardless of whether a given NGO dealt with human rights, education, the environment or indigenous issues, all of these were considered pawns for outside interests to manipulate local actors.

NGOs have risen on a worldwide scale, picking up the slack where governments have retreated or been remiss in providing goods and services for their respective populations. While some governments have both praised and encouraged the establishment of NGOs to fill in the gaps left open as a result of budgetary cutbacks or a policy reorientation, some countries, such as Brazil, have not received the expansion of NGOs well. This backlash came primarily from conservative members of Brazil's Congress and local interests who claimed that outsiders had long been meddling in internal affairs and were stalling the country's own efforts to achieve autonomy with its development plans. Some even feared that there were conspiracies to internationalize the Amazon region.

Brazil's Energy Scenario and Belo Monte Dam

Since the Kararaô outcry, the Brazilian government has continued its attempts to solve its energy problems by redesigning plans for the middle and lower Xingu region. Eletronorte resubmitted plans for a new hydroelectric dam on the Xingu in 2000. The new project, Belo Monte dam, would flood an area significantly smaller than the Babaquara and Kararaô dams. On paper, the new project appeared attractive, especially in light of Brazil's energy crisis in 2001, when many Brazilian cities and states suffered blackouts and were forced to ration electricity. At that time, the administration of then president Fernando Enrique Cardoso openly admitted that it had not anticipated this scenario, which was brought on by a series of unexpected droughts and poor planning. Several energy experts claimed that without new energy sources the country would undergo an even more severe shortage by 2010.

The project was presented as ecologically and politically correct because an area of only 400km^2 would be inundated compared with the 7500km^2 reservoir anticipated for the previous complex of dams. According to the new calculations, the only indigenous area which stood to be affected by the Belo Monte project would be a small indigenous reserve inhabited by the Juruna Indians (Paquiçamba Reserve), downriver from Altamira, where the Xingu takes a big turn (*Volta Grande do Xingu* or the Xingu's Big Bend) before proceeding northwards and emptying into the Amazon near the town of Porto de Moz. Currently, about 80 people live on this reserve of 4348ha.

Eletronorte planned to take advantage of this area's natural topography and redirect the Xingu across its big bend, thus forming a small reservoir just north of the Paquiçamba reserve. This relatively small man-made lake would feed the 20 turbines at the future power plant near Belo Monte and generate approximately 11,000MW of energy. The hydroelectric project would be the second largest in Brazil and the fourth largest in the world. Energy production would also be very efficient. Officials estimated that Belo Monte would generate 27,500kW of energy for every square kilometre of flooded land (Goodland, 1995).

According to Eletronorte, the Juruna's reserve would not suffer any direct impact from the dam. The dam's main engineering feat would be to divert the Xingu River through two canals to the power generation plant at Belo Monte. The reservoir would be small and far enough away from the Juruna reserve not to cause any problems. Eletronorte claimed that the only people to be displaced would be a small group of non-indigenous landholders settled by the Instituto Nacional de Colonização e Reforma Agrária (INCRA), the Brazilian Land Settlement and Agrarian Reform Institute. But diverting the Xingu would affect all riverine communities downriver from the dam, particularly in the region of the Big Bend. The river's volume around the Big Bend would be greatly reduced as most of the Xingu's water would be diverted into the energy-generating canals feeding the power plant, and ecological cycles attuned to rainy and dry season hydrology would be altered, invariably affecting the livelihoods of many riverine dwellers along this stretch of the watercourse.

Eletronorte commissioned an impact study from the Universidade Federal do Pará (UFPA) and the Goeldi Museum to evaluate the social and environmental impacts of the proposed dam. The company allocated approximately US$1.3 million to conduct these studies and hired Fundação de Amparo ao Desenvolvimento da Pesquisa (FADESP), the University's Grant Administration Office, to administer this contract. The funds represented about one-fifth of the University's annual budget. The final details of the contract were negotiated with FADESP in October 2000 and Eletronorte convened all researchers to a meeting where the guidelines for the impact studies were presented. The assessment team included researchers from the natural and social sciences in addition to ad hoc consultants. Eletronorte briefly described the project to researchers who were, in turn, expected to explain it to local communities in and near Altamira as they evaluated the potential impacts of the dam. The power company had shown general aspects of the project to local politicians, businessmen and the regional elite, but not to the general population.

Surprisingly, no formal contract was drafted between Eletronorte and individual assessment team members. The power company claimed that the general provisions of its contract with FADESP were all-encompassing and committed researchers to the stipulations laid down in the project's terms of reference. Researchers submitted individual research proposals indicating the equipment and financial resources they needed to conduct their studies. This situation created the impression among many researchers that they were working under an informal agreement that would only be honoured if they followed the power company's guidelines. After the impact studies were under way, researchers were expected to periodically meet with Eletronorte officials and present progress reports. Later, Eletronorte directed researchers to submit periodic reports to its review committee and ad hoc consultants who would edit research findings and draft the final report. In turn, this report would be submitted to the Secretaria de Ciência e Tecnologia da Amazônia (SECTAM) of Pará state, its Environmental Secretariat. The Environmental Secretariat would then assess the magnitude of social–environmental impacts the dam would incur and act as the de facto licensing agency if it deemed the project feasible.

In an attempt to further control the research, members of Eletronorte's staff periodically accompanied researchers to their study sites to monitor fieldwork and provide advice, as they claimed to have amassed vast experience from prior impact studies. Later, the power company asked researchers to provide them with their raw data. Many researchers refused to comply with this request for ethical reasons. Researchers working with human subjects wanted to maintain the anonymity and confidentiality of the communities where they worked. During periodic meetings, researchers asked the company to provide more information about the dam, as full disclosure of the project's details had yet to occur. For example, a number of researchers asked for details regarding the possibility of building a complex of dams instead of only one, as planned (see Figure 2.1). Eletronorte officials politely skirted these questions, although the company later renamed the project *Complexo Belo Monte* (Belo Monte complex), dropping its previous name of *Usina Hidrelétrica Belo Monte* (Belo Monte hydroelectric plant). Renaming the project implied that the power company was planning to build a series of dams on the Xingu. As many of the researchers were from the Amazon region, they asked whether any of the energy generated from the dam would be provided to local communities. The research staff expected that most of the energy would go to Brazil's large urban centres in the south, but wanted the company to know that only 60 per cent of the rural households in Pará state have electricity, much of this being very unreliable.

Researchers were also irked by Eletronorte's eagerness to conclude the impact studies in a hurried manner. The terms of reference stipulated that the impact studies should run the course of one full year, to accurately portray the Xingu's biophysical and social characteristics during the wet and dry seasons, yet the power company ignored this clause and urged researchers to finish their tasks before the prescribed time. Eletronorte representatives justified their actions by claiming that the one-year period should incorporate both the research and submission of the final report. In truth, the power company wanted the impact studies concluded as quickly as possible in order for SECTAM to fast-track approval of their project. The political climate was ripe for Eletronorte to have its project approved, as it was one of the federal government's top priorities. Eletronorte also wanted to secure approval before the upcoming presidential elections in October 2002.

Belo Monte Encounters Resistance

Approval for this project seemed almost certain until many leaders of local social movements along the lower Xingu united and organized protests against the dam. Eletronorte had repeated its earlier error with the public. No public hearings were held to discuss the project and people were angry about being left out of the decision-making process. The power company tried to placate the public, stating that hearings would be held after the impact studies were concluded. Evidently, the power company was confident their project would be approved despite its frequent violations of rules and regulations. Discontent mounted among the local communities of Altamira and elsewhere. Political opposition to Fernando H. Cardoso's administration was increasing as presidential elections approached, and many researchers participating in the impact study were dissatisfied with the power company's procedures. The concerns of anthropologists working on the project grew in proportion to Eletronorte's negligence and unwillingness to comply with requests to treat the indigenous question with more seriousness and transparency. What followed was a series of meetings among Altamira's local NGOs, who in turn contacted the *Ministério Público Federal* (MPF, or Federal Prosecution Service) to suspend the impact studies. Local indigenous groups were informed of the power company's manoeuvres and were advised to merge with other social movements to protest the project.

Pará state's federal prosecutors then submitted legal arguments to a local district court demonstrating that Eletronorte had not followed appropriate procedures for conducting the impact study:

1 The power company did not bid out the contract for administering the impact study. Rather, the power company had informally approached the university, freely awarding the contract to FADESP. Given the international backlash of the Kararaô outcry of 1989, the power company wanted to draw as little attention as possible to its new project in hopes of avoiding controversy. Eletronorte countered the charge, claiming that it wanted to give regional Amazonian institutions a fair chance to participate in the impact studies. They also claimed that the university and FADESP were non-profit public institutions with the competence to handle the job of executing the impact studies.

2 The MPF claimed that since the Xingu River runs through two Brazilian states, Mato Grosso and Pará, it should necessarily be considered a 'national' river and not a single state river. Accordingly, it would be incumbent upon the Instituto Brasileiro do Meio Ambiente e dos Recursos Naturais Renováveis (IBAMA), the Brazilian government's federal environmental agency, to assess the impact study and decide the Belo Monte dam's feasibility, not SECTAM in Pará state. To this, Eletronorte had little to say other than countering that the impacts of the dam would only occur in Pará state. Despite adopting this position, it quickly urged all their impact study team researchers to register their names with IBAMA.

3 The MPF of Pará State also argued that that since indigenous communities would be impacted by the dam, congressional approval was necessary to authorize any development project that would affect their lands. This is mandated in Article 231 of the Brazilian constitution. Eletronorte's legal staff refuted this argument by claiming that the impact studies should not be interpreted as part of the project, and they added that their calculations showed the Juruna's reserve would only be indirectly impacted. Nowhere in Eletronorte's terms of reference were any of the indigenous communities studied considered to be directly impacted, despite the fact that more than nine ethnic groups reside in this general region, not to mention the urban Indians residing in Altamira, which according to estimates stood to be partially flooded.

In view of these arguments, the local district court ordered the immediate suspension of the impact studies. A legal battle ensued in the Brazilian court system over the next year. Eletronorte appealed the lower court's decision and won. Pará state's Federal Prosecutor's office then took its arguments to a higher court which reordered the suspension of the impact studies. Again, Eletronorte appealed, thereby placing the final decision upon the Supremo Tribunal Federal (STF, Brazil's Supreme Court), as what was in question were constitutional directives open to interpretation. While the impact studies were suspended, neither Eletronorte nor the research team was allowed to discuss the project or resume work. In an effort to get around this order while awaiting the court's decision, the company hired independent consultants to finish the study, many of which were unfamiliar with the peoples or natural characteristics of the region. This move was illegal as many of these ad hoc consultants acted primarily as 'yes men' to the power company and swiftly performed their surveys in a manner that would conform to their project. In this way, once the STF gave its decision their studies would be completed and work could proceed according to schedule. Much to its dismay, however, the STF sustained the arguments of Pará's federal state attorneys. In November 2002, the Brazilian Supreme Court's chief justice decided to maintain the lower court's ruling and Eletronorte was faced with a tough choice. The power company could appeal this decision, which was reached solely by the deliberations of the STF's chief justice. An appeal would force the issue to be examined by all 11 justices of Brazil's Supreme Court, which would have entailed a long and drawn out court process before arriving at a final decision.

Belo Monte and New Directions in Energy Policy

Eletronorte's quiet and elusive strategy for obtaining approval of the Belo Monte project backfired without the wide-scale national and international attention it feared. Instead, the power company encountered solid resistance from local social movements, indigenous peoples, NGOs, researchers and Brazilian state attorneys. Although comparatively little media attention was given to the fight over this project, the legal framework laid down by Brazil's constitution of 1988 was used to suspend the project and force the power company to once again redraft its future plans for the Xingu region. Thus, while outside interference could be cited by Brazilian nationalists as the culprit for undermining development efforts in the Amazon, as was the case of the Kararaô outcry, the Belo Monte project was strictly challenged by the state's internal mechanisms. The political opposition to Fernando H. Cardoso's neo-liberal policies also contributed towards cancelling the Belo Monte enterprise, as many leftist parties regarded his approach to development as environmentally harmful and bereft of social responsibility. In response, President Cardoso said that he would transfer the political decision about whether or not to build Belo Monte to the incoming administration. In November 2002, Cardoso's Social Democratic party lost the presidential elections to *Partido dos Trabalhadores* (Brazil's Worker's Party). The new Brazilian President, Luíz Inácio Lula da Silva (Lula), promised to develop alternative energy strategies during his presidential election campaign, yet recently his administration resurrected Belo Monte. In newer developments, the government approved prior studies, although not thoroughly completed, and has commissioned another group of researchers to perform the complementary impact studies needed to conclude project approval.

Ecologist Phillip Fearnside (2000) once remarked that many large-scale Brazilian development projects are akin to vampires. If a stake is not driven through their hearts they will certainly resuscitate, cloaked in a new language to come back to haunt the environment and politically disenfranchised communities that stand in their way. Members of the social movements that stalled Eletronorte's plans in Altamira are apprehensive about the newly elected government's about-face in regard to Belo Monte. While their victory against the power company may only be temporary, local people are gaining confidence and growing increasingly formidable. In the event the project is reinstated, they will have the juridical knowledge to contest any untoward action taken by the power company as evinced in their quiet yet solid victory against Eletronorte. Many of these communities now have their own associations and represent themselves in political negotiations, often without the use of mediators. Thus, any future project contemplated for the region will have to defer to their decisions and incorporate them in the impact studies.

The Vampire Reawakens

These two standoffs against Eletronorte were remarkable and showed the resolve and organization of indigenous peoples in the face of unbridled development. Certainly, as many critics of these movements have maintained, their victories would not have been accomplished without the collaboration of interlocutors and sympathizers to their cause. And it is interesting that while globalization in the main has been criticized for overwhelming indigenous peoples, its cross-currents have actually helped put them on the map. Mass media has been critical in giving indigenous peoples a platform to voice their concerns and the increased use of the internet has also provided them with another avenue of expression. Yet at the same time many apparent faults were exposed in the procedures for social–environmental impact assessments and forced the Brazilian government to rethink its posture and methods in proposing projects to the public. The much-lauded informed prior consent so common in development discourse seemed to be nothing more than a façade by the power company going through the motions yet making no serious effort to enlist the participation of stakeholders.

Participation in development projects by all interested parties requires much discussion, transparency and political will. During Brazil's political reopening, a weekly news magazine (*Veja*) once remarked that democracy entails a painful process (*a democracia doi*): that is, engaging stakeholders in discussions, creating participatory models and following through with them is a painstaking task. It requires a long and drawn-out dialogue and mechanisms that ensure that no one is slighted in the decision-making process. It is difficult for disenfranchised people to articulate their concerns and interlocutors have to be attuned to this deficit in order to succeed in representing stakeholders and guaranteeing that their opinions, participation and, sometimes their votes, are secured. Evidently, there is no final word as to how this policy should be enacted and development facilitators have been (and continue to be) struggling with this issue. This development creates a tyranny in both directions – one from project proponents and government officials, while the other hails from stakeholders and their interlocutors (Cook and Kothari, 2001).

The current administration of President Lula has received mixed reviews, both internally and from abroad, in terms of its human rights record and policy towards the environment. Previously hailed as the people's party, advocating equal rights for the downtrodden and oppressed, some observers claimed that the administration's performance in relation to Brazil's indigenous peoples and the environment has not delivered up to expectations. His administration was lauded for appointing Marina Silva, a native Amazonian, as Minister of the Environment, yet her effectiveness in terms of formulating policy has been largely neutralized by other ministries deemed more important in terms of the government's political and economic agenda. A heavy development agenda proceeded, drawing protests from internal and overseas critics. The government agency charged with administering indigenous affairs, Fundação Nacional do Índio (FUNAI), went through two directors during Lula's first term in office, and currently has a third director in place. And as proposed plans for large-scale development proceeded, the government openly criticized the lengthy procedures that the Ministry of Environment tasked developers to embrace in applying for permits and conducting environmental impact assessments.

Government officials claimed that approval for proposed projects incurred excessive bureaucratic legwork, undermining any hopes of expediting much needed development. Furthermore, the Brazilian government found that

IBAMA had too much authority and autonomy in dealing with development projects. As a result of this frustration, Lula's administration attempted to gut IBAMA and diminish the powers of the Ministry of Environment. It was indeed ironic that the workers' party, that had for so many years championed the cause of under-represented people and the environment, and criticized previous governments for imposing authoritarian rule, now positioned itself as railroading projects it formerly had challenged.

In view of all of these developments, the Lula administration is reviving the Belo Monte project. Before coming to power, the workers' party was one of the main opponents of this dam yet now that it can score political points through growth and development it has advocated implementing several projects in the Amazon region. Again, picking up on the discourse of growing energy demands, the Lula government echoed the alarming warnings of the previous administration by stating that if it does not act quickly enough in anticipating shortages Brazil will experience countrywide blackouts by 2010 and some observers claim that such a scenario could occur even earlier (Anonymous, 2007).

In the meantime, Eletronorte worked steadily at courting the community of Altamira. Although it was handed a second defeat when it proposed the Belo Monte dam, the power company campaigned heavily to sway the residents of Altamira and neighbouring communities. It launched educational campaigns, held town meetings, lobbied the political and business elite of the community and lavished the city with donations in an attempt to show that it was acting in its better interests. Not until the current administration proposed reviving the project did it openly mention the prospects of building the dam. It set up a local headquarters in Altamira and fanned out in many directions, even setting up a display for passersby to walk through a model of the Belo Monte Project (see Figure 2.2). Meanwhile, local politicians did not reject any subsidies donated by the company and relied on its presence to bolster local infrastructure and services.

This approach of seducing the community would make it hard to reject any proposal Eletronorte would eventually have in reiterating its request to build Belo Monte. In the meantime, local indigenous groups were also being induced to accept Eletronorte's presence and donations. This divided some of the members of the greater Altamira indigenous community, a schism the power company had hoped to achieve. Earlier, during the impact studies, Eletronorte's person in charge of the company's indigenous affairs, a former FUNAI agent, also urged that any meetings with the indigenous communities should be conducted separately, among each individual ethnic group. His claim was that if all indigenous communities were brought in for a face-to-face meeting to discuss the benefits and impacts of the Belo Monte dam that this encounter would only inhibit some groups, and old rivalries would resurface, thus undermining any hopes of achieving a consensus. However, this was a manipulative strategy urged by the power company to individually induce each group to accept the project. The ideal meeting agenda would have been to meet

Figure 2.2 Eletronorte model of Belo Monte project on display for passersby

separately and then collectively, to exchange ideas and voice concerns and desires, independent of Eletronorte's presence. This approach backfired however, as many of the indigenous stakeholders found it lacking in transparency. They and their interlocutors responded by stating that meeting with each group on an individual basis would only stir feelings of suspicion and fuel distrust among them. Yet, unfortunately, some indigenous people were eventually enticed into accepting the power company's donations and advocated for the construction of the dam. Others still accepted donations but never made any promise of whether they would lobby in favour of it. This latter group was actually advised by interlocutors to proceed in this way in order to receive the maximum amount of benefits possible, without committing to any particular agenda, so that they would not come under any obligation to compromise themselves or their communities. Yet continued favours and presents are creating a potential for a relationship of dependency as the Indian Service does not have the resources to furnish all of the necessary goods and services to local indigenous communities.

Indian Identity and State Recognition in the Face of Development

While a stalemate was achieved by the indigenous communities of Altamira another issue still looms large and keeps a number of Brazilian Indians from achieving full recognition as legitimate citizens. Indigenous identity in Brazil is as complicated as the way Brazilians apply racial and ethnic classification. This is an ambiguous domain and anthropologists have been wrestling with the issue for many years. Contrary to the method of classifying people in the US, in Brazil there is little consensus in racial and ethnic categorization (Harris, 1970). Although Harris' data were later shown to have more orderliness than previously imagined (Byrne and Forline, 1997), there is so much ambiguity in this domain that people can negotiate their own identities according to different social situations.

It has been observed that Brazil has lots of racial harmony yet the ideal of having light-coloured skin still prevails (Chasteen, 2006). While the US may be criticized for its rigid form of classification, Brazil's system of categorization, for its part, has been considered a form of forced assimilation. It is interesting that it is on this point where the US and Brazil converge. The US has, indeed, been largely observed to have an inflexible classification system for people of African descent, but it applies another standard for classifying indigenous peoples. As indigenous peoples are the first inhabitants of the New World, their just claim to land has always been disputed by the US government. The Bureau of Indian Affairs has often demanded a blood quantum for American Indians to prove their indigenity. Therefore, for someone to be considered an Indian, that person needs to demonstrate that at least one of their grandparents belongs to a federally recognized tribe.

Before Brazil's current constitution was enacted in 1988 indigenous peoples were steadily induced and coerced to integrate into mainstream society. Integration in this vein meant relinquishing one's identity either passively or forcibly and accommodating to the agenda of the nation state. Invariably, this assimilationist policy eliminated indigenous status for many Brazilian Indians living outside of reservations. In this situation, most lose their Indian status and find themselves trapped on the lowest rung of the Brazilian socio-economic ladder. Thus, they become hostage to total disenfranchisement and cannot participate in decision-making processes imperative to their well-being, even facing the larger challenge of being rightly recognized as indigenous citizens and stakeholders.

On the other hand, Brazil's indigenous peoples located in reservations, although recognized as legitimate Indians, are still considered wards of the state. In this regard, they are under a system of tutelage that still does not consider them full-fledged citizens capable of making enlightened and rational decisions. Given this status, their affairs are managed under the auspices of FUNAI. The positive aspect of being recognized as an indigenous citizen in Brazil is that there are special rights and privileges that accrue. These rights include land, health care, subsidized productive activities such as farming, differentiated education and a basic infrastructure to maintain their livelihoods. Yet, in the main, all of these benefits are administered through FUNAI.

In contrast to other countries of Latin America such as Peru, Guatemala, Bolivia and Ecuador, Brazil does not have a large indigenous population. In all, Brazilian Indians comprise less than one per cent of Brazil's total population of 193 million people. Yet after reaching an all-time low of 100,000 people at the turn of the 20th century, they rebounded as a result of increased immunity to illnesses and disease, health care and better representation in advocating for their rights. However, there is still an ongoing battle in determining Brazil's indigenous population and recent demographic figures are disputed. For its part, the Instituto Brasileiro de Geografia e Estatística (IBGE, Brazil's Institute for Geography and Statistics) claimed that in the last census it counted over 700,000 Indians. FUNAI, however, claims that approximately 400,000 is a more accurate number.

FUNAI has resisted acknowledging the high count reported by the IBGE stating that these numbers are inflated and that many people enlisted on tribal rolls are mixed bloods and opportunists trying to gain the advantage of Indian status. Many of the people not counted by the state's Indian Service are indigenous peoples living in urban areas. FUNAI has resisted acknowledging urban Indians as the agency cannot keep track of people with mixed ancestry and also finds them more difficult to deal with as many of them are more aware of their rights and are more aggressive in asserting them. But increased pressure from social movements has led FUNAI to recognize more of these people as legitimate Indians and the agency began investigating the possibility of creating reserves in urban areas, some of these traditionally occupied by indigenous players prior to colonization, settlement and later the establishment of the Brazilian state. As previously mentioned, many current Amazonian cities were formerly indigenous communities until they were forced to succumb to the forces of colonization. This scenario takes us back to Altamira, home to a number of indigenous ethnic groups now residing in many neighbourhoods of this city.

Eletronorte and FUNAI reluctantly agreed to acknowledge urban Indians of Altamira as legitimate indigenous people, and more frequently referred to them as *caboclos* or the common class of peasant groups encountered in the Amazon (Nugent, 1993; Parker, 1985; Ross, 1978). In these terms, they were not considered full-fledged indigenous citizens but a mass of detribalized Indians of mixed blood. The same held true for the Juruna of the Paquiçamba reserve. Together, these two groups became perhaps the most vociferous in airing their concerns about the Belo Monte dam. Both groups were able to regain their indigenous status through the help of anthropologists and the Conselho Indigenista Missionário (CIMI, Brazil's Indigenous Missionary Council). This represented a lengthy struggle for both groups as FUNAI and Eletronorte resisted acknowledging their status as Indians. But the pressure paid off and the Juruna's Paquiçamba reserve was demarcated and officially registered to them in 1994. This group represents the remnants of a community that split up during the 19th century, one segment remaining in the mid-lower courses of the Xingu, currently residing on the Paquiçamba reserve, while the other travelled upriver and settled in what is now the Parque Indígena Xingu, a large land area shared with other indigenous groups located near the headwaters of this river.

Since the 1980s both the urban indigenous residents of Altamira and the Juruna of Paquiçamba have slowly been able to regain their indigenous status.

Their pressure on FUNAI, along with lobbying from church groups, anthropologists and other sympathizers assisted them in this process of due recognition. Only a handful of them were able to deal with the Indian Service prior to this period, often achieving this through negotiation, friendship in the form of patron–client relations with FUNAI, outsider assistance or through outright hostility in demanding their rights. In the 1980s surveys were conducted to begin delimiting the Juruna's Paquiçamba reserve, primarily through CIMI's assistance, and later in 2003 studies were undertaken by one of the authors (Forline) to examine the possibility of revising this area to include neighbouring river islands and other areas containing Juruna cultural landmarks. Similarly, other studies were conducted to ascertain the feasibility of establishing a tract of land in the actual city of Altamira that would be assigned to its indigenous residents.

Towards the Future

In light of all of these events, it is worth pausing and questioning the irony of the adage that Brazil is a racial democracy, once lauded as a racial paradise. As David Treece (2000) aptly pointed out, in the history of Brazilian literature indigenous peoples went through an interesting set of phases where they were first considered exiles, next allies and then rebels. In recent years, they have gained a place for themselves under the sun, both through their own efforts, and with the collaboration of interlocutors such as church groups, anthropologists and other sympathizers. A significant portion of these gains were achieved in the face of the development projects that stood to affect them, particularly hydroelectric dams. This, of course, has extended to them a stepped-up role in project participation. The plurality of Brazil's fledgling democracy may still be criticized for now stressing a more rigid structure in classifying people, akin to the US system of categorizing different racial and ethnic groups. While full recognition as legitimate indigenous actors in Brazilian society may have made some significant headway, total recognition is still being constructed and articulated. There is a fuller exposition of Brazilian Indians in both the national press and international arenas. While some press given to indigenous peoples detracts from their plight, most of the coverage given to the Indians of Brazil deals squarely with issues pertaining to their rights, security, culture and well-being. Charismatic indigenous personalities have been portrayed and some of their best advocates in mainstream society have been celebrity figures in Brazil, such as the late senator and anthropologist Darcy Ribeiro. Advocacy and increased exposure have also taken the realm of indigenity beyond the broad-brush category of generic Indian and a number of people in mainstream society are beginning to realize that there are many different ethnic groups among Brazil's indigenous peoples. This, naturally, returns to the dilemma of classifying people since the days of mercantilism and, later, colonialism. In the world arena, since colonial times the concept of tribe has been deeply flawed. This misnomer was frequently used by colonial authority to divide, segregate, merge and manage conquered peoples. Now that these issues are being dealt with in a globalized context, there is an increased awareness of indigenous concerns. As previously observed, the cross-currents of globalization have provided indigenous peoples with a better opportunity to represent themselves and garner support through increased travel, communications and contact. Brazil has recently ratified the International Labour Organization's Convention 169, which extends to indigenous peoples legitimacy in representation and prior consultation in issues dealing with their well-being. The United Nations Declaration on the Rights of Indigenous Peoples in 2007 also fine-tuned previous resolutions, urging countries to duly respect and honour questions of sovereignty, intellectual property rights and informed prior consent vis-à-vis development projects. It is noteworthy that Kring Kaingang, an indigenous Brazilian woman, was responsible for drafting part of this text.

Looking towards the future an interesting battle promises to unfold as the Lula administration insists on full scale development to implement his social programmes. If his policies have not fulfilled expectations in regard to indigenous peoples and the environment perhaps this is because most of his political base stems from labour unions, working class people and urban intellectuals. His rural constituency would have more affinities with peasant leagues and landless people in need of basic social services, all of these often squarely pitted against indigenous interests. And while he promises to revive and expand the nearly moribund Indian Service, it is anticipated that these measures will be done with the objective of instituting his particular policy and staffing FUNAI with his personnel. In the meantime, Brazil's indigenous organizations will be tested in the face of the upcoming package of projects under review.

Note

1 See also MAB's website at www.mabnacional.org.br/

References

Anonymous (2001) 'Pará corre risco de ter energia racionada', *O Liberal*, article, May 5

Anonymous (2007) 'Dinheiro', *Folha de São Paulo*, editorial, December 11

Byrne, B. and Forline, L. (1997) 'The use of emic racial categories as a tool for enumerating Brazilian demographic profiles: a re-analysis of Harris's 1970 study', *Boletim do Museu Paraense Emílio Göeldi (Antropologia)*, vol 13, no 1, pp3–25

Chasteen, J. (2006) *Born in Blood and Fire: A Concise History of Latin America*, Norton, New York, NY

Cook B. and Kothari U. (2001) *Participation: The New Tyranny?*, Zed Books, London

Davis, S. (1978) *Victims of the Miracle*, Cambridge University Press, Cambridge

Fearnside, P. (1999) 'Social impacts of Brazil's Tucuruí dam', *Environmental Management*, vol 24, no 4, pp485–495

Fearnside, P. (2000) 'O Cultivo da Soja como Ameaça para o meio ambiente na Amazônia Brasileira', Paper presented at the international symposium *Amazônia 500 Anos: Lições de História e Reflexões para uma Nova Era*, Goeldi Museum, Belém, Brazil

Forline, L. (2004) 'Uma avaliação sócio-ambiental da uhe curua-una: primeiros diagnósticos e reflexões', Report submitted to Centrais Elétricas do Pará (CELPA)

Goodland, R. (1995) 'Ethical priorities in environmentally sustainable energy systems: the case of tropical hydropower', in M. Di Lascio, V. Di Lacio and L. da Paz (eds) *Energy Policy for the Sustainable Development of the Amazon Region,* UNB Grupo de Planejamento Energético, Brasília, pp15–36

Goodland, R. (2005) 'Evolução histórica da avaliação do impacto ambiental e social no Brasil: sugestões para o complexo hidrelétrico no Xingu', in A. O. Sevá Filho (ed) *Tenotã-Mõ: Alertas Sobre as Consequências Dos Projetos Hidrelétricos no Rio Xingu*, International Rivers Network, São Paulo, pp175–191

Harris, M. (1970) 'Referential ambiguity in the calculus of Brazilian racial identity', *Southwestern Journal of Anthropology*, vol 26, pp1–14

McDonald, D. (1977) 'Food taboos: a primitive Environmental Protection Agency, South America', *Anthropos*, vol 72, pp734–748

Moran, E. (1991) 'Human adaptive strategies in Amazonian blackwater ecosystems', *American Anthropologist*, vol 93, pp361–382

Nugent, S. (1993) *Amazonian Caboclo Society: an Essay on Invisibility and Peasant Economy,* Berg, Oxford

Parker, E. (1985) 'Caboclization: the transformation of the Amerindian in Amazonia, 1615–1800', *Studies in Third World Societies*, vol 29, ppxvii–li

Ramos, A. (1984) 'Frontier expansion and Indian peoples in the Brazilian Amazon', in M. Schmink and C. Wood (eds) *Frontier Expansion in Amazonia*, University of Florida Press, Gainesville, FL, pp83–104

Ross, E. (1978) 'The evolution of the Amazonian peasantry', *Journal of Latin American Studies*, vol 10, no 2, pp193–218

Sponsel, L. (1989) 'Farming and foraging: a necessary complementarity in Amazonia?', in S. Kent (ed) *Farmers as Hunters,* Cambridge University Press, Cambridge, pp37–45

Treece, D. (1987) *Bound in Misery and Iron: The Impact of the Grande Carajás Programme on the Indians of Brazil,* Survival International, London

Treece, D. (2000) *Exiles, Allies, Rebels: Brazil's Indianist Movement, Indigenist Politics, and the Imperial Nation-State*, Greenwood Press, London

Part II

Participation and the Dynamics of Gender in Water Management

3

Gender and Social Participation in a Rural Water Supply Organization in Rajasthan, India

Kate A. Berry

> Water is essentially a woman's issue. Men are not really bothered about it. They just wash their hands and sit down for food. It's the woman who has to arrange water for all day. Women need the water. And if there is no water in the house, the man will take a stick in his hand and ask – you didn't get water? It's the women who have to pay the price. It's the woman who needs water for the household work and to sustain the family. It is a woman's resource. (K. Bai, Rajasthani activist and grandmother, quoted in Parmar, 2004)

Introduction

India is one of the three largest water users in the world, the others being China and the United States. While irrigation is the largest water-using sector in the country, water use for domestic and industrial purposes is low, approximately 59m^3 per capita, less than half that of China (Sharma, 2006). Throughout the country an estimated 40 million households in rural areas do not have a safe or accessible source for drinking water (UNDP, 2008). In the state of Rajasthan in northwestern India fewer than 20 per cent of the households are reported to have water within the house, with the primary sources for water being outdoor handpumps and wells (Census of India, 2001).[1] As suggested by the opening quotation, women bear primary responsibility for ensuring there is adequate water for rural household use. With 77 per cent of people in the Indian state of Rajasthan living in rural communities, the population of the central western region, the Marwar, is estimated at more than 20 million, making it the most densely populated, rural, arid zone in the world (Census of India, 2001; Institute of Development

Studies, 2005). Consequently, working through the challenges of addressing women's participation in water management is of particular significance in Rajasthan.

This chapter examines how social participation has become connected to the articulation of gender within the work of a non-governmental organization (NGO) working on rural water supply development in the Marwar region of western Rajasthan. Instead of examining women's participation in an NGO as a gendered project of modernity associated with commodification of water and women (O'Reilly, 2006a), my argument is that ideologies and activities associated with social participation shape the gendered nature of water management – both within an NGO and through its work in rural communities in the Marwar region. In other words, gender is dynamic and gender norms can be altered when matters as significant as social participation are negotiated and water management is put into practice. Extending beyond the notion that gender as a form of cultural capital may be deployed to political and economic ends (Williams, 2004), this chapter argues the inverse: that social participation in water governance may reshape gender. As a result, the focus is on examining the knowledge, performance and process associated with social participation and water management to reveal the dynamic nature of gender. The approach adopted here considers the institutional relationships, customary practices of men and women and organizational orientation that frame participation in water management and, in turn, reshape experiences of what it means to be a woman or a man in India.

This project came about as a result of an invitation to do research with the Jal Bhagirathi Foundation (JBF) based in Jodhpur, Rajasthan. One of the three specific objectives JBF has is 'facilitating women's empowerment and ensuring their major role and participation in the creation and management of drinking water supply systems and services' (Pastakia, 2008, Table 1). The organization establishes women's self-help groups, works with micro-enterprises, ensures women's role in drinking water management and explores opportunities for augmenting women's livelihoods through the provision of water services. My involvement with the organization was from February to April 2008 and included:

- numerous discussions with board members, managers, fieldworkers and volunteers within JBF;
- several field visits to rural communities where projects were either ongoing or completed;
- assisting with an organizational workshop on gender equity.

This chapter begins with an overview of the links between gender and water development in South Asia. Attention then turns to the evolving role of NGOs as they have adopted participatory frameworks that concerned themselves with gender and water matters. JBF is then examined with respect to three aspects that link social participation with gender and water management:

- interactions and relationships with institutional donors and agencies;
- customary practices and context of the communities;
- orientation of the organization.

The chapter concludes with a discussion about the ramifications of social participation in local NGOs as they reshape gender through their work on water matters.[2]

Gender and Water Management in South Asia

Geographically speaking, gender is uneven terrain, becoming significant within the spaces in which power is expressed. While there are asymmetries between women and men, gender cannot be taken for granted as a defined set of roles or a static identity. Fundamentally relational and historically situated, gender intersects with other identities and is expressed in a variety of ways in different places (Lahiri-Dutt, 2006). Some are advantaged by gender while others may suffer its effects because not all individuals have the same degree of influence, wealth, concerns, rights or ability to meet their needs. Even for a single individual, gender is not static since different points in one's life have different gendered outcomes. Gender inequity can be seen as rooted in social stratification of roles and responsibilities assigned to men and women and the differential valuation of these roles (Gupta and Yesudian, 2006). Patriarchal structures that perpetuate gender inequities extend from the private sphere into public sphere (and vice versa), yet the lack of homogeneity results in different outcomes for initiatives aimed at women.

Perhaps the most apparent connections between gender issues and water matters in South Asia are between women's responsibilities to fetch domestic water and how they stand to gain by positive changes to water supplies (Cleaver, 1998a). As suggested by the introductory quotation, fetching water for household use and ensuring that it is usable is considered women's work; in rural India, women are responsible for collecting water for household use 86 per cent of the time (Bhatia, 2004). Consequently, women disproportionately experience the effects of changes in environmental or social conditions that make accessing household water in reasonable quantities and quality more difficult. They may need to travel longer distances, experience reductions in water supply availability, alter their access to water or spend time storing and preventing water from being contaminated. Something as simple as erratic power supplies, for example, may pose significant challenges in women's daily routines since this can delay water collection for hours until power is available for pumping (Shah, 2006).

Yet there is more that connects gender with water management because water is used and controlled in a variety of gender-specific ways at the household level as well as through broader scale institutions and sociopolitical structures. Consequently, men and women develop different perceptions, experiences, values and priorities about water and their actions vary accordingly.

The household is a key locus for shaping of gender relations as well as the primary site for water use, positioned to connect gender and water through intra-household dynamics as well as through relations between households and broader forms of collective action (Cleaver, 1998b). Beyond domestic uses of water for cooking and washing, men and women often have different household roles with respect to ensuring water is available for bathing, washing clothes, watering vegetable gardens, irrigating fields, watering livestock and house construction and repair (Cleaver, 1998a; Shah, 2006). Responsibilities for household work in conjunction with the need for economic endeavours may put a woman in a situation where she is unable or unwilling to participate in water management initiatives. Interacting with household dynamics, local culture can also be oppressive as a consequence of customs that reinforce gender inequities (Cleaver, 2001). For example, in a Nepalese village the decreased social standing of women resulted in them being portrayed as uneducated, unknowledgeable about official matters and poor in accounting. This, in turn, resulted in constraints to their participation in formal water user association meetings as few women attended and little was said by those who were there (Bhushan-Udas and Zwarteveen, 2005).

Men face challenges that influence their roles in water management and capacity to participate as well. Disproportionate emigration of men has created changes in household water matters and imposed new gendered roles and responsibilities (Lokur-Pangare and Farrington, 1999). As an increasing number of Indian men under the age of 60 migrate in search of wage labour for income to sustain their families, rural areas have become redefined as the primary residence of females, the young and the elderly. While women and girls migrate for marriage and family reasons, young and middle-aged Rajasthani men move for employment, business or educational opportunities, outnumbering women in such migration by ten times (Census of India, 2001). Many young and middle aged men circulate from town to country seasonally or at periodic intervals to maintain their social obligations within their home communities, but are not present for the daily work of water matters or to participate regularly in water management activities.

Formal institutions associated with water management have rules, resources and activities that influence gendered relations. Institutions that launch new water technologies, for example, influence men and women differentially because women are often seen as lacking relevant skills, partly due to lower access to education. In many places, women are absent or poorly represented within managerial, engineering and policy-making institutions (Maharaj et al, 1999).[3] The distribution of benefits in water projects may be differentiated by gender as well. For example, the introduction of technical expertise and funds in water infrastructure development may be accompanied by new water rights titling criteria and procedures that affect women and men differently (Van Koppen, 2000). Gendered differences in how sociopolitical structures are engaged and information is exchanged are related to shifting balances of power. When faced with formidable barriers to institutions or political discourse, women tend to rely on informal forms of participation, including less established sociopolitical hierarchies and informal, inclusive networks (Mohan and Hickey, 2004).

Time and circumstances also alter gender as it is expressed in water matters. Gender norms may vary with caste, race, class, religion, age or family position in ways that shape the experiences of individuals in particular settings and influence their access to information and power as decision-makers and participants (Bell and Franceys, 1995). Bhushan-Udas and Zwarteveen (2005) provide an example of different expectations between high caste women in Nepal, who were unlikely to be involved in construction or maintenance work, and lower caste women, who were often involved in such physical work on water conveyance structures. Caste may also influence the reasons why women support water supply development in their villages as upper caste women may participate because of their interests in the constancy and quality of water supply, whereas lower caste women may be more interested in the convenience of a new supply (O'Reilly, 2006b). With already busy schedules, poor women's participation in water management may simply add to the workload without any clear advantages, particularly if power relations are not in their favour. South Asian women's perspectives may also vary with age and position within a household as mother, daughter or daughter-in-law.

The Work of NGOs

This is the era of NGOs. The incredible increase in the numbers of NGOs in recent decades has been matched by augmented functions and amplified powers as well as increased responsibilities and roles with respect to the state. When the World Bank redirected focus to support private development as a means of public assistance in the 1970s, NGOs got a strong nudge and they have continued to grow in numbers, funding and missions ever since. These local NGOs tend to be action-oriented organizations, many of which consciously work to improve situations for vulnerable communities and marginalized individuals, including poor or rural women. These NGOs have been praised for:

- their ability to deliver services at lower costs, particularly in rural areas;
- their rapid and innovative responses as needs emerge in communities;
- their familiarity in working with their target groups;
- their ability to emphasize the processes through which people gain control over their lives;
- their less bureaucratic structures and formalities;
- their adoption of local technology, resources and skills. (Basu, 2004)

NGOs focusing on gender equity began to emerge in large numbers during the 1980s. In recent years India has become part of this worldwide trend with a growing number and greater diversity of local NGOs that play a major role in expanding spaces for public participation for women (Sekhon, 2006).

Discussions concerning the involvement of NGOs in gender and participatory development have largely focused on whether and how NGOs meet their goals of empowering women through participatory water management. The discussion has become somewhat polarized. On the one hand, NGOs are seen as being well positioned to establish settings where women can present their requirements and meaningfully contribute in community organizations. They may organize women's groups, address men's reactions or support organizations in electing women who can conveniently attend meetings and are motivated. Moreover, NGOs may influence perceptions that women can serve as a family or village representative (Shah, 2006). Within this vision of NGOs, participation is idealistically seen as a means to transform the ways women assert themselves, providing opportunities to articulate their beliefs, experiences and ideas through meaningful engagement in water management activities and decision-making (Cornwall, 2004).

On the other hand, concerns have emerged about the ends to which participation is used and who stands to benefit. When considered as social engineering, participation within development initiatives may rescript people's options in ways that are less than benign. If individuals lose their significance except as they provide legitimacy to decisions and actions, participation may fall prey to becoming a tool for engineering consent for projects whose framework has already been determined (Hildyard et al, 2001). Others wonder what benefit marginalized people receive from sitting on committees or individually speaking at meetings. How can participation be transformative rather than merely tokenistic when, for example, women are simply required to be represented in particular numbers or percentages in water user groups (Bhushan-Udas and Zwarteveen, 2005; Cleaver, 2001; Hickey and Mohan, 2004)?

In the backdrop of this debate, however, is a rather static notion of gender. Missing from these discussions are questions about whether and how gender norms are being transformed (intentionally or inadvertently) by NGOs deploying participatory water management. The first perspective suggests NGOs contribute to the well-being of women without recognizing that through participatory practices, changes may be advocated to traditional gendered roles. Neither does the second perspective make explicit the interests that specific women (as well as men) have in altering gender norms or, in other cases, preventing such changes. Most who have engaged the topic have not clarified how women may be agents in their own right, with options to resist, feign compliance or participate for tactical or self-interested reasons and, in so doing, alter the dynamics of gender.

In approaching this, Agrawal and Gibson's (1999) advice is worth heeding: pay attention to how people interact and the politics through which their interests emerge along with the institutions that shape these political processes. More specifically, Hailey (2001) suggests that an NGO is shaped partially by its relationships with government agencies and aid donors; partially by customary practices and community context; and partially by its own aspirations and management style. What follows is a close look at how these three aspects influence the Rajasthani NGO, Jal Bhagirathi Foundation (JBF) as it develops its interests in and capacity to shape gender practices and norms through social participation in water management.

Institutional Relationships

There has been an extensive history of governmental initiatives aimed at improving political and economic conditions for Indian women, yet many essentialist notions propagated in national political discourse have been rooted in ideas that women's reproductive roles constrain their productive capacities. The first constitution of newly independent India in 1950 stressed gender equality in all phases of life. Starting in 1952 government-run programmes incorporated women's issues in rural development, including self-help groups to serve as village-level forums for organizing women in economic development. Since 1975 there has been broader recognition that women constitute an overwhelming number among the poor in India and that development plans often structurally favour men. Consequently, India's Five Year Plan in 1980 included a chapter that signalled a shift from providing women with welfare to emphasizing developing women's potential. Numeric targets have also been promoted; for example, in 1993 a national law went into effect requiring the reservation for women of 33 per cent of elected positions in village assemblies. In the 1990s a shift in political discourse emphasized empowerment of women, in which females become equal partners and participants with males in development processes (Gupta and Yesudian, 2006; Khullar, 1997; Raju, 2006; Sekhon, 2006). This is not to suggest that the state has been fully supportive of gender equity. Legal provisions have often been interpreted in ways that are counter to gender equity and many initiatives were never fully implemented, while others failed. As an example, national legislation frequently establishes numerical targets, but cannot guarantee that women participate effectively. Preparation, the ability to act independently of one's spouse and confidence in the setting are also factors that contribute to one's capacity to participate (Lokur-Pangare and Farrington, 1999; Sekhon, 2006).

However, it is through interactions with international agencies and foreign donors rather than directly with the state that institutional relations are brought into focus for JBF. Promoting gender equity has international cachet to the extent that not only is this one of the United Nations Millennium Development Goals (MDGs), but achieving this has been closely tied to the other seven goals. Consequently, international agencies and foreign donors often expect gender equity initiatives to be undertaken in conjunction with participatory water development projects. The World Bank is arguably the best known international organization promoting participatory development, with the idea that transparent and inclusive decision-making will lead to accountable systems that allocate water to the highest value use and provide equitable allocation and service provision (World Bank, 2007). While not directly involved with JBF, the World Bank has been active in Rajasthan.[4]

The United Nations Development Programme (UNDP) is another high profile international organization promoting women's empowerment and participatory water initiatives that has been actively involved with JBF since soon after its inception. In India, UNDP is concerned with making governance more participatory, gender-balanced, transparent and accountable. Rajasthan has been identified as one of seven states in India where, due to low rates of human development, great gender disparity and high proportions of scheduled castes and tribes, programme initiatives in the future are to be focused, including collaborations on sustainable water management and gender-sensitivity strategies. Increasingly UNDP emphasizes partnerships, including those with local NGOs and private sector organizations (UNDP, 2007, 2008). In addition to assisting JBF in securing funding from the Italian government for recent project work, UNDP also provides consultation and evaluates project accomplishments through a biennial evaluation.

A UNDP-sponsored evaluation in early 2008 identified progress made by JBF. The evaluation noted the reduction of walking distance and decreased time commitment for women collecting water (previously up to three or four kilometres and often five to six hours daily) because of JBF's assistance in constructing community wells and tanks closer to homes (Pastakia, 2008). Another accomplishment noted was the rise in the number of women's self-help groups from 17 in late 2005 to 53 in early 2008, with total membership of 645 women. About one third of the self-help groups had established linkages with banks and loans that could leverage their savings. Nevertheless, criticism was levelled and suggestions made about the lack of convergence between water initiatives undertaken by mainstream village institutions dominated by men and the women's self-help groups (Pastakia, 2008). The management of JBF views these accountability reviews as useful because they assist in problem identification and suggest ideas and approaches to empower women given the 'semi-feudal social relations in the project area' (Pastakia, 2008, p19). UNDP and JBF concur that, while traditional Marwar gender relations are enduring, progressive reform is needed. Increasing women's participation in water management and self-help groups have become a means to incrementally transform gender norms, to change how men and women act and make decisions on a daily basis.

The dependence that NGOs develop on international donors who financially support them has drawn considerable criticism in recent years. NGOs have been identified as pawns in a universalizing drive for participation; in other words, readily co-optable. With this in mind, NGOs would be expected in the end to do what the donors want, so participatory processes become a versatile but insidious means by which international donors move forward on neoliberal reforms (Cooke, 2004; Sengupta, 2000). While there may be reason for concern about the motives and authenticity of participation associated with international donors, it is too simplistic to portray NGOs simply as the wards of international donors who, as guardians, are promoting participation solely for their own purposes. NGOs recognize the potential for using these relationships to their own advantage.

JBF, for example, has responded to UNDP's earlier advice to get more actively and extensively involved in promoting gender equity, yet much of the impetus, decision-making and implementation resulted from the organization's own approach to gender issues. At this point, UNDP is supportive of JBF's initiatives and sees its own role as contributing to their success in helping marginalized rural Marwar residents. The UNDP evaluation report spells out JBF's successes, as well as work that will reshape gender norms:

- bring together men's and women's village groups where possible;
- provide additional opportunities for women's village groups to work on water structures;
- strategically time and structure village group formation so as to facilitate women's learning within the group;
- prioritize gender equity with JBF itself;
- offer capacity building camps for women, sensitization camps for men and guidance on micro-economics for both;
- have JBF staff and managers visit Indian women's empowerment work in similar settings (Pastakia, 2008).

The terms of collaboration between NGOs, such as JBF, and international donors and agencies, such as UNDP and the Italian and US governments, cannot be static. These terms change as the NGO evolves and in conjunction with broader political and economic forces. While NGOs are not independent of government agencies and aid donors, relationships between them might better be characterized as functional and interactive, rather than as politically captive and uni-directional. While NGOs clearly depend on international donors and states, there are many ways to make projects appealing and many different types of donors looking to assist the marginalized people that NGOs, such as JBF, work with and represent.

JBF has recently expanded into new alliances. One example is Project Undercurrent, supported by the Gates Foundation, Acumen Foundation and IDEO, whose goal is to:

> improve the transportation and storage of water for household use to impact the health and quality of life of low-income communities in the developing world. (IDEO et al, 2007, p2)

JBF participated in phase 1 of Project Undercurrent and raised the visibility of women's issues with water in the process. Another project piloting new technology into Rajasthan, the WaterPyramid – a hybrid system of rainwater harvesting and solar distillation – is a collaboration between JBF, the state government and the private sector producer (Bhasipol et al, 2007). Such alliances offer JBF the potential to forge different sort of ties with foundations and private firms and in the process develop their own social capital.

While new alliances in most instances directly or indirectly provide alternate sources of funds, these collaborations are also significant in the transmission of ideas, technologies and approaches. Such collaborations are often associated with greater publicity that raises the visibility of JBF and, in so doing, changes expectations and communicates perspectives about how to link women with participation. Moreover, institutional contacts with representatives from the media, international agencies, foundations or the government are one means by which alternate gendered norms (or expressions of men's and women's roles from outside the Marwar) may reinforce JBF's ideas or come into contact with customary practices within the context of local communities.

Customary Practices and Community Context

In India, states are designated as primary water managers and decision-makers yet the central government remains significant, particularly with regards to major development initiatives and infrastructure construction that crosses political boundaries.[5] Throughout the country the water rights system is not truly riparian because during British colonial times water rights evolved differently throughout the country. Since independence, users do not possess statutory rights over sources. Water rights tend to be poorly defined, generally based on a combination of traditional use and customs, evolving colonial and then republican bureaucratic decisions and the precedence of case law (Sengupta, 2000).

Some villagers believe that the government has the intention to provide community water supplies but has been unable to do so. Others in rural Rajasthani communities feel that government does not have their best interests in mind when it comes to water supplies. Much of the scepticism dates back to the colonial period when local self-sufficiency was eroded. When formed in 1863, the state of Rajasthan Public Work Department laid claim on all water infrastructure, including those built and maintained by households or communities. Colonial efforts to expand on water technologies also reconfigured social relations and advanced British governance. The confiscation and loss of control marked the end of interest of many communities in water supply and has continued to affect perceptions about water management through the decades following India's independence. As centralized approaches to water management were perpetuated, capacity for local management and participation dwindled (Birkenholtz, 2008; D'Monte, 2005; Parmar, 2004). Suspicion about the government's motives and approaches to water matters has not evaporated, exacerbated by failed government projects, rifts created by party politics, breakdown of leadership and community organization, commercialization that expands the gap between rich and poor and changing population dynamics that include out-migration (Krishna, 1997). Despite calls to enhance people's participation in water resource management, under such conditions building collective participation has proven challenging, if not impossible, for both the central and state governments.

Concerns about government officials' attempts to assimilate material resources and disparage local knowledge have led to a gap in which other approaches and techniques are emerging. The revitalization of water traditions in Rajasthan has been applauded, as customary water harvesting structures and management techniques are promoted as simple, sustainable and appropriate environmental technology in conjunction with minor redesign elements for new conditions or materials (D'Monte, 2005; Parmar, 2004). Groundwater for example has long been an important source of water supply in arid Rajasthan, and the Beldar caste, which is identified with well digging expertise, provides traditional sources of knowledge as do Hindu water diviners who identify the best location for and timing of well construction (Birkenholtz, 2008).[6] Given the monsoonal nature of precipitation throughout India and the aridity of the Marwar region, the ability to store water for later use has been important enough that, in recent years, traditions of capturing the scant rainfall have been revived. Common water harvesting structures include simple reservoirs made of soil and rock, designed to collect runoff during the monsoon and allow water to percolate so as to recharge groundwater or create surface storage. Modern twists on these small reservoirs can be found in the ways that a catchment area is prepared. Another water harvesting technique with a long tradition in arid Rajasthan is collecting water from rooftops, which involves directing rain that falls on an impervious surface into gutters and from there into a storage tank for future use (Jain and Jain, 2000). Thus, mixing the appeal of tradition with the modern in ways marked distinctively as Rajasthani, coupled with the relative ease and small scale of development marks this as community water management that tends to steer clear of direct involvement by government or from abroad.

As a result, water harvesting and similar water approaches in Rajasthan have been linked with participation, where they have been promoted as the people's business. According to Parmer (2004) who analysed gender issues within a NGO in northeastern Rajasthan, water harvesting also involves the public in water management 'making water management everybody's business'. NGOs fit into the gap in governance through work with village organizations that activate community members and stimulate participation. JBF views itself as an advocate for community rights over common property resources such as water, so its primary interface with government has been in the policy arena, and it advances participatory water development and management initiatives to enhance community capacity. Viewing rural water supply security as fundamental to its Gandhian-style strategy, JBF attempts to facilitate decentralized rural development within the context of contemporary community capacities and needs. The NGO supports its work with villagers through mutual understandings about the limitations of government-driven water initiatives and shared appreciation for traditional water approaches. JBF has tried to build momentum for participation across a wide spectrum through its position as an advocate for the potential of communities.

With revitalizing rural communities as the goal, JBF recognizes that women's involvement is central to the community ideal; indeed, mobilizing women is an organizational mandate (JBF, 2008). As well as being active sites of solidarity, however, communities may also be sites of conflict that encompass shifting alliances, power and social structures, defined to be exclusionary as well as inclusive (Cleaver, 2001). And in Rajasthan's rural communities women's issues are particularly thorny. Rajasthani women have a low degree of autonomy, even within their own households. Gupta and Yesudian (2006) found that women's participation in household decision-making was lower in Rajasthan than in any other state in India. Other aspects of women's status in rural Marwar are problematic as well:

- restricted mobility;
- low education levels;
- poor health and nutrition;
- limited access to land, capital and technology;
- practices of women's veiling and isolation;
- preference for sons over daughters. (Institute of Development Studies, 2005).

Aware of the challenges of working within this patriarchal setting, JBF has to tread cautiously into the arena of women's empowerment.

In this context, the capacity of JBF is in part contingent upon the quality and types of interactions configured within communities. A pivotal means through which rural communities are engaged with gender equity, participation and water management is through interaction with fieldworkers, which in the parlance of JBF includes staff designated as project animators, community motivators, community organizers, project surveyors and project coordinators (JBF, 2008). Fieldworkers are significant points of transmission between households and NGOs on matters of gender, since they are active agents in implementation who make choices about what to emphasize during their interactions, in essence defining the meaning and experience of participation on the ground (McKinnon, 2007; O'Reilly, 2006b). Given this situation, fieldworkers face contradictions of gendered norms within an NGO while simultaneously trying to serve as agents of change. UNDP evaluators viewed the shortage of women fieldworkers in JBF as a significant challenge; in early 2008 JBF had only seven women on its staff out of a total of 48 and nearly half of these were in management positions (Pastakia, 2008). Using a model that assigned fieldworkers their own area of responsibility and tasks within a village cluster, JBF provided each with basic guidelines and training; if a fieldworker needs help, technical or other assistance is generally offered within the organization. Less experienced fieldworkers may be assigned to work with those who are more experienced. Some are originally from rural Marwar and even work in their home village, while other fieldworkers are from a city or from another state.

During a JBF-sponsored workshop in April 2008, fieldworkers, as well as other staff, managers and volunteers gathered for two days to examine gender equity issues within the organization as well as within villages where they worked. The workshop was designed for all individuals to consider their own roles along with the organization's capacity and limitations to define terms that encourage women's participation in water management within the context of the villages they work in. At the start of the workshop, 41 participants completed a simple survey; two of the survey's five questions are relevant here as they were directed at their experiences and strategies in working with women and men. While many respondents had no challenges to report, many mentioned constraints similar to those described in other parts of Rajasthan:

- women were often not able to work out of the house and sometimes did not attend meetings because of social norms;
- isolation was viewed as a problem as was women's reserve, modesty or shyness around men
- illiteracy was a constraint in getting village women involved.

Positives were mentioned as well. Many viewed village women as honest, hardworking, quick learners and full of ideas; and a number of respondents saw their own work with village women as contributing to building women's self-confidence and equal rights. The vast majority saw increased education and training as being fundamental to altering gender norms; several mentioned that encouraging village women's efforts was an important part of their work and a few reinforced the significance of work with women's self-help groups as well as the need to recruit more women fieldworkers within the organization. Group discussions during the workshop reinforced how lack of education, poverty and caste division further complicated women's involvement in water management. Also underscored were the needs to work on achieving mutual respect and to enlist village men as well as women, as these might otherwise try to undermine initiatives on gender equity.

O'Reilly (2006a, 2006b) found that within an NGO in northern Rajasthan, the dynamics of gender were complicated by the perceptions of other staff because some support women fieldworkers' involvement and women's participation in water management generally, while others denigrate these individuals and initiatives. In the case of JBF, conflicts were apparent between individuals over how gendered norms should be structured within the organization, yet the mandate to empower rural women seemed to be taken seriously, despite the lack of clarity on how this could best be accomplished. The conundrums of gender equity were tough issues that many faced individually as fieldworkers, given the cultural and political contexts of villages they worked in. Yet it is precisely because individuals in JBF were positioned to struggle over the introduction of alternate gender norms that were sanctioned organizationally – introducing ways to accommodate water supply needs while engaging new participatory roles for women within villages – that gender challenges and contradictions have not proven easy to resolve.

Organizational Orientation

Founded in 2002, JBF has taken many cues from the experiences of other NGOs in Rajasthan that are linked to rural water supply development. The experiences of Tarun Bharat Sangh (TBS), an NGO operating in northeastern Rajasthan, have been particularly instructive as the two organizations share many of the same approaches, along with two trustees who have established the vision for both organizations. While TBS did not start out with the objective of empowering women, their initiatives acted as a catalyst by getting women involved in selecting sites for water features, arranging voluntary labour and allocating water and other project benefits (Parmar, 2004). Both NGOs emphasize combining voluntary participation in water management with integrated multi-sectoral programmes that sponsor small-scale enterprise and promote participation of women from village organizations in the development, construction and financing of water infrastructure. TBS supported Rajasthani villagers who mobilized in opposition to state interference with community-built reservoirs and worked as a liaison to communicate their concerns (D'Monte, 2005). Marches and massive public gatherings, sometimes involving hundreds or thousands of people, are other means of collective mobilization that both NGOs used to generate awareness and get people together, while simultaneously increasing visibility of issues and engaging participation. In the case of JBF, the organization was associated with marches and public gatherings both in February and in April 2008, which were designed to bring attention to the plight of the Luni River (the only perennial river within JBF's project area) and pressure government officials to enforce water quality protection laws against polluting industries.

One of the trustees shared by both NGOs is the charismatic Rajendra Singh, known throughout much of India as the Water Man. His reputation for promoting the rights and responsibilities of Indian villagers on water matters and community self-sufficiency has been at the core of protests, the centrepiece of news stories and the target of international water documentaries (D'Monte, 2005; Parmer, 2004). The other trustee shared in common is Prithvi Raj Singh, who as Managing Trustee of JBF plans, manages and executes the objectives of the organization, overseeing all aspects of its administration. The two other trustees of JBF, Maharaja Gaj Singh (who also serves as chairperson) and his wife Maharani Hemlata Rayje, are the traditional rulers and custodians of the Marwar. While no longer having official political roles, the Maharaja and his wife continue to be powerful philanthropic leaders in western Rajasthan, concerned with the social, economic and environmental issues of the region. Leadership by these four individuals has been influential in steering the course of JBF as well as developing relationships of trust with villagers, many of whom respect the trustees' longstanding commitment to the land, water and people of Marwar. Notably as one of the four trustees, the Maharani is a woman of high social position and caste, known for being attentive to educational and family issues in Rajasthan, and her position as trustee of JBF raises the profile of women's issues and connects this to water matters.

Women are emerging in high positions within JBF as well, changing how gender equity is acknowledged and addressed. The most prominent position within the NGO is the Project Director, who as second in command under the

Managing Trustee, supervises all employees at the main office in Jodhpur and administers all project work. Since 2006, this position has been held by Kanupriya Harish, a well-educated woman from Uttar Pradesh. She is well aware of gender equity issues within the organization as well as the implications of women's issues in their work in the villages. As the leader of the staff, her work on a daily basis highlights women's skills and competencies to those within the NGO as well as creates a high profile as she engages directly with villagers, ensuring they understand JBF's terms of agreement and providing assistance for water projects and self-help groups. In the process she is, through example, displaying a much broader set of skills than many villagers have acknowledged for women. In addition to the Project Director, other management positions have been filled by women including a newly established position, the Program Communications Officer based in the Jodhpur office, and the Financial Controller based in the Jaipur office. Highlighting women trustees and incorporating competent women as leaders fits the JBF logic of unobtrusive engagement with gender issues, preferring to show through examples that there are alternatives to patriarchal traditions. In an understated approach, gender is being re-negotiated, not only through the positions and actions of these women, but also by the individuals who promoted these changes (the trustees), individuals within the NGO who are supervised by these women (other managers, fieldworkers and volunteers) as well as the villagers themselves.

Striving to bridge the customary and modern, JBF strategically combines native elements with imported. As a relatively new, idealistic NGO it relies on visionary spokespeople and involved villagers and supports old labour-intensive technology constructed in novel ways with capital from the outside. By moving away from what is viewed as a pervasive system of entitlements based on Nehruvian handouts and towards a model that is by design more modern, JBF encourages village self-sufficiency and community investment in their own water supply projects. They have successfully targeted water projects that help a single caste as well as those that incorporate multiple castes within the project development, design and implementation phases. Within the NGO, personal relationships in which key individuals mediate informal consultation processes and facilitate shared decision-making are significant, yet there is also a formal management framework in place that requires villages to match funds and work within administrative guidelines. With a founding vision that embraces contemporary self-sufficiency while simultaneously incorporating Rajasthani traditions, JBF is steadily gaining latitude to work under conditions marked as its own. National and international media coverage, such as articles in *Newsweek* and Japanese newspapers and film coverage by the BBC, University of Rome and NDTV, provide the leverage needed to ensure they can continue to chart a direction consistent with their ambitions, style and experience (Pastakia, 2008). Sympathetic press coverage that is influenced by a NGO can be particularly important in forging networks of support (Jenkins and Goetz, 1999). JBF takes seriously the work of maintaining strong public relations and developing high visibility through the media and in national and international development circles. One way this has been done is by hosting prominent guests, such as HRH Prince Charles of the United Kingdom. Website development is also important for the NGO. Raising the profile of the NGO serves to:

- capture the attention of donors;
- update the domestic and international press;
- provide more organizational autonomy;
- attract and retain good staff and volunteers.

This is not to suggest that JBF has moved beyond the gender issues that surround the organization. As pointed out in one of the group discussions at the April 2008 workshop, 'JBF is still as patriarchal as the rest of society' (Berry, 2008). Nevertheless, many ideas were generated at the workshop and hope was expressed that positive change was possible. The following suggestions about JBF's role in promoting women's equity through participation in water management were generated in the group discussions at the workshop:

- Women and men alike should be involved in discussions on water matters as women's concerns about water should be respected and their suggestions taken seriously. JBF should take the lead to encourage women as equal and active partners in water management.
- Education and training affects women's ability to participate in water management. Similarly customary practices, opinions of family members and divisions within society can hinder women's participation. These matters should be addressed by both men and women.
- Supporting women's economic activities associated with water management and starting a federation of women's self-help groups could promote empowerment.
- The governance system is already in place at the community level for better representation of women and could be leveraged to encourage more women to serve as policy makers.
- Addressing women's roles as equal partners with men has human rights implications as well as backing within Hindu spiritual traditions.
- Gender balance and equity issues should be tracked and worked on within the organization itself.

These are perceptual and substantive divides that JBF staff, managers and volunteers felt need to be negotiated if more women in rural Marwar are to achieve active and meaningful involvement in managing water, a resource in which they are already highly invested. Each of the points made relates directly to transforming the experiences of rural women and suggests ways that JBF could frame participatory development and water management activities as they reshape gender practices and norms.

Conclusions

Many at JBF are aware that negotiating participation with and for women cannot be readily accomplished on demand or by using a heavy-handed approach imposed from the outside, and are mindful not to create a gender paradox in which the institutional mandate to empower women is at odds with villagers' ideas. Acknowledging contradictions and identifying points of convergence with traditional approaches to water management involves commitment of resources, funds, time and creativity by the NGO. JBF is now at the point where it will choose how to act on the suggestions made at the workshop and how to approach the UNDP evaluator's recommendations on the gender equity. In this vein, the suggestions of Lokur-Pangare and Farrington (1999) may be useful: support from men and women who are resistant to changing traditions may be easier to generate as these individuals come to understand women's work burden, if lower caste and poor individuals can be encouraged to recognize similar processes of oppression and if equal rights to participate in development processes become broadly appreciated.

This is not simple terrain to negotiate as McCusker and Oberhauser (2006) point out in a study of South African development projects. They found that South African women had to negotiate power with their male counterparts in a culture steeped in traditional gender relations, in which value of women's activities was determined by men and these activities were often characterized negatively. These women straddled two contradictory processes: on the one hand, traditional customs and patriarchal practices governing social relations bent on maintaining a position of privilege and, on the other hand, broader structural forces dedicated to transforming rural economies through market-based, neoliberal reforms, each vying for hegemony. Traditional understandings of gender came into conflict with those promulgated from the outside, and in the process women stood to gain little from projects that mandated gender equity and participation on the one hand while reifying traditional cultural practices on the other (McCusker and Oberhauser, 2006).

Since its inception in 2002, JBF has enhanced water supplies for rural communities in arid western Rajasthan. The NGO's work with villages has resulted in nearly 200 small water projects being built, restructured or managed in the past three years; in the process, social capital has developed through enhanced community skills, resources and capacities (Pastakia, 2008). In working with villages on water management, women's issues are inevitably raised, putting the NGO in the middle of competing gender norms. On the one hand are ideas and approaches on gender equity originating from international donors, state and national governments representatives as well as the experiences of trustees, managers and staff from outside Marwar. On the other hand are the customary practices and lived experiences of men and women within the context of rural communities. Much of the organization's impetus to work towards gender equity originates from outside rural Rajasthan, not from within the range of experiences and interests of villagers in the Marwar, yet JBF serves as a facilitator and mediator. The NGO has consciously chosen to promote women's empowerment within rural villages as part of the new water management schemes, trying to enhance the prospect that women will find support. Nevertheless, the NGO is cautious about challenging traditional power structures. Their hope is that participation in water management may influence the ways other goals are achieved when women are better positioned to influence other decisions about themselves, their households and their communities. JBF is beginning to encounter challenges and contradictions as their initiatives re-shape the gendered terrain they are working within.

Acknowledgements

This project was made possible through the financial support of a Fulbright Middle East, North Africa, South Asia Research Grant. I appreciate the assistance of the US Educational Foundation in India, in particular Dr Girish Kaul, as well as the cooperation and generosity of the Jal Bhagirathi Foundation, in particular Prithvi Raj Singh, Kanupriya Harish and Nilima Hembram.

Notes

1 Within the arid Marwar region of western Rajasthan there are very few streams and the average annual rainfall of 10–40cm is associated with the monsoon period of July to September (Government of Rajasthan, 2008).

2 This analysis does not consider the work of multinational NGOs with mandates in various parts of the world. For more on these organizations see Jasanoff (2003).

3 Indian women still make up only 20 per cent of the professional and technical workers, about 2 per cent of the administrators and managers and 8 per cent of the national legislature (Basu, 2004).

4 In the 1990s the World Bank and Government of India spent US$70 million to form the Rajasthan Department of Watershed Development and Soil Conservation with the intent to promote participatory development but no funds were dedicated specifically for this purpose (Krishna, 1997). In 2002, the World Bank approved funding of US$140 million to reinforce the State's capacity to develop water resources, increase water availability for irrigated agriculture, and form the Rajasthan Water Planning Department; according to their press release, the project was designed to promote the empowerment of the rural poor and women through the formation of community groups (World Bank, 2002). Yet at the end of 2007, of the 83 separate private sector contracts on this project 95 per cent were for construction or maintenance work on canals and dams.

5 In northwestern India, for example, the Indira Gandhi Canal is a project of the central government that transports river water from northern Punjab to arid lands in western Rajasthan.

6 A variety of wells were also customary in western Rajasthan and have different names and meanings depending upon their construction type, depth, ownership and religious significance (Jain and Jain, 2000).

References

Agrawal, A. and Gibson, C. C. (1999) 'Enchantment and disenchantment: the role of community in natural resource conservation', *World Development*, vol 27, no 4, pp629–649

Basu, S. (2004) 'NGOs in women's economic development in India: an evaluation', in S. Hasan and M. Lyons (eds) *Social Capital in Asian Sustainable Development Management*, Nova Science Pub, New York, NY, pp71–100

Bell, M. and Franceys, R. (1995) 'Improving human welfare through appropriate technology: government responsibility, citizen duty or customer choice', *Social Science & Medicine*, vol 40, no 9, pp1169–1179

Berry, K. A. (2008) 'Report on the strategic workshop on gender equity in water management', unpublished report, Jodhpur, India

Bhasipol, B., Foran, M., Kita, K. and Salmon, M. (2007) 'Attacking the water crisis in Rajasthan using the Water Pyramid innovation', unpublished report, MIT Sloan School of Management Global Entrepreneurship Lab, Cambridge, MA

Bhatia, R. (2004) 'Role of women in sustainable development – a statistical perspective', in *Proceedings of National Seminar on Gender Statistics and Data Gaps,* New Delhi, India: Central Statistical Organisation, Government of India, February, pp233–241

Bhushan-Udas, P. B. and Zwarteveen, M. (2005) Prescribing gender equity? The case of Tukucha Nala irrigation system, central Nepal', in D. Roth, R. Boelens and M. Zwarteveen (eds) *Liquid Relations: Contested Water Rights and Legal Complexity*, Rutgers University Press, New Brunswick, NJ, pp21–43

Birkenholtz, T. (2008) 'Contesting expertise: the politics of environmental knowledge in northern Indian groundwater practices', *Geoforum* vol 39, pp466–482

Census of India (2001) '(08) State of Rajasthan housing profile and demographic files', Data Dissemination Wing, Office of the Registrar General, New Delhi, India

Cleaver, F. (1998a) 'Choice, complexity and change: gendered livelihoods and the management of water', *Agriculture and Human Values*, vol 15, pp293–299

Cleaver, F. (1998b) 'Incentives and informal institutions: gender and the management of water', *Agriculture and Human Values*, vol 15, pp347–360

Cleaver, F. (2001) 'Institutions, agency and the limitations of participatory approaches to development', in B. Cooke and U. Kothari (eds) *Participation: The New Tyranny?*, Zed Books, London, pp36–55

Cooke, B. (2004) 'Rules of thumb for participatory change agents', in B. Cooke and U. Kothari (eds) *Participation: The New Tyranny?*, Zed Books, London, pp42–55

Cornwall, A. (2004) 'Spaces for transformation? Reflections on issues of power and difference in participation in development', in B. Cooke and U. Kothari (eds) *Participation: The New Tyranny?*, Zed Books, London, pp75–91

D'Monte, D. R. (2005) 'Water management in Rajasthan, India: making a difference', in J. Velasquez, M. Yashiro, S. Yoshimura and I. Ono (eds) *Innovative Communities: People-Centred Approaches to Environmental Management in the Asia-Pacific Region,* United Nations University Press, Tokyo, pp292–308

Government of Rajasthan (Water Resources Department) (2008) Rajasthan rainfall patterns, www.rajirrigation.gov.in/1rainfall.htm, accessed 26 June, 2008

Gupta, K. and Yesudian, P. P. (2006) 'Evidence of women's empowerment in India: a study of socio-spatial disparities', *GeoJournal*, vol 65, pp365–380

Hailey, J. (2001) 'Beyond the formulaic: process and practice in South Asian NGOs', in B. Cooke and U. Kothari (eds) *Participation: The New Tyranny?*, Zed Books, London, pp88–101

Hickey, S. and Mohan, G. (2004) 'Relocating participation within a radical politics of development: insights from political action and practice', in B. Cooke and U. Kothari (eds) *Participation: The New Tyranny?*, Zed Books, London, pp159–174

Hildyard, N., Hegde P., Wolvekamp, P. and Reddy S. (2001) 'Pluralism, participation and power: joint forest management in India', in B. Cooke and U. Kothari (eds) *Participation: The New Tyranny?*, Zed Books, London, pp56–71

IDEO, Acumen Fund and Bill and Melinda Gates Foundation (2007) 'Project undercurrent: phase 1 summary', unpublished report, September

Institute of Development Studies (2005) 'Mid-term evaluation report of JBF-UNDP project on vulnerability reduction through community empowerment and control of water in the drought prone areas of Marwar Regions of Rajasthan', unpublished report, Jaipur, India

Jain, K. and Jain, M. (2000) *Architecture of the Indian Desert,* AADI Centre, Ahmedabad, India

Jasanoff, S. (2003) 'NGOs and the environment: from knowledge to action', in D. E. Lorey (ed) *Global Challenges of the Twenty-first Century: Resources Consumption, and Sustainable Solutions,* Scholarly Resources Inc, Wilmington, DE, pp269–287

JBF (2008) 'Management system manual', revision January, unpublished manuscript, Jodhpur, India

Jenkins, R. and Goetz, A. M. (1999) 'Accounts and accountability: theoretical implications of the right-to-information movement in India', *Third World Quarterly*, vol 21, no 3, pp603–622

Khullar, M. (1997) 'Emergence of the women's movement in India', *Asian Journal of Women's Studies*, vol 3, no 2, pp94–129

Krishna, A. (1997) 'Participatory watershed development and soil conservation in Rajasthan, India', in A. Krishna, N. Uphoff and M. J. Esman (eds) *Reason for Hope: Instructive Experiences in Rural Development,* Kumarian Press, West Hartford, CT, pp255–272

Lahiri-Dutt, K. (2006) 'Introduction', in K. Lahiri-Dutt (ed) *Fluid Bonds: Views on Gender and Water,* Stree, Calcutta, ppxiii–xl

Lokur-Pangare, V. and Farrington, J. (1999) 'Strengthening the participation of women in watershed management', in J. Farrington, C. Turton, and A. J. James (eds) *Participatory Watershed Development: Challenges for the Twenty-First Century*, Oxford University Press, Oxford, pp118–158

Maharaj, N., Athukorala, K., Garcia Vargas, M. and Richardson G. (1999) 'Mainstreaming gender in water resource management: why and how', unpublished report, World Water Vision and World Water Council

McCusker, B. and Oberhauser, A. M. (2006) 'An assessment of women's access to natural resources through communal projects in South Africa', *GeoJournal*, vol 66, pp325–339

McKinnon, K. (2007) 'Postdevelopment, professionalism, and the politics of participation', *Annals of the Association of American Geographers*, vol 97, no 4, pp772–785

Mohan, G. and Hickey S. (2004) 'Relocating participation within a radical politics of development: critical modernism and citizenship', in S. Hickey and G. Mohan (eds) *Participation: From Tyranny to Transformation?*, Zed Books, London, pp59–74

O'Reilly, K. (2006a) '"Traditional" women, "modern" water: linking gender and commodification in Rajasthan, India', *Geoforum*, vol 37, pp958–972

O'Reilly, K. (2006b) 'Women fieldworkers and the politics of participation', *Signs*, vol 31, no 4, pp1075–1098

Parmar, A. (2004) 'Ocean in a drop of water: empowerment, water and women', *Canadian Woman Studies*, vol 23, no 1, pp124–128

Pastakia, A. (2008) *Meeting the Challenge of Drinking Water Security in Marwar Region of Rajasthan*, mid-term evaluation study of the project 00038717: Vulnerability reduction through community empowerment and control of water in the drought prone areas of Marwar Region (Rajasthan) for UNDP and Italian Development Cooperation, New Delhi, India

Raju, S. (2006) 'Contextualising gender empowerment at the grassroots: a tale of two policy initiatives', *GeoJournal*, vol 65, pp287–300

Sekhon, J. (2006) 'Engendering grassroots democracy: research, training and networking for women in local self-governance in India', *NWSA Journal*, vol 18, no 2, pp101–122

Sengupta, N. (2000) 'Negotiation with an under-informed bureaucracy: water rights on system tanks in Bihar', in B. R. Bruns and R. S. Meinsen-Dick (eds) *Negotiating Water Rights,* ITDG Publishing, London, pp137–161

Shah, A. (2006) 'Women and water: perceptions and priorities in India', in K. Lahiri-Dutt (ed) *Fluid Bonds: Views on Gender and Water,* Stree, Calcutta, pp172–184

Sharma, B. R. (2006) 'Water demand growth trends: projects for the world and India', in *Proceedings of International Workshop on Water Saving Technologies*, Fulbright Indo-American Environmental Leadership Program, New Delhi, India, February, pp51–58

UNDP India (2007) 'UNDP country programme for India (2008–2012)', unpublished report, New Delhi, India, July

UNDP (2008) MDGs – UNDP India's Role, www.undp.org.in/index.php?option=com_content&task=view&id=73&Itemid=157, accessed 26 February 2008

Van Koppen, B. (2000) 'Gendered water and land rights in rice valley improvement, Burkina Faso', in D. Roth, R. Boelens and M. Zwarteveen (eds) *Liquid Relations: Contested Water Rights and Legal Complexity,* Rutgers University Press, New Brunswick, NJ, pp83–111

Williams, G. (2004) 'Towards a repoliticization of participatory development: political capabilities and spaces of empowerment', in S. Hickey and G. Mohan (eds) *Participation: From Tyranny to Transformation?*, Zed Books, London, pp92–107

World Bank (2002) 'India: World Bank supports two water sector projects in rural India in the amount of US$289.2 million', press release (2002/211/SAR) 19 February, http://web.worldbank.org/WBSITE/EXTERNAL/PROJECTS/0,,contentMDK:20034387~menuPK:64282137~pagePK:41367~piPK:279616~theSitePK:40941,00.html, accessed 23 June 2008

World Bank (2007) *Making the Most of Scarcity: Accountability for Better Water Management Results in the Middle East and North Africa,* World Bank, Washington, DC

4

Gendered Dynamics of Participation in Water Management in Nepal and Peru: Revisiting the Linkages between Membership and Power

Margreet Z. Zwarteveen, Pranita Bhushan Udas and Juana Vera Delgado

Introduction

In this chapter, we use examples and illustrations from Nepal and Peru to critically rethink conventional gender wisdoms and strategies to either empower women or achieve more gender equity in water management. By now, it is quite well documented that water decision-making and management in irrigation (at least the formalized and public parts of it) are, almost everywhere in the world, dominated by men (Arroyo and Boelens, 1997; Bustamente et al, 2005; Meinzen-Dick and Zwarteveen, 1998; Shyamala and Rao, 2002; Vera, 2005, 2006a; Zwarteveen and Meinzen-Dick, 2001). Male dominance is not limited to collective action in water, but has also been documented for other participatory natural resource management organizations (Agarwal, 2000, 2001; Sarin, 1995). The near absence of women in management organizations is hard to reconcile with the rhetoric of accessibility, democracy and participation that characterizes thinking about and acting on user-based water management groups.

The most common approach to make sense of the non-participation of women from irrigation management decision-making processes follows conventional theories about the determinants of well-performing institutions, epitomized in the work of Ostrom (1990). These theories consider user organizations for water resource management as the main (and sometimes single) forum for decision-making, the location where rules and regulations, punishments, and rewards for collective action in water are generated. The theory goes that it is through these institutions that the contributions from members necessary for operating and maintaining the infrastructure are mobilized and channelled, and the ways in which water is acquired and distributed are regulated. When seen from the perspective of these theories, the non-involvement of women in the organization is a problem when women are among the users of the system or are among those affected by the operational rules. Their non-participation would negatively affect their possibilities for meeting their water needs and concerns. It would also imply difficulties for the scheme's management to enforce its rules on women: because they are not official members, it becomes difficult to punish them for free-riding or other types of non-compliance (Zwarteveen, 2006).

In line with this theoretical model, conventional gender and development policy wisdom tends to interpret the absence of women from management organizations as an indicator of gender inequity in water. Proposed remedies focus either on the identification and removal of formal entrance barriers for women or on quota systems that stipulate that organizations need to have a minimum number of female members. The assumption that inclusion and participation will lead to (or implies) equalization (with men) or even empowerment (Cleaver, 1999) informs many statements at international water conferences and guiding principles for the water sector (GWA, 2003).

Our aim in this chapter is to critically discuss such conventional policy wisdoms about the linkages between membership, participation and power. More specifically, we examine two central assumptions about the functioning of water user organizations on which the current model of participatory water management and conventional gender policy wisdoms are based:

1 That it is possible for participants in user-based irrigation organizations to bracket status differentials and power inequalities and to deliberate as if they were social equals. The assumption, in other words, that water user organizations are relatively insulated social and political domains in which the behaviour of members is primarily determined by rules and laws that are internal to these domains.
2 The assumption that water powers and decisions predominantly happen within this formally designated domain of the water user organization. Women's lack of participation in such formal bodies and meetings have been interpreted as women's exclusion from water management decisions: they are not where researchers and policy makers tend to assume management powers are situated.[1]

Our analysis engages with, and is informed by, the emerging literature criticizing an over-optimistic faith in participation and inclusion in user groups. Rather than just focusing on institutional form and membership, this literature suggests the need to focus on institutional processes and patterns of inclusion and exclusion, and on the dynamic and negotiated nature of institutional evolution (Cleaver and Franks, 2005; Mehta et al, 2001).

We use examples and illustrations from Nepal and Peru to propose an alternative understanding of the linkages between membership, participation and power in water. The information from Peru is based on field research by Juana Vera in 2005 and 2006 in the Colca Valley in the occidental Southern Andes, focusing on the community of Coporaque. This is an area with little rainfall (354mm/year according to sequential data of 100 years), which is concentrated in only two to three months around February to March. To cultivate crops, farmers need to use irrigation water from August until March. Water is one of the main limiting factors of production; on average farmers only have an irrigation turn of one day once every 40–70 days. A little extra irrigation would immediately translate into higher productivities.

The irrigation system of Coporaque consists of an interconnected network of two principal and three secondary canals that lead water to different reservoirs. One of the main canals called Malqapi has been in existence from before the Inca period. This canal captures water from the snow-capped mountain Willcaya (5250m above sea level) and conducts it through a system of interlinked rivers and channels called Sawara-Qantumayu to two very old reservoirs, Santa Rosa and Mallkuqocha. The canal only carries water between August and January, in quantities varying between 40 and 80l/sec. Until 2003, this canal was the only source of water that all 320 water users made use of. The second main canal, the Coporaque, only came into use in 2003. It takes water from the Colca river, and although it has a maximum capacity of 280l/s it usually only carries about half of this flow. This is because large amounts of water from the Colca river are diverted to a large modern irrigation system much lower in the plains. The canal Coporaque also feeds the Mallkuqocha and Santa Rosa reservoirs. The big advantage of this canal is that it carries water throughout the year, allowing farmers to crop two harvests a year. At the time of the study, the canal Coporaque was the most important source of water. The secondary canals Chilliwitira-Wallallik'ucho, Qachule and Ch'aqere are historic and used respectively by 100, 120 and 25 users.

The data from Nepal are based on field research in Eastern Nepal led by Pranita Udas in 2004 and 2005 focusing on Udayapur district. The average annual rainfall of the district is 2152mm, but is very erratic. Of the total rainfall, 80 per cent rain falls during the monsoon. Because of spatial and temporal variation in rainfall, irrigation is important as a buffer in periods of drought. Two of the three water user associations discussed in this chapter are located in the plains of the Terai, and one is in the hills:

1 The Lama Khola irrigation system was initiated by farmers. After a recent rehabilitation with government support the total command area at the time of the study was 200ha. The project area is accessible by road during the dry season, however in rainy season the road becomes a drain.
2 At the time of the study, the Bhusune Asari Irrigation System had 281 members, of which 22 were female (nine per cent). The Bhusune Asari irrigation system was constructed with a loan from the Asian Development

Bank to the government of Nepal. The project began in 1995 and was completed in 1996, when the water user association, represented by 11 male committee members, took charge of the system. The main canal of the Bhusune Asari system is 6km long, and it has a gross command area of 144ha, and its headwork is a side intake scheme. The fact that water hardly ever reaches the tail end has weakened relationships among the users. The committee was blamed by most of the users for mismanagement of the funds, and the mistrust between the users and the committee has led to difficulties in mobilizing labour for repairs and maintenance.

3 The Baruwa irrigation system is located in the plains and has field canals on either side of the main canal. Initiated by local Chaudhari farmers, the canal was rehabilitated in 1991 with support from the government irrigation office and, although old, the command areas of the system increased after rehabilitation. A first rehabilitation was started in 1991 and completed in 1993, when the system was handed over to users. The gross command area as officially registered is 150ha, and the net command area developed was 70ha. The total canal length was 1750m with some 400m of idle canal, and the canal discharge at that time was 250l/s. The system was again rehabilitated in 2003, when the intake was improved and canal linings were done. The total population benefiting at the time of the handover was 125 households. At present the total number of users has increased to 215, partly because the canal was extended to new villages.

Before going into the details of the Peruvian and Nepali case studies, we first briefly describe the theoretical origins of models of participatory irrigation management that are most commonly used and referred to in irrigation policies and guidelines. We focus on how these conceptualize the linkages between membership, participation and power.

Participatory Irrigation Management: Theories and Policies

Policy attempts to promote participation in irrigation management have theoretical roots in different schools of thought, among which thinking about common property management and new institutionalism are probably most prominent. Thinking about common property management stems from the continuing debate about the tragedy of the commons (Hardin, 1968; Bromley, 1992) which has been revived by growing concerns about the environment. It tends to have a rather legal focus, and examines the prospects of community ownership and management of resources in the light of environmental sustainability. New institutionalism primarily focuses on the creation of financial incentives by markets and institutions (often conceptualized as organizations, Cleaver, 1999, p600) so as to improve the efficiency of resource use. Both schools of thought share a strong belief in local institutions for increasing irrigation management performance. This belief is based on the expectation that institutions help formalize mutual expectations of cooperative behaviour, allow the exercise of sanctions for non-cooperation and thereby reduce the costs of individual transactions (Granovetter, 1992). Although these theoretical premises and assumptions have been subject to criticisms and doubts (Cleaver, 1999, 2000; McCay and Jentoft, 1998; Mosse, 1999), they continue to guide water policies and projects.

Following the above logic, most efforts to increase participation in irrigation are first and foremost inspired by hopes of increased efficiencies (both in water use and in the use of public funds) and productivities. Participation is promoted primarily because it is expected to help managers, planners and engineers meet their objectives of making irrigation systems perform better, as the following citation neatly illustrates:

> The success of irrigation projects depends largely on the active participation and cooperation of individual farmers. Therefore, a group such as a farmers' association should be organized, preferably at the farmers' initiative or if necessary with initial government assistance, to help in attaining the objectives of the irrigation project. Irrigation technicians alone cannot satisfactorily operate and maintain the system. (Asian Development Bank, 1973, p50, cited in Ostrom, 1990)[2]

Pioneers of participation in irrigation considered water users and irrigators a kind of software that was essential for making more effective and efficient use of the hardware of physical structures for the capture, conveyance, distribution and drainage of water (Uphoff, 1986).

Participatory irrigation policies are often cloaked in the rhetoric of empowerment (Cleaver, 1999), but not much attention is usually given to their precise meaning, nor are participatory approaches in irrigation normally evaluated in terms of empowerment (that is, for instance in terms of whether they enhance the capacity of individuals to improve their own lives, or in terms of whether processes of social change have been set in motion to the advantage of marginalized groups). Indeed, as Sagardoy contends in discussing empowerment efforts in irrigation 'there is often no real intention of dedicating the efforts and resources that such programmes will require' (Sagardoy, 1995, p40). Instead, participation in itself is considered empowering (Cleaver, 1999), a thought that is likely to stem from the widespread belief that chances of fairness in access to and sharing the benefits of a resource are greatest when it is managed under a common property regime.[3] This belief is fuelled by the fact that many people share (the use of) a resource, and by the (assumed) existence of traditional community values of solidarity, cooperation and independence.

Efforts to promote participation in irrigation usually do not devote much attention to who are or should be the participants of water user organizations. Instead, the group of farmers or irrigators is referred to as a group that is already existing and easily identifiable: those people who are served by a common irrigation facility. Participation is about the involvement of this group in the project or system of another group, the engineers or irrigation managers who are part of a state irrigation bureaucracy. Implicit in the theory that informs creation of water user associations (WUAs) is that sharing the irrigation facility is what binds users together and what motivates them to meet and collaborate. Also in theory, for water user organizations to effectively and successfully undertake the tasks they have been assigned, users are expected to set aside their non-irrigating roles and identities. The ideal model of a water user association thus presupposes the absence of social inequalities other than those ordained by the physical layout or division of the irrigation system, or by the organizational division of irrigation related functions (Zwarteveen, 2006).

Such institutional theorizing reflects a belief that can be characterized as politically liberal. The water user organization and the irrigation management decision-making processes that go on within it are seen as (and indeed ideally need to be) insulated from what are considered (in liberal terms) non-political or pre-political processes, those characteristic for instance of the economy (or the market), or the family (the household) (Fraser, 1997). The theory not only assumes, but also actively proclaims the possibility and desirability of a power-free domain of deliberative interaction in which social differences and inequities are effectively set aside, and in which participants speak to another as if they are social and economic peers. This is for instance reflected in phrases like 'social divisions should not be so serious as to disrupt communication and decision making between farmers' (Vermillion and Sagardoy, 1999, p72). Importantly, the success of irrigation management institutions is seen to be, in part, a function of the success with which they have achieved this insulation of water management from the rest of the world.

The social context in which community organizations function is not entirely ignored, but is seen in the light of this problem. The ultimate concern of much theorizing and thinking about participation in irrigation is to unravel the determinants of well-performing irrigation management institutions. What good performance means has already been decided, based on universal laws of human nature and behaviour, and mostly expressed in rather narrow technical, productionist and economic terms. Real situations are thus described and judged on the basis of whether or to what extent they follow, or can be made to follow, the ideal model. The existing social relations of power and the existing culture and norms are loosely treated as the raw material from which institutions can be crafted, 'the institutional resource bank from which arrangements can be drawn which reduce the social overhead costs of co-operation in resource management' (Cleaver, 2000, p365).

Increasing the Participation of Women

In spite of the rhetoric of inclusion and democracy that pervades participatory management thinking and policies, most existing irrigation organizations consist of mostly or only men (Boelens and Zwarteveen, 2003; Meinzen-Dick and Zwarteveen, 1998; Zwarteveen, 2006). This is also true for Peru and Nepal. In Peru, participation of women in irrigator organizations is generally much lower than that of men, but there is some variation throughout the country. In the Southern Andean region, characterized by more indigenous influences and by smaller irrigations systems with a longer history, participation of women is generally higher than in the Northern coastal area with more mestizos and larger, more modern irrigation systems. In the studied mountain village of Coporaque, female participation in the formal water organization was 30 per cent; 90 of the 300 registered members were women. This is high when compared to for instance the coastal Chancay-Lambayeque system, where no women at all were registered in spite of the fact that 41 per cent of the land is joint property of husband and wife (Kome, 2002). It is also high when compared to most other mountain systems, such as that of Llullucha near Cusco, in which women were even officially prohibited from being members of the water user organization (Vera, 2004, 2005).

In Nepal, irrigation organizations generally are almost exclusively male domains. A study conducted by the Water and Energy Secretariat of 15 irrigation systems revealed that only three WUAs had 18 per cent female representation in their committees, while the rest were all-male committees (WECS, 1998). Likewise, an analysis of the implementation of the Second Irrigation Sector Project revealed that out of 249 WUAs, only 43 per cent had at least one woman on their committees. Twenty-five per cent of these had 20 per cent participation of women (Subedee, 2001). Data from the National Federation of Water Users' Association show a similar pattern. Analysis of 588 WUAs registered with the Federation from 31 districts in 2006 indicates that only in five districts did all the WUAs have at least one woman on the committee. Only 58 women assumed office bearer posts, either as president, vice president, secretary, joint secretary or treasurer. Twenty districts out of the 31 had an average of less than 20 per cent female participation on committees, whereas only four districts had more than 20 per cent women on committees, on average.

That there are few women in irrigation decision-making is something that is increasingly recognized, and most irrigation projects and policies include specific gender components or propose measures to increase and improve the participation of women in water decision-making and management. Quota systems that stipulate a minimum participation of women in user organizations are strategies frequently used towards this goal. Nepal is a good example here. Influenced by global discourse on gender, and because of donor conditionality, the government of Nepal has made female participation in water user associations mandatory. The irrigation regulation of 2000 stipulated the compulsory presence of at least two women members in executive committees of at most nine members (Bhushan Udas and Zwarteveen, 2005), while the irrigation regulation of 2003 further stipulated that irrigation committees should have at least 33 per cent of women, if available.

However, and as the above given figures show, the results of this proactive gender policy have been mixed at best. In one of the systems studied, the Lama Khola irrigation system, most of the users have a *magaloid* ethnic background,[4] a community with relatively egalitarian gender relations (Acharya and Bennett, 1983). The relatively high percentage of female members in the WUA reflects this: at the time of the study the WUA record indicated that 64 out of 358 registered members were women, or 18 per cent. The WUA was considered as one of the best-performing in the district. During the 1999 committee elections, and complying with government regulations to have at least 20 per cent women in the committee, three women were chosen to sit on the 12-member committee of the WUA. When the community was visited in 2005, however, the committee consisted of men only. One of the female committee members had migrated out of the village and was not replaced; the other two women had both resigned in 2003. The male committee members explained that the female members never attended WUA meetings, and their ability to serve on the committee was therefore in question. The female ex-committee members indicated that they found it difficult to find the time to attend meetings because of their other activities.

The few successes in measures to promote female participation can be partly attributed to their top-down character: they are dictated from the highest policy levels. Implementers at different levels often do not attribute a high importance to gender policy objectives, while female irrigators themselves are often not aware of the policy, and few see inclusion in water users' organizations as a priority need (Bhushan Udas and Zwarteveen, 2005).

The situation in Peru is somewhat different. Here, water laws and regulations do not make specific mention of women or gender, on account of the fact that these are supposed to be gender neutral (Vera, 2006b). The only formal requirement to become registered as a member of a water user organization is to be a landholder and to pay yearly the established water tariff. This makes many women eligible for membership, since the Peruvian Civil Code gives equal rights to sons and daughters in relation to land inheritance, and customary laws in Andean communities also stipulate that both sons and daughters should inherit parental land. In spite of the relatively high number of female landholders in Peru, however, several studies show that the number of female members in water user organizations is lower than the number of female land-title holders (Kome, 2002). This was also true for Coporaque. When asked why they did not use their land rights to obtain membership of the water user organization, women replied that only one representative of each household can register as a member, and that official water authorities usually consider men, as the heads of households, to be best suited for this role. Hence, in spite of the formal eligibility of most farm women to become members of a water user organization, most women (with the exception of widows) remain outside of formal water decision-making domains and also do not enjoy the formal possibility of owning water or voicing water concerns. Some Coporaque women experienced difficulties when applying for loans or legal assistance, because they did not formally exist as water users. The formal title of membership was often required to be eligible for certain financial or legal support.

At the same time, many in Peru consider it desirable to have some minimum female participation in water user committees. When new authorities are elected, for instance, the users are always asked to elect at least one women board member. Women have a reputation of being more honest and trustworthy than men, which makes them suitable as secretaries or treasurers. Women can also be helpful during the many social events that WUA committees are expected to organize. When receiving important official guests, or when organizing training sessions or the water feast, the female committee member is the one in charge of the drinks (*chicha*, a special beverage made of maize) and food.

When comparing the two countries, what attracts attention is that the formal barriers to female inclusion in water user organizations are being removed or do not exist. In spite of this, female participation remains much lower than that of men (in the case of Peru), or almost negligible (in the case of Nepal). Analysing participatory forest organizations, Agarwal states:

> women's limited participation in decision-making ... means that they have little say in the framing of rules on forest use, monitoring, benefit distribution, etc., with implications for distributional equity and efficiency. (Agarwal, 2000, p286)

Most analyses share this assumption of positive linkages between participation and:

1 power to (co-) decide on how natural resources are used, maintained and distributed;
2 access to the resource and to information about or benefits stemming from the use of the resource.

While we wholeheartedly agree on the importance of concerted efforts to more fully involve women in water decision-making and management, we use the rest of this chapter to shed some critical doubts on the assumed linkages between membership, power and benefits.

Water Management as a Masculine Domain

As noted above, the assumption of positive linkages between membership and participation in decision-making is based on an ideal model of water user organizations, in which it is to be an arena in which participants set aside such characteristics as differences in birth, wealth and gender and speak to one another as if they are social and economic peers. The operative phrase here is *as if*. In the reality of many irrigation organizations, the social inequalities among participants appear difficult to eliminate or even just bracket. In terms of gender, it is clear that women's freedom to publicly interact with men is not just a function of removing formal entrance barriers, but is constrained by social practices and norms that define what sorts of interaction are permissible, with which men, in what contexts and using defined modes of conduct (Agarwal, 1994).

Observations at meetings of the water user organization of the Bhusune Asari irrigation system in Nepal serve to illustrate that what happens inside the domain of the water user organization reflects what happens outside of it; water user organizations are not insulated from the rest of society, but are on the contrary embedded and constitutive of it. In Bhusune Asari, membership of the water user organization is open to anyone who cultivates irrigated land, but restricted to one person per household. Most women do not own their own land, and men are considered heads of households, which is why they usually register as members. At the time of the study, the Bhusune Asari Irrigation System, which was rehabilitated in 2004 and 2005, had 281 members, 22 of whom were female (nine per cent). Female members were those women who had land in the command area. Fifteen per cent of the total cost of the rehabilitation project was to be met through contributions from farmers, of which five per cent was to be made in form of cash deposited with the bank, while the rest was to be contributed in labour. During the rehabilitation the amount of work to be done to comply with this fifteen per cent rule was identified and the WUA leaders called for equal participation of labour from all users, irrespective of the size of landholding. The work included removal of soil and digging at the main intake, about one and a half hour's walking distance from the tail end. A few people from the tail-end present at the meetings complained about the amount of time it would cost them just to be present for the work, and among those protesting was a female committee member. She also expressed her concern about the ability of poorer households to contribute labour without direct compensation, as their livelihood depended on working for a daily wage. She especially emphasized the plight of female headed households. The concerns of the tail enders were, however, overruled by some rich male landowners and a male committee member, who argued that all have to work together in order to bring water to the tail end. 'If we do not all collaborate, water will not reach the tail-end.' The male committee members and the larger landowners also successfully argued against the suggestion that labour contributions should be proportional to landholdings and instead proposed that all contribute equally, irrespective of the size of land. It was difficult for the poorer tail-end households to disagree with the large landowners, because many of them depended on these landowners for access to land (as tenants) and for work as wage labourers. This example shows how not just gender, but also wealth and location in the system (at the head end or in the tail end) together influence a person's leverage in meetings.

Similar observations about what happens at irrigators' meetings have been done in other studies in Nepal, which note that it is exceptional that women speak at all. In Tukucha Nala, for instance:

> many women, and also many men, did not perceive the meeting as a place where they could voice their concerns and opinions. Because of their lower social status and dependency on the bigger and more vocal men, most people remained silent at meetings and just listened to what the more literate and articulate people had to say. (Bajracharya, 2000; Bhushan and Zwarteveen, 2005, p41; Poudel, 2000; Pun, 2000; Schaaf van der, 2000)

It has already been noted that in Peru only a few of the women eligible for membership of the water user organization actually become members. When women were asked why they did not use their land rights to become a member of the water user organization, many answered that according to the official water authorities (Administración Técnica de Riego or ATDR) membership is limited to the head of the household, or to one representative of each family. Men are normally considered as the heads of households, and as the most suitable representative. At the same time, some women admitted that not becoming a member of the Comisión de Regantes, (the formal water organization, recognized by the Irrigation Department and officially registered) made it easier for them to evade the collective and compulsory work parties to clean and maintain the canals (faenas), while it also exempted them from the obligation to attend official meetings of the WUA. They considered going to such meetings a loss of time in two ways: nobody listens when a women speaks and they could use the time more beneficially to do things at home or even work in the fields.

Observations at the meetings of the Comisión de Regantes in Coporaque showed that those women who attended meetings barely used the meetings to voice their concerns or to articulate their opinions. Only in two of ten meetings attended did a woman formally ask the Assembly for the opportunity to speak. In one of those two cases, the woman who wanted to say something experienced great difficulties catching the attention of the authorities. Another woman present stood up and acted as intermediary, demanding that the turn to speak be given to the women who had raised her hand. In the few instances that women did speak, they made their point rather quickly, in contrast to some of the men who easily spoke for more than ten minutes. At these formal meetings, most women just whispered among themselves, nevertheless making so much noise that the authorities had to ask them to keep quiet.

Many studies of irrigation systems in the Andes report very similar findings (Arroyo and Boelens, 1997; Bastidas, 2004; Gutiérrez and Cardona, 1998; Lynch, 1991). A woman in the Peruvian Llullucha community in Paucartambo commented:

> we are too shy and too afraid to voice our ideas about water management in community meetings; sometimes we talk to our husbands, but we do not have experience to talk to outsiders.

Men from the same community added:

> Llullucha women do not have any experience in public meetings, and they cannot talk in public. Women lack character in making decisions, and most are shy and illiterate. (Vera, 2005, p109)

That women, like many men, are not very successful in getting their opinions and needs across in meetings is partly because the social dynamics at meetings reflect wider social hierarchies and power relations. Influence in decision-making and the authority to make decisions are not just derived from criteria internal to the organization, but the result of historical processes of investments and of social relations of power that are coloured by ethnicity and gender.

There is more at stake, however. The observations during water management meetings in Udayapur and Coporaque also suggest that discursive interactions within the public domain of the water user organization are governed by protocols and styles of decorum that are themselves correlates and markers of gender inequality (Fraser, 1997). Both in Nepal as in Peru, women said they felt diffident about articulating their opinions at meetings, referring both to their illiteracy (or lack of Spanish language skills in the case of Peru), but also to a prevailing gender ideology that discourages women's public performance in meetings. To be outspoken and opinionated are markers of masculine distinction, ways of defining and reconfirming masculinity and male superiority, evaluated positively when found in men but negatively when found in women. Since such markers are attached to membership and participation in the water user organization, the water user organization itself becomes a space to define and reconfirm existing gendered norms and practices, of performing masculinity, and of distinguishing a separate male public domain from a female domestic and private domain.

Both in Peru as in Nepal, women's lack of participation and influence in formal meetings was not a reflection of their overall lack of interest or participation in water management or irrigated agriculture. For instance, women's involvement in the construction works in Bhusune Asari, one of the systems in Udayapur, was almost equal to that of men. In Coporaque as well, women played visible and prominent roles in canal cleaning and construction, in the traditional water feasts and in irrigated agriculture. Women's invisibility in the water management organization and its meetings reflects a gender division of labour that delegates the task of public decision-making to men, and marks it as a masculine domain.

This association of water management with masculinity is reinforced and reasserted through irrigation agency officials, most of whom are male engineers. This shows up most strongly in Coporaque, although it is also true for Udayapur. Irrigation engineers working with government agencies have been socialized during their training to delineate their professional identity through a clear gender demarcation: the attributes and skills that are seen as typical characteristics of good irrigation professionals, such as technical competence, physical strength, mathematical skills, being in command, self-confidence and rationality, tend to be seen as characteristics more commonly found in men (Zwarteveen, 2006). Engineers working in the communities often displayed a clear preference for interacting with male farmers, and when asked admitted that they considered men as the real and better irrigators and water managers. As one woman explained:

> in the faenas they don't accept women. When we go to these works, they always ask us 'where is your husband?' He is the entitled member and not the woman ... They think that women cannot work like men, they consider our work as a half of the men.

One example illustrating the preference of engineers for male farmers comes from a neighbouring community of Coporaque, Lari. Here, a development organization aimed to increase irrigation efficiencies through an improvement in irrigation methods and a reduction of the frequency of irrigation. They presented their proposal in a meeting of the water user organization and everybody agreed. The development organization proposed what they considered a truly participatory method of training and learning, by which peasant irrigators teach each other how best to irrigate. Expert irrigators, called kamayoq, were invited from other villages that already had an efficient water management system. All kamayoq were male farmers, and they were only introduced to other male farmers, even though women are the ones more involved in irrigation in Lari. In one incident, a woman observed that the kamayoq and her husband were doing some designs on the plot where she planned to irrigate. She approached them, asking 'what are you doing here, just now that I must start to irrigate?' The husband tried to explain to her that the kamayoq was teaching him how to improve their irrigation method. The wife replied 'but I am the irrigator, and I have my own way how to deal with water, please could you remove all these designs?' The husband had to give in, knowing quite well that she was the one responsible for irrigated farming. The example is illustrative in that even though most engineers working at field level know that many women are involved in farming and irrigating, few would identify female farmers as knowledgeable or as the people to interact with about (improved) farming and irrigation methods.

Organizational and Institutional Plurality

Policies and interventions often view water user organizations as the single, or at least the most important, domain in which irrigation decisions are taken, and as the place where irrigation powers are concentrated. Such a view informs the idea of women as excluded from management decisions: they are not where researchers and policy makers tend to assume management powers are located. Gender, then, is seen and analysed primarily as a form of exclusion and social disadvantage, suggesting a rather unitary notion of power in which the included are powerful and the excluded are powerless. Our observations from Coporaque and Udayapur, instead, suggest that irrigation management powers are uneven, uncertain and struggled for rather than concentrated in the hands of the included (Jackson, 1998). In this section we suggest that another reason why the linkages between participation (or inclusion) in water user organizations and powers over water are less direct than received policy wisdom assumes, is that the formally designated institutional space for water management decision-making only plays a minor role in actual water distribution or resource mobilization.

In Udayapur, Nepal, the main task of the water user organizations is not so much the organization of water distribution schedules and mobilization of resources for maintenance, but maintaining good relations with outside officials. An important, if not the most important, task of the water user organization president is to direct flows of money and resources from outside sources to the village. It is important in this respect that the water user organization maintains an image of credibility and legitimacy in the eyes of outsiders. This can be done by diligently keeping all written and financial records, since these are the evidence most outside organizations look for in search of indicators of organizational performance. It is also important that the presidents of community organizations portray themselves as trustworthy and respectable; they represent the face of the organization. It is telling in this respect that the chairman of the construction committee of the previously referred to Bhusune Asari Irrigation system was also the president of the Baruwa irrigation system further downstream, which had been rehabilitated a few years earlier, in 2003. He was clearly a person who inspires confidence with outside agencies.

This president claimed that the success of the Baruwa rehabilitation effort was mainly due to his skills in mobilizing labourers, and because of his transparent and democratic leadership. He did indeed succeed in calling the attention of the District Irrigation Office officials, who accepted the request of the users to appoint him as a chair of the construction committee, even though he himself was not a user of the Bhusune Asari system. He was also quite successful in mediating between the irrigators and the engineers in solving disagreements about the design of the new system. The irrigators wanted the older siphons replaced with an aqueduct, because the siphons were difficult to clean and therefore proved a nuisance to use. Yet, the engineer refused because an aqueduct would cost too much. The chairman nevertheless succeeded in saving so much money during the earthworks that the aqueduct could be built.

While successful at the construction stage and in dealing with outside organizations, the chairman of the Bhusune Asari construction committee neglected his duties as president of the water user organization of the Baruwa system in terms of water management. Instead, water management continued to be arranged and decided by the traditional Tharu village leader, called the jewar. The jewar, who is always a man, is responsible for managing all village affairs, including water management. He also plays an important role in social ceremonies such as weddings and funerals. The gorait is his assistant, a messenger who passes information from one house to another and from one hamlet to another. Whenever there was a need to clean the system, a farmer who planned to irrigate first asked the gorait to convey his request to the jewar, who could then plan and announce a date and time for all villagers to come together and clean the canal. There was no clear water distribution plan to indicate the time of each irrigator's turn. When water was scarce, people needed to wait in the fields for their turn, and negotiate with their field neighbours about the order in which different people would receive water. One's ability to actually receive water at the preferred time also depended on one's bargaining powers and skills, which in turn reflected larger social hierarchies. Indeed, without much status and power, waiting for water could become quite time-consuming. One woman for instance experienced a sharp decline in the productivity of her fields after her husband left to work outside the village because she was not able to spend long hours in the fields to wait for water, especially at night and also did not manage to convince her field neighbours to allow her to take water first.

That the formal water user organizations for water management are not crucial when it comes to actual water management is also illustrated by the fact that many irrigators in the three systems studied hardly knew the members of their water management committees. In the Upper Baruwa Irrigation System (located upstream) more than 44 per cent of the users did not know any of the committee members. In the Lama Khola Irrigation System, more than 81 per cent replied that they did not know any WUA committee member. Only in the Baruwa irrigation system did 14 per cent of the users reply that they did not know any member of the water user organization. Here, almost all knew the president, who is quite famous and who also is the president of other village groups, such as the drinking water organization.

The actual function of the formal water user organizations in the three systems studied was related more to bringing resources from outside to improve the system structure, than to performing day-to-day activities like arranging water distribution, mobilizing resources, labour for maintenance or solving water related conflicts. The degree and nature of different people's participation in the water user organization, therefore, had few direct linkages to their power and rights related to water.

Observations in Coporaque, Peru, are strikingly similar. Officially, Coporaque's irrigation waters and systems fall under the control of the ATDR. At community level, the official water management body is the Comisión de Regantes. In practice, however, many important water management duties continue to be assumed by the traditional local organization, the regina, and authority, the regidor. The regidor is responsible for distributing water, but also for the maintenance, cleaning and improvements of the irrigation infrastructures. In addition, a very important duty of the regidor is the organization of the yearly water feast[5] and all other water rituals. In Coporaque there are two principal regidores and three minor regidores.

Most regidores are men, but women can also become regidoras. In everyday reality, both spouses receive the charge. As one of the regidores explained:

> the charge falls on both of us, it would be impossible only for one man to fulfill it ... The woman is most important for organizing the water feast. She must look after the preparation of all the beverages and meals so that the whole community can be entertained during three days ... Also when I am busy in the fields controlling and distributing the water to the users, my wife is in charge to carry out all the responsibilities at home, she has to see how to manage the irrigation and agricultural activities, she must look after the livestock. That is why my children complain ... What can I do, this duty is a sacrifice, but we must comply and provide the service to our community.

The function of regidor is a rotating one, and everyone in the community is expected to serve as regidor at least once, reflecting the community service principle that is extremely important in many Andean communities. Someone who does not serve the community is considered as an allku (dog, petty slacker). Serving the community confers status and prestige, which in turn grants rights and powers (Moffat et al, 1991; Vera, 2004). Women who occupy the position of regidora, directly or indirectly (as the wife of the regidor) acquire special status, respect and enhanced self-esteem. Women are largely responsible for organizing the water festival and establishing networks of contacts, alliances and reciprocal relationships among families and relatives who will help prepare food and beverages to offer the community work party members.

The fact that women are hardly involved in the Comisión de Regantes is all the more striking given their active engagement in the reginas. In terms of numbers, most of the time there were as many women as men in the reginas, and women frequently raised their voices and spoke out when their water entitlement was at stake or when someone who stole water had to be punished. A few reginas were led by women, and female regidoras were seen to perform quite well. In reginas, the language spoken is Quechua, the local language, which removes one important hurdle for many women to participate when compared to the Comisión de Regantes, where Spanish is the main language.

As in Udayapur, the Comisión de Regantes seems to be important primarily because it lends legitimacy to the community's water-related demands in the eyes of outsiders, while also making it possible for the community to request assistance and support from the government. This is truly important; it is for instance through the formal water user organization that negotiations about who owns and can make use of upstream water resources are carried out. Yet, for water management inside the community, and even for arranging water distribution and resource mobilization questions with neighbouring communities, the irrigators do not normally make use of this formal organization but instead rely on traditional institutions that are embedded in everyday community life and politics, and reinforced by rituals and parties. Women's involvement in these traditional institutions is much higher, and much more visible, as compared to their participation in the formal Comisión de Regantes.

In Udayapur, as in Coporaque, many important water management decisions are also taken in the fields (along the canals) and in households. That decisions over whose turn it is to have water are made in the fields has already been mentioned for the Baruwa system in Udayapur. In Coporaque, this is also true. Agreements on water distribution schedules are renegotiated in the fields, where water turns may be exchanged among field neighbours. Husbands and wives, and other family members, often discuss and negotiate agricultural decisions among themselves within the domain of the household.

Conclusions: Are Women Excluded From Water Management?

The ideal model of a water user association presupposes the absence of social inequalities. In the reality of many participatory irrigation programmes, the social inequalities among participants are not eliminated, but only bracketed.[6] But can they be effectively bracketed? Our observations from Nepal and Peru suggest not; discursive interactions within the public domain of the water user organizations are governed by protocols and styles of decorum that are themselves correlates and markers of gender difference and inequality (Fraser, 1997). Women said they felt uncomfortable with the aggressive tone in which matters were discussed at meetings, while they also referred to the prevailing gender ideology that discourages outspokenness in a woman to explain their reluctance to play more prominent roles in irrigation decision-making.

Gendered norms and divisions thus function informally to marginalize women and to prevent them from participating as peers. Skills and abilities deemed necessary to function well in meetings, or to engage in other irrigation management tasks, are thought to come more easily to men than to women. What is more, they tend to be evaluated positively when found in men and negatively when found in women. The professional culture of irrigation engineers helps to further sanction and institutionalize the association of masculinity with water management expertise.

Some quite important efforts have been undertaken by projects and through policy measures to increase and improve the participation of women in water management organizations. Unfortunately, most studies have a tendency to take the very fact that there is female participation (or that there are female members) as the most important or only indicator of success in gender terms. But is it indeed true that a redefinition of membership rules, or a removal of the entrance barriers for women, leads to irrigation systems that are better geared towards meeting the interests and needs of women? Do these improve women's access to and control over water? The assumption that this will happen is crucially based on the above-mentioned image of the water user organization as the single, or most important, domain of water decision-making and as the place where water powers are concentrated. This is an image that corresponds poorly with our observations from Nepal and Peru.

Long observed that:

> the question of non-involvement should not be interpreted to imply that nonparticipants have no influence on the constitution and outcomes ... On the contrary, they can as 'backstage' actors, have a decisive influence on strategies and scenarios. (Long, 1989, p240)

Our studies in Nepal and Peru illustrate that power over water is not just where the men are, even though women's own direct water interests may well be better served by maintaining and reproducing the portrayal of irrigation as a male domain. Delegating the costs and responsibilities of accessing water to men also allows women to remain relatively invisible and thus freer in their irrigation behaviour. In other words, women's control over irrigation may be located partly due to their invisibility and absence from the formal water user organization. This carries fewer costs and allows women to remain within the boundaries of what is considered appropriate female social behaviour. It also implies that women may be much more involved and have greater influence and a greater voice in decisions than is suggested when just looking at their participation in formal water management domains. When it comes to deciding about water turns or about the mobilization of labour and funds for maintenance women often play important roles.

At the same time, one may wonder how strategic these important roles are in terms of wider social hierarchies and power relations. Our observations from Peru and Nepal indicate the water user organizations' main function is not just the management of water, but that of an intermediary between formal state and donor agencies from outside the community with the water users. Actual water management in both the Peruvian and the Nepali case studies was largely done in other domains, and through other means. Formal water user organizations thus existed for the sake of satisfying donor and government demands and were maintained by water users because their existence provided access to outside resources and established their legitimacy in formal laws.

In Udayapur and Coporaque, government interference with and interventions in the irrigation systems were relatively recent and partly inspired by desires to manage water more efficiently so that it could be made available for other communities or purposes. Some water management decisions that used to be taken at community level are increasingly taken at other levels of decision-making by formal water agencies. Water management is no longer just an internal community affair, but increasingly implies having to deal with the outside world, especially for legal recognition, and for obtaining funds and technical support. The effect of this is, as Lynch argued for irrigation management in Peru that:

> the scarce resource for water management is money for repair and maintenance of infrastructure, not water per se ... Control over access to resources no longer occurs at the interface between the community and the household (between the canal and the farm field), but at the interface between the community, as represented by its formal irrigation organization, and the funding agencies. (Lynch, 1991, p49)

Our studies suggest that there is a danger of this process of externalizing water management powers being accompanied with a new gender division, with (some) men becoming increasingly engaged and important at extra-community levels and women remaining active at community levels. The process of externalization may become one of masculinization; because men have had more exposure to the outside world from a young age, they also have more experience and are considered as more able to deal with government and project officials. Men in general also have more education and are more literate, which increases their confidence in dealing with outsiders and their ability to discuss technical and administrative matters. In addition, outside officials (extension officers and project staff), most of whom are men themselves, tend to contact men more readily than women partly because they only recognize men as irrigators and water managers (Lynch, 1991, p47).

In sum, our studies in Nepal and Peru suggest that understanding the linkages between powers over water and gender equity involves looking beyond the boundaries of formal water management organizations. And, while admitting that these boundaries are flexible and permeable, a gendered analysis of water management would be well served by critically investigating how the social construction of these boundaries itself becomes a symbolic and political act of delineating a specific domain for performing gender and gendered power. Our analysis also suggests that achieving real and lasting changes in unequal gender relations, or fostering changes that lead women themselves to actively demand rights and a voice in water management, cannot be expected to happen just as a mere effect of increasing the number of women in water user committees. More structural changes in gendered power relations require more general development efforts explicitly aimed at the empowerment of women, most of which would require broadening the scope of what is normally considered to belong to the area of water management. Literacy and education for women and preventing young girls from dropping out of school might be more effective entry points for such efforts than water. Further strengthening water user associations and possibly expanding their range of activities to make them less male-oriented and more attractive to women are others.

Acknowledgements

We would like to thank the Dutch Science Council NWO (respectively the programmes 'Water for Society' and Science for Global Development

'WOTRO') for financing the two PhD studies of Pranita Bhushan Udas and Juana Vera Delgado.

Notes

1 This analysis is based on and inspired by the discussion of Fraser (1997) about gender and democracy.
2 Michael Goldman, in a critical review of the commons literature, observes that the questions asked by commons scholars 'reflect the search for the holy grail of successful common models'. He concludes that 'their prescriptions are meant for the ubiquitous professional-class "we", recommending that development professionals get investment portfolios right, for the benefit of the development's alleged clients, the world's commoners' (Goldman, 1997, p2).
3 The reasoning is that private property is more efficient and common property more just. Baland and Platteau (1996, pp36–46) provide a review and discussion of this reasoning.
4 People in Nepal can be broadly divided in two categories. The first are people with Indo-Aryan origins, who are known as *nonmatwali* and who consider drinking liquor as a crime. The second are *magaloid* with Tibeto-Burman origins, known as *matwali*. In the *matwali* group drinking of liquor is important in rituals related to birth and death (Acharya and Bennett, 1983).
5 The water feast is a traditional religious celebration of ceremonies and rituals related to water. It is managed by the two main water regidores. In former times they were known as *Mallku Kamayoq* (in the former local language this meant the person in charge to service the god of water). In Coporaque the water feast is called Yarqa Haspiy and takes place during the whole month of August, entailing a series of ceremonial rites for the god of water, hills and land and offering food and drink to all community members who take part in the maintenance of the irrigation infrastructure.
6 It needs to be noted here that while the ideal of a water user organization as a power-free arena of deliberation and interaction is shared by many irrigation professionals, there are important distinctions between how different scholars and policy makers analyse and propose to deal with social and political differences between members. The major disparity is between the relatively influential group of sophisticated social engineers, and those (many from NGO or activist backgrounds) who believe in social organization as a way towards (partly) reducing social differences. The first group recognizes social differences but either believes that these are not important or thinks that there are successful ways of circumventing them. The second group explicitly questions social differentiation and believe that empowerment of some social groups is possible through facilitated processes of social organization.

References

Acharya, M. and Bennett, L. (1983) *Women and the Subsistence Sector: Economic Participation and Household Decision Making in Nepal*, World Bank Staff Papers 526, Washington, DC

Agarwal, B. (1994) *A Field of One's Own. Gender and Land Rights in South Asia*, Cambridge University Press, Cambridge

Agarwal, B. (2000) 'Conceptualising environmental collective action: why gender matters', *Cambridge Journal of Economics*, vol 24, pp283–310

Agarwal, B. (2001) 'Participatory exclusions, community forestry and gender: an analysis for South Asia and a conceptual framework', *World Development* vol 29, no 10, pp1623–1648

Arroyo, A. and Boelens, R. (1997) *Mujer Campesina e Intervención en el Riego Andino: Sistemas de Riego y Relaciones de Género, Caso Licto, Ecuador* (*Peasant Women and Intervention in Andean Irrigation: Irrigation Systems and Gender Relations, the Case of Licto, Ecuador*), CAMAREN, Riego Comunitario Andino, CESA-SNV, Quito, Ecuador

Bajracharya, P. (2000) 'Gendered water rights in the Hile Khola Kulo irrigation system, Shakhejung VDC, Ilam', in R. Pradhan, F. von Benda-Beckmann and K. von Benda-Benckmann (eds) *Water, Land and Law. Changing Rights to Land and Water in Nepal*, FREEDEAL/WUR/EUR, Kathmandu, Wageningen, Rotterdam, pp129–146

Baland, J.-M. and Platteau, J. P. (1996) *Halting Degradation of Natural Resources. Is There a Role for Rural Communities?*, Clarendon Press, Oxford

Bastidas, E. (2004) 'Women and water in the northern Ecuadorean Andes', in V. Bennett, S. Dávila Poblete and M. Nieves Rico (eds) *Opposing Currents. The Politics of Water and Gender in Latin America,* University of Pittsburgh Press, Pittsburgh, PA, pp154–169

Bhushan Udas, P. and Zwarteveen, M. Z. (2005) 'Prescribing gender equity? The case of the Tukucha Nala irrigation system, central Nepal', in D. Roth, R. Boelens and M. Zwarteveen (eds) *Liquid Relations: Contested Water Rights and Legal Complexity,* Rutgers University Press, New Brunswick, NJ, pp21–43

Boelens, R. and Zwarteveen, M. (2003) 'Water, gender and "Andeanity": conflict or harmony? Gender dimensions of water rights in diverging regimes of representation', in T. Salman and A. Zoomers (eds) *Imaging the Andes: Shifting Margins of a Marginal World,* CEDLA Latin America Studies 91, Aksant/CEDLA, Amsterdam, pp145–166

Bromley, D. (1992) *Making the Commons Work: Theory, Practice and Policy*, Institute for Contemporary Studies, San Francisco, CA

Bustamente R., Peredo, E. and Undeata, M. E. (2005) 'Women in the "water war" in the Cochabamba valleys', in V. Bennett, S. Dávila Poblete and M. Nieves Rico (eds) *Opposing Currents: The Politics of Water and Gender in Latin America*, University of Pittsburgh Press, Pittsburgh, PA

Cleaver, F. (1999) 'Paradoxes of participation: questioning participatory approaches to development', *Journal of International Development*, vol 11, no 4, pp597–612

Cleaver, F. (2000) 'Moral ecological rationality, institutions and the management of common property resources', *Development and Change*, vol 31, pp361–383

Cleaver, F. and Franks, T. (2005) *How Institutions Elude Design: River Basin Management and Sustainable Livelihoods*, BCID research paper 12, University of Bradford, Bradford

Fraser, N. (1997) *Justice Interruptus: Critical Reflections on the 'Postsocialist' Condition*, Routledge, New York, NY

Goldman, M. (1997) '"Customs in common": the epistemic world of the commons scholars', *Theory and Society*, vol 26, no 1, pp1–37

Granovetter, M. (1992) 'Economic action and social structure: the problem of embeddedness', in M. Granovetter and R. Swedberg (eds) *The Sociology of Economic Life*, Westview Press, Oxford, pp53–81

Gutiérrez Z. and Cardona, S. E. (1998) *Las relaciones de género en la gestion de aguas en sistemas de producción intensivas: estudios de Caso Combuye, Vinto, Cochabamba*, PEIRAV, Cochabababamba, Bolivia

GWA (2003) *Tapping into Sustainability: Issues and Trends in Gender Mainstreaming in Water and Sanitation*, Background document for the gender and water session, 3rd World Water Forum, Kyoto, Japan, GWA Secretariat, IRC, Delft, The Netherlands

Hardin, G. (1968) 'The tragedy of the commons', *Science*, vol 162, pp1243–1248

Jackson, C. (1998) 'Gender, irrigation and environment: arguing for agency', *Agriculture and Human Values*, vol 15, no 4, pp313–324

Kome, A. (2002) 'La copropiedad de la tierra, el derecho de uso de agua y el derecho de asociación de las organizaciones de osuarios del norte del Perú' (Co-ownership of land, the water rights and the right to membership in water user associations in northern Peru), in M. Pulgar-Vidal, E. Zegarra and J. Urrutia (eds) *Perú: El problema agrario en debate* (*Peru: the agricultural issue debated*), SEPIA IX, Lima, Peru, pp379–397

Long, N. (ed.) (1989) *Encounters at the Interface: A Perspective on Social Discontinuities in Rural Development*, Wageningen Studies, no 27, Wageningen Agricultural University, Wageningen.

Lynch Deutch, B. (1991) 'Women and irrigation in highland Peru', *Society and Natural Resources*, vol 4, pp37–52

McCay B. J. and Jentoft, S. (1998) 'Market or community failure? Critical perspectives on common property research', *Human Organization*, vol 57, no 1, pp21–29

Mehta, L., Leach, M. and Scoones, I. (eds) (2001) 'Environmental governance in an uncertain world', *IDS Bulletin*, vol 32, no 4, pp1–9

Meinzen-Dick, R. and Zwarteveen, M. (1998) 'Gendered participation in water management: issues and illustrations from water users' associations in South Asia', *Agriculture and Human Values*, vol 15, no 4, pp337–345

Moffat, L., Geadah, Y. and Stuart, R. (1991) 'Two halves make a whole: balancing gender relations in development', Canadian Council for International Co-Operation (CCIC), MATCH International Center, and Association Québécoise des Organismes de Coopération Internationale (AQOCI), Ottawa

Mosse, D. (1999) 'The symbolic making of a common property resource: history, ecology and locality in a tank-irrigated landscape in South India', *Development and Change*, vol 28, pp467–504

Ostrom, E. (1990) *Governing the Commons. The Evolution of Institutions for Collective Action*, Cambridge University Press, Cambridge

Poudel, R. (2000) 'Farmers' laws and irrigation: water rights and dispute management in the hills of Nepal', PhD thesis, Wageningen University

Pun, S. (2000) 'Gender, land and irrigation management in Rajapur', in R. Pradhan, F. von Benda-Beckmann and K. von Benda-Benckmann (eds) *Water, Land and Law. Changing Rights to Land and Water in Nepal*, FREEDEAL/WUR/EUR, Kathmandu, Wageningen, Rotterdam, pp195–216

Sagardoy, J. (1995) 'Lessons learned from irrigation management transfer programmes', in S. Johnson, D. L. Vermillion and J. A. Sagardoy (eds) *Irrigation management transfer: selected papers from the international conference on irrigation management transfer in Wuhan, China*, Colombo: IIMI and Rome: FAO,Saguneeti Sangam, WALAMTARI, Hyderabad, India

Sarin, M. (1995) 'Regenerating India's forests: reconciling gender equity with joint forest management', *IDS Bulletin*, vol 26, no 1, pp83–91

Schaaf, C. van der. (2000) 'Land, water and gender in Rupakot village, Nepal', in R. Pradhan, F. von Benda-Beckmann and K. von Benda-Benckmann (eds) *Water, Land and Law. Changing Rights to Land and Water in Nepal*, FREEDEAL/WUR/EUR, Kathmandu, Wageningen, Rotterdam, pp169–194

Shyamala, C. V. and Rao, S. V. (2002) 'Role of women in participatory irrigation management: a study in Andhra Pradesh', in R. Hooja, G. Pangare and K. V. Raju (eds) *Users in Water Management, The Andhra Model and its Replicability in India*, Rawat Pub, New Delhi

Subedee, S. K. (2001) 'Initial socioeconomic assessment: community managed irrigation sector project in Central and Eastern Basins (Third Irrigation Project), Nepal', Routledge, London

Uphoff, N. (1986) *Improving International Irrigation Management with Farmer Participation: Getting the Process Right*, Studies in Water Policy and Management, no 11, Westview Press, Boulder, CO

Vera, D. J. (2004) '"Cuánto más doy, más soy ...": discursos, normas y género: la institucionalidad de las organizaciones de riego tradicionales en los Andes del sur Peruano' ('I am as much as I give'... discourses, norms and gender: the institutionalization of traditional water organizations in the southern Andes of Peru), in F. Peña (ed) *Los Pueblos Indígenas y el Agua: Desafíos Del Siglo XXI* (*Indigenous People and Water: Challenges of the Twenty-First Century*), Colegio de San Luís, WALIR, SEMARNAT, IMTA, San Luís Potosí, pp17–35

Vera, D. J. (2005) 'Irrigation management, the participatory approach, and equity in an Andean Community', in V. Bennett, S. Dávila and M. N. Rico (eds) *Opposing Currents. The Politics of Water and Gender in Latin America,* University of Pittsburgh Press, Pittsburgh, PA, pp109–122

Vera, D. J. (2006a) 'Género, etnicidad, normas e institucionalidad de los derechos de agua' (Gender, ethnicity, norms and the institutionalization of water rights), WALIR Studies, vol 8, WUR-IWE and CEPAL, Wageningen, The Netherlands

Vera, D. J. (2006b) 'Derechos de agua, etnicidad y sesgos de género: un estudio comparativo de las legislaciones hídricas de tres países Andinos (Water rights, ethnicity and gender divisions: a comparative study of water legislation in three Andean countries), in R. Boelens, D. Getches, and A. Guevara (eds) *Políticas hídricas, derechos consuetudinarios e identidades locales* (*Water policies, customary rights and local identities: water law and indigenous rights*), Instituto de Estudios Peruanos (IEP), Lima, Perú, pp385–405

Vermillion, D. L. and J. A. Sagardoy (1999) *Transfer of Irrigation Management Services. Guidelines*, FAO Irrigation and Drainage Paper no 58, FAO, Rome

WECS (1998) 'Study on performance of participatory management in irrigation', Dip Consultancy, Kathmandu, Nepal

Zwarteveen, M. and Meinzen-Dick, R. (2001) 'Gender and property rights in the commons: examples of water rights in South Asia', *Agriculture and Human Values*, vol 18, pp11–25

Zwarteveen, M. Z. (2006) 'Wedlock or deadlock? Feminists attempts to engage irrigation engineers', PhD dissertation, Wageningen University, The Netherlands

Part III

Participation and River Basin Governance

5

Social Participation in French Water Management: Contributions to River Basin Governance and New Challenges

Sophie Allain

Since the 1992 Rio Declaration, the development of social participation has been strongly anchored in the environmental policies of many countries, with the purpose of improving both the efficiency and legitimacy of policy-making. However, in France, social participation has existed for longer in the field of water management as a key element associated with the establishment of a river basin governance system. Beyond words and expectations, what does social participation actually bring to water management? The aim of this chapter is to provide some answers to this question. To the extent that social participation covers a variety of situations, I will begin by giving an overview of it in the field of water management. This will be followed by an examination of the contributions of social participation to water management through the analysis of collaborative planning procedure called the water management plan. As many of these experiences are still in progress, because they may sometimes reach deadlock in spite of the potential they present, and because the forms of social participation themselves change, it will also be important to pay attention to the challenges that social participation faces today.

Social Participation: A Strong Orientation In French Water Management

Social participation was introduced in the French water policy by the 1964 Water Act in relation to the creation of the water agencies (*agences de l'eau*), but has been greatly extended by the 1992 Water Act through the introduction of collaborative planning procedures. While it represents a key feature of the

French system of river basin governance, it is important to note that participation has also been associated with the development of contractual procedures, and may be locally boosted in the framework of ad hoc procedures.

Social participation as a key feature of the establishment of river basin governance

French water agencies are financial bodies levying fees and granting subsidies to public or private owners for the protection of the water resources. Six water agencies were created, covering the seven large metropolitan watersheds. Like any other state agency, a water agency is led by a board of directors (an organ that decides) and a director (the executive), but the innovative point here is that the board of directors is a multipartite body, composed not only of public representatives but also of private stakeholders and non-governmental organizations (NGOs). Those members are chosen by and among a political body, the basin committee (*comité de bassin*), which brings together representatives from three groups:

1 local elected authorities;
2 state agencies;
3 water users and environmentalist or consumer NGOs.

The board of directors prepares multi-year intervention programmes and submits them to the basin committee, which plays a consultative role and gives a formal opinion about the rates and fees that fund the water agency's participation. While in the beginning, the basin committees and boards of directors were composed of equal parts from each group, state agency representation has decreased in each body, especially after the 1982 Decentralization Laws and today represents 20 per cent of the total. Table 5.1 presents the recent composition of the basin committees stemming from the decree of 15 May 2007.

Table 5.1 Basin committee membership

Basin committees	Local elected authorities	Water users and NGOs	State agencies	Total
Adour-Garonne	54	54	27	135
Artois-Picardie	32	32	16	80
Loire-Bretagne	76	76	38	190
Rhin-Meuse	40	40	20	100
Rhône-Méditerranée	66	66	33	165
Seine-Normandie	74	74	37	185

Alongside those official bodies, geographical commissions (*commissions géographiques*), which work at smaller scales of the Water Agency's territory, represent more open assemblies. Specific commissions are also set, which may collect new members in order to obtain a wider range of opinions.

Extension of social participation in the framework of collaborative planning procedures

The 1992 Water Act has clearly strengthened the place of social participation in river basin governance by introducing two instruments of collaborative planning:

1. The master water management plan (*schéma directeur d'aménagement et de gestion des eaux* or SDAGE), which aims at determining quantitative and qualitative objectives of management and general lines of action at the level of a Water Agency's territory.
2. The water management plan (*schéma d'aménagement et de gestion des eaux* or SAGE), which has to verify those objectives and define rules and plans of management at the level of a smaller hydrographical basin in a manner compatible with the master water management plan.

Whilst this latter plan is defined by the basin committee, the law provides for the creation of a specific multipartite body, the local water commission (*commission locale de l'eau* or CLE) to draw up a water management plan. This specific commission is composed of three different groups as in the basin committee. The 1992 Water Act defined precise rules of selection to compose a local water commission, such as fixed proportions for each group (50 per cent from elected authorities, 25 per cent from water users and NGOs and 25 per cent from state agencies) and many guidelines to guarantee representation from a large variety of interests and viewpoints, such as local elected authorities coming from places both upstream and downstream and from both rural and urban areas. The 2006 Water Act, however, gives more leeway to the prefect, the representative of the state authority, in composing a CLE. As both plans obtain legal value after the final step of approbation by the prefect, social participation here becomes an ingredient of the rule-making process itself.

Development of social participation and extension of contractual procedures

Whereas the establishment of a river basin governance system represents a key component of French water policy, the development of 'contractualization' in water management, which also implies social participation, is more generally in line with French environmental policy. Pierre Lascoumes (1991) explained that this practice was put forward by the Ministry of the Environment from its very creation in 1971 as a pragmatic way to introduce measures of environmental protection in the face of powerful organized interests. Jean-Pierre Gaudin (1999) stresses that today it covers many kinds of procedures with a variety of names such as contract, charter or agreement.

In water management, an initial kind of contract was one signed bilaterally between the state and economic interests (Lascoumes, 1991). The Ministry of the Environment then settled for different kinds of branch contracts (*contrats de branche*) with polluting industries in the 1970s, in which the state accepted that it would defer controls and sanctions in return for industry's commitments to decrease pollution. Today, contracts have expanded greatly. Contracts have also been signed between the state and irrigating farmers in many areas in order to reduce irrigation water usage in instances of water shortage. Instead of being subject to a prohibition of irrigation during lengthy, fixed periods, farmers committed themselves to withdraw not more than the volume of water previously agreed upon, so that they may use water more flexibly (Allain, 2000). Although social participation is mainly restricted to the inclusion of private stakeholders in the public process in these situations, it may sometimes be enlarged to incorporate other stakeholders. In the Charente département, for example, other users of water and NGOs also participate in meetings devoted to the definition of the available volumes of water for farmers' irrigation.

A very different kind of contract is the river contract (*contrat de rivière*), created by a 1981 ministry circular, which aims to mobilize public subsidies for financing actions and works in favour of water resources. The definition process of such a contract is close to the water management plan to the extent that it relies on a multipartite body bringing together local elected authorities, state agencies, water users and NGOs. But unlike the collaborative planning procedure, a river contract does not have legal significance. Similarity among these procedures is not a coincidence; in fact, the river contract procedure was considered as a way to develop planning after the failure of introducing planning in the framework of the 1964 Water Law. However, in the context of preparing the 1982 Decentralization Law, the decision was made to avoid any compulsory instrument, and therefore to introduce the river contract by a simple ministry circular and to put emphasis on the contractual dimension. After the creation of the water management plan in 1992, the role of each procedure was somewhat unclear for a while, but today procedures tend to be articulated pragmatically in the context of concrete situations (Allain, 2004a). For example, a river contract may initiate dialogue among stakeholders and handle some problems, after that a water management plan may define the establishment of a more complete system of management, and a new river contract may then be signed to implement some of the decisions made in this plan.

Social participation deeply rooted in local initiatives and will

Whilst those first descriptions of social participation in French water management might suggest that this practice was initiated by the state, many examples, on the contrary, show that it was fostered by local initiatives and will. The idea of developing social participation in water management clearly emerged from the national consultation organized in 1990 by Prime Minister Michel Rocard in order to build a new water policy and was at the heart of legislative debates during the vote for the 1992 Water Act. According to one senator, 'I hope that, between concertation and regulation, the former shall prevail and that we shall be able to set up tools for local and contractual water management.' Another is quoted as saying, 'The enforcement of this law must allow every user of the water to present their opinion and to participate fully in the definition of the decisions concerning the management and the preservation of the water resources.'[1] In fact, local elected authorities had demanded more responsibilities in the management of water resources for many years. Several public bodies have been created along these lines, for example by local elected bodies to finance interdepartmental investments concerning water (such as dams) even before 1964, but above all after the 1982 Decentralization Laws. Some of them have encouraged social participation such as the Charente River Institution (*Institution du Fleuve Charente*), which initiated a kind of collaborative planning for quantitative management before the 1992 Water Act. This model of plan was taken up by the Adour-Garonne water agency, in which the Charente River Basin lies, to settle a specific kind of plan, the water shortage management plan (*plan de gestion des Étiages* or PGE) devoted to quantitative management. Defining such a plan was then recommended by the master water management plan adopted in 1996 in southwest France, where irrigation expanded greatly for corn production and where water shortages represent the main water management problem. The Charente River Institution's initiative also inspired the model of the water management plan (SAGE), which, however, deals with a wider variety of issues and has legal significance. Moreover, dialogue was initiated several times by water users, in particular by farmers confronted with increasing conflicts with other stakeholders. One of the important experiences is the operation, Dividing the river water (*Rivière, partage de l'eau*), aimed at promoting dialogue among users of the same river.

Concertation among institutionalized stakeholders

In France, concertation refers to both the idea of consultation of the stakeholders before making a decision and to the idea of coming to an agreement about a collective plan. In the first case, the emphasis is put on the collection of opinions and nothing is said about the conditions of decision-making, which may not necessarily take place in the framework of the concertation process. In the second case, concertation corresponds to a collaborative process aimed at concluding an agreement or collective action. Social participation, then, has been highly rooted in French water management for many years and along different axes. Participation mainly occurs through concertation among local elected authorities, state agencies and representatives from water users and NGOs and, in this respect, social participation in French water management represents a real openness in the public decision-making process. Yet participation is restricted to institutionalized stakeholders, far from ordinary citizens. The degree of involvement in the decision-making process also often remains unclear, as the notion of concertation is equivocal.

Analysing what social participation brings to water management, then, requires examining the process itself. The case of river basin concertation planning will be investigated, and particularly the water management plan. Whilst the definition of master water management plans is compulsory and remains a rather formal process, water management plans have to emerge from local initiatives and may be considered as learning processes to the extent that they are defined by ad hoc, multipartite bodies that have a great deal of leeway and few references to organize themselves.

Learning from the Water Management Plan Procedure

Since the 1992 Water Act, 148 water management plans have been initiated covering today 40 per cent of the metropolitan territory in France. To the extent that the definition of most of them has often begun much later than the promulgation of the law and is still in progress (of those 148 plans, 45 have been approved, but most of them quite recently and few have yet been implemented), it is difficult to analyse what such kind of participatory procedures brings in terms of concrete improvements in the state of the water resources. So instead this is approached by examining what was intended by those who initiated the process and how the process has been oriented because of a collective dynamic in order to appreciate the contributions of social participation to water management. Therefore, a qualitative analysis conducted on the definition of ten water management plans is relied upon (Allain, 2001, 2002, 2005b). This begins with a description of the theoretical framework of such an analysis.

Analysing the concerted planning process through negotiation

Several scholars have stressed the political dimension of planning processes (Forester, 1989; Wildavski, 1979). Along these lines, concertation planning procedures tend to emphasize the integrative dimension of planning processes. The water management plan is then presented in the 1992 Water Law as an instrument to handle the water resources as a common heritage. In that way, the procedure defines several elements designed to guarantee a more democratic process and intended to avoid the selection of some issues to the detriment of others: the openness of the process, the demand to handle the water resources according to all their dimensions (water shortages, pollutions, wetlands, for example) and the demand to follow a step-by-step process that relies on a careful diagnosis of the situation in the river basin. However, at the same time that public room for direct interactions is organized so as to develop collaboration in water management, the procedure is also a new element likely to change the social regulation conditions, which may be considered either as an opportunity or as a threat by local stakeholders. Furthermore, with the variety of interests at stake, the aim of making decisions and defining legal rules implies that collaboration cannot be taken for granted as regards the definition of the plan and its scope. The implementation of the water management plan may then be subject to a negotiation process from its very initiation. This notion of negotiation involves an exchange of arguments and behaviours aimed at reaching an agreement between stakeholders who do not share the same interests or viewpoints.

Acknowledging negotiation as a generic social activity is in line with Anselm Strauss' (1978) interactionist perspective that sees such activities stemming from the strategic behaviour of actors situated as interdependent as they seek to pursue their own lines of action. But whereas negotiation is often understood as bargaining among selfish actors, it is important to emphasize a key feature of negotiation, which leads us to understand this concept somewhat differently: any negotiation presents the double ability of reciprocal influence and production of meaning.[2] Therefore, stakeholders involved in a negotiation are permanently in tension between a competitive behaviour aimed to influence the other ones and a cooperative behaviour aimed at collective problem-solving and they may change their behaviour during and through the negotiation process (Walton and McKersie, 1993). Considering the planning process as a negotiation may then provide a means to analyse it in progress and to identify at any moment whether it turns towards distributive or integrative social regulation (Allain, 2004b). Following Walton and McKersie (1993), distributive social regulation may be defined as pure bargaining processes, in which stakeholders defend their own interests and one stakeholder's gain is a loss to the other. Integrative social regulation may be defined as pure joint problem-solving processes, in which stakeholders collaborate to find solutions that benefit all the parties.

Strategies underlying the initiation of a water management plan

The initiation of a water management plan generally responds to specific strategies directed to make a specific issue of social regulation negotiable in the field of water management. It may, however, sometimes be considered within the framework of an integrative strategy.

Making some specific issues negotiable

In this situation, a water management plan is considered as a strategic instrument to produce new rules, in order to:

1 remedy legal intricacies;
2 question some rights;
3 avoid constraining public decisions;
4 promote new modes of management.

Let us examine these four issues:

1 Finding remedies to legal intricacies. Although the legal framework offers more and more measures in favour of water resources, many of them remain unimplemented. Beyond the opposition of local stakeholders, the intricacies of the legal framework themselves may explain such situations, sometimes resulting either in conflicting rules and scopes of activities, or in gaps in the law concerning the definition of responsibilities. A water management plan may then be an opportunity to establish a consistent, legitimate framework for breaking regulation stalemates at the level of a river basin. It is the deliberative process within a multipartite body established by the prefect, which gives legitimacy to the rules of water management stemming from the plan, beyond their legal power.

   Such a motivation played a role in the initiation of the Beauce groundwater management plan to improve the division of water among users in a large area of intensive agriculture where the pressure of irrigation withdrawals is very high. Although contracts of the same type as those described previously have been signed between the state and the farmers to limit the irrigation in case of water shortage, the Ministry of the Environment greatly pressured local stakeholders to go further and collectively define rules for the division of water among all the users. Such rules would thereby allow state representatives to grant the authorizations (compulsory since 1992) for water withdrawals within a clear legal framework. Indeed, the power of state services comes up against rules of private law in that situation, to the extent that it falls to the judicial judge to reconcile riparian owners' interests in accordance with the French civil code of 1804, yet not to establish general rules. Such general rules depend on the state as the authority responsible for the public interest, whilst this one is not allowed to establish individual rules for the division of water. Therefore, the greater the number of interests and the area of the river basin at stake, the more the situation becomes muddled as regards identifying the responsibilities of regulation. The farmers withdrawing Beauce groundwater finally committed themselves to the definition of a water management plan in order to work in the direction proposed by the Ministry of the Environment. Whereas the process is still in progress, the recent 2006 Water Act will further strengthen this option. It specifies that a water management plan will now have to include rules likely 'to define priorities among the uses of water as well as the division of total volumes of withdrawn water per use' (article 77).

   Another example is the Loiret water management plan. Close to the city of Orléans, the small Loiret River appears as a succession of reaches because of the presence of many old mills today transformed into nice residences. In this area, the river is subject to many conflicts between the riparian owners defending their access to private banks and the city dwellers striving for areas of water sports and leisure to row or go canoeing. The former put forward their right of closure as their property right while the latter claim a right of free movement on the river in accordance with the commons status of water resources. Whereas the 1964 Water Act had given the prefect the authority to regulate the movement of outboard motors (article 25), no clear rule had been defined for non-motorized leisure boats. However the 1992 Water Act came out in favour of the free movement of this type of boat 'in the absence of an approved water management plan' (article 6), and that is the reason why the Loiret riparian owners formed a property owners' syndicate that was one of the local stakeholders greatly promoting the initiation of a water management plan.

2 A water management plan may be seen as a means to question established rights. Along these lines, the purpose is to readjust powers among stakeholders in situations where one actor is considered to be using water resources to the detriment of the other users or in a way that jeopardizes the preservation of water resources, but where this actor is in a strong position to defend his/her own interests because of rules established previously or because of longstanding practices. This type of powerful stakeholder includes public companies, such as the state electricity producer or the national shipping company, which were granted concessions allowing them to keep a given rate of flow in the river, whilst other users would require higher or lower flows. Farmers, fishermen and shipping companies may then complain to the state electricity producer that they would like more water in summer when the rate of flow is low, whilst water is stocked by the company in upper dams for producing electricity in winter. Another example would be local elected authorities claiming the use of shipping canals to drain water off in case of a rise in river water levels or floods. In other situations, local stakeholders may ask that the rate of flow in ship canals is decreased in order to keep higher rates of flow in the rivers connected to those canals. The definition of a water management plan may then offer an opportunity for the local stakeholders to negotiate new rules to the extent that each member of a local water commission has a vote and one powerful member may not override all others. In the Ain Low Valley, located between the city of Lyon and the Jura mountains, famous for trout fishing, the initiation of a water management plan was considered by the local stakeholders, especially fishermen's associations, as a way to force the state electricity producer to accept new rules of management for its upstream dams in order to let more water into the Ain River in summer. In the same way, the Largue water management plan was promoted by local stakeholders in part to get the national shipping company to keep more water in the Largue River in a situation where this river could suffer from very low rates of flow due to withdrawals to supply water to the Rhine–Rhône ship canal.

3 A water management plan may also be seen as a way to oppose public plans or measures in order to avoid constraining public decisions. The

Drôme water management plan, the first to be approved, was initiated partly to handle severe problems of quantitative management due to an important development of irrigation and to look for another option to the solution put forward by the state agency that relied on building an upstream dam. In the same vein, the motivations of one of the local stakeholders in defining a water management plan in the Upper Tarn River Basin was local opposition to a state measure that would grant a specific status of protection to this area, rich in beauty spots, and which would introduce heavy constraints for the users of the river. The local stakeholders, already aware of the area's high tourist value, preferred to come to an agreement together to define the measures and rules likely to preserve the river rather than to be subjected to those imposed by the state.

4 A water management plan may be seen as a means to promote ecological modes of management. Such a motivation led the water agency to sustain the initiation of a water management plan in the Ain Low Valley. This stakeholder aimed to make local authorities adopt more natural ways of managing the river, in order to allow it to wander more, and therefore to maintain the specific type of landscape generated by the river meanders. In the same vein, the Largue water management plan was initiated by a local authority devoted to the management of the river for the same type of reason, and also to develop the use of environmental technology for sewerage. A water management plan, in this instance, is seen as a mean of legitimizing ecological plans and practices.

Looking for river basin cooperation

A water management plan can hardly be considered as an integrative instrument likely to organize cooperation among the stakeholders from the start. Stakeholders may, however, agree on the necessity of cooperation from the outset and rely on the water management plan to carry out this function. This happens when an issue of water management is already acknowledged as a collective stake, including such issues as the protection against flood risks or drinking water supplies. The first issue motivated the initiation of the Lys Valley water management plan, which straddles two departments in northern France in an area that has been severely affected by floods many times. Consumption of water was a key motivation in defining the Vilaine water management plan, which covers a large area in Brittany (9000km^2), where intensive agriculture causes severe pollution problems of surface waters, and where there is little accessible groundwater.

Whereas different types of strategies to initiate a water management plan have been distinguished, it is important to note that in practice several of them may be present in the same situation. Next, attention is turned to the outcomes that can be obtained at the end of the planning process.

A step in building river basin governance

While water management plans have legal significance, most of the plans present themselves as complex systems of measures including not only juridical rules – namely, rules having legal significance – but also plans of works, modes of management and cognitive actions.

The Ain Low Valley water management plan, for example, combines these four types of measures. Some of the juridical rules aim at controlling the behaviour of private users, especially quarry owners' activities, but most are directed at making other public decisions more consistent with the orientation of the water management plan:

1. The Ain Low plan introduces measures concerning town planning for example, such as limitations on building in floodplains.
2. Measures corresponding to plans of works, such as those aimed at improving the efficiency of sewerage systems or establishing fishways.
3. Measures directed at the design of information systems, which provides a means to control the implementation of the water management plan or promote new modes of management concerning the river itself (for example, restricting dredging) or land occupancy that affects water resources (for example, cropping systems that are likely to limit runoff).
4. Measures that are cognitive actions directed to increase the knowledge concerning the water resources or heighten public awareness of concrete problems, such as floods or risks generated by hydroelectric plants.

For some stakeholders, the limited place of juridical rules is a sign of inefficiency in the concertation planning. State representatives and environmentalists hold this view, but the strategies underlying the initiation of a water management plan show this procedure is subject to a variety of expectations that cannot be satisfied by defining only juridical rules; a water management plan is fundamentally a planning instrument designed to organize. Moreover, until the 2006 Water Law, the juridical significance of the water management plans was limited to the extent that plans could not be opposed by private stakeholders *(opposé au tiers)*. The heterogeneity of the measures has to be seen as an expression of the negotiated feature of concertation planning and of the nature of the agreement that has been reached at a given moment to improve the management of water resources. Next, attention is turned to examining the nature of agreements stemming from social participation.

Priority to integrative solutions

Despite the heterogeneity of measures that compose a water management plan, it is striking that most of them encompass integrative solutions, namely those likely to be beneficial for the river basin community without spoiling specific interests or solutions that encourage water users to act in ways that allow preservation of water resources. Such solutions may include using equipment in ways that minimize negative impacts, soft modes of management or awareness campaigns. Plans rely more on contractual instruments than on juridical rules to make private behaviours change. When important installations are planned, there are relatively small reserves of water or pipelines rather than dams, and such installations clearly aim to improve social regulation and not to develop a specific use like irrigation. Finally, juridical rules are directed at organizing the implementation of public decisions more often than to produce new norms.

On the contrary, distributive solutions that force private water users to change their activities are not developed very often. Such solutions concern very few users; for instance, banning quarrying in specific places on a river. Or, alternately, they must be considered in a framework of integration that relies on mutual compensation; for instance, reducing irrigation in compensation for developing new reserves of water that are likely to secure irrigation.

Construction of a river basin community for managing water resources

By bringing together stakeholders who are most of the time barely aware of their interdependencies and their interests in cooperating, social participation may contribute to shaping a river basin community. Such a community is first and foremost a community of management, directed at coordinating actions and finding solutions. One should not consider that issues of water management, then, are only the business of public stakeholders. This is because dialogue among public stakeholders takes place within an enlarged body involving private stakeholders and allows multilateral debates, so that deeper questions of coordination may be handled and new options discovered. This may result in agreements going beyond usual public practices.

To what extent does social participation contribute to shaping a political community, in which local stakeholders collectively become more aware of their common heritage? The planning process itself contributes to weaving links among local stakeholders and to shaping a feeling of solidarity within territories often subjected to social splits. The Loiret water management plan, which is still in progress, has provided the opportunity to initiate dialogues among rural and urban populations and bring them together around the idea of collaborating through acknowledging their mutual interests. In the Drôme water management plan, a common representation of development around the idea of sustainable tourism arose from debates among upstream and downstream representatives. However, the planning process remains managed according to an instrumental logic. Most documents are written in a very technical way and little attention is paid to the political justification of the orientation(s) selected or of the decision(s) made. The Vilaine water management plan is one of the rare documents clearly expressing that the main stake in this watershed is the preservation and the restoration of water for human consumption and that this stake must be considered as the underlying basis in all decisions made concerning water management. Finally, whereas concertation planning procedure, such as the water management plans, contributes to raising the political dimensions of water issues through shaping a community of management dedicated to a watershed, it must be considered as only a step in the construction of a political community.

The Challenges of Collective Negotiations

The analysis of the implementation of a water management plan showed ways in which social participation is shaping collective negotiation processes in the political regulation of water management. Improving the contributions of social participation to water management should take this dimension into account. This issue is addressed from three angles:

1 Improving mediation of negotiated planning.
2 Negotiating in situations of public participation.
3 Negotiating local water governance.

Improving mediation of negotiated planning

Because the initiation of a water management plan responds to specific strategies, which most of the time aim to make certain issues negotiable, the planning process tends to be framed at the inception by the expectations of the coalition working on it. Whereas this step of initiation may be seen as an unmanaged pre-negotiation process, establishment of a local water commission marks the openness of a new round in the negotiation process by bringing new interests and viewpoints around the table. It also allows stabilization of the domain of issues likely to be handled by defining the territorial area covered by the water management plan. Furthermore, somebody is often employed to manage the day-by-day process, and this individual is not concerned by the outcomes of the process and may be seen as a neutral party. However, the way to manage the negotiated planning process in order to come to an agreement is far from being well controlled. An in-depth analysis of ten planning processes regarding water management plans (Allain, 2001, 2002, 2004c) has revealed three ways of planning that could lead and have sometimes led to deadlocks or unsatisfying compromises. Specifically, these include ways of planning that are:

- managerial;
- expert;
- collaborative.

1 In terms of managerial ways of planning, a small committee may, for example, take control over the process in order to make sure that progress will be made in the direction that it wishes. This is not to suggest that such a committee intends to manipulate the local water commission in order to defend specific interests. Instead, its control over the process is designed to organize matters so as to resolve severe and well-identified problems. Such

a situation was involved in controlling the definition of the Drôme water management plan. Bringing together representatives from the state working at the local level and elected local authorities primarily from the downstream part of the river-basin, a small committee concentrated on two main issues that were not handled by previous actions. One of them, previously described, involved the improvement of downstream water quantity management in a context of heavy irrigation pressure. The other main issue concerned drastic problems of incision caused by a continuous development of gravel quarrying between 1950 and 1980. River downcutting led many riparian stakeholders to complain about the weakening and damaging of hydraulic structures (bridges and dikes) and the drainage of groundwater that diminished water availability. The small committee chose a specific way to handle each of those issues:

1. The committee insisted on bilateral dialogues with the farmers to make them accept some reduction of their withdrawals in case of shortage while simultaneously looking for the possibility of creating new reserves of water.
2. The committee called on well-recognized researchers to assess the situation and make proposals and then gave much publicity to the results among local elected authorities.

In such a process, the local water commission was more akin to an approval chamber than a deliberative assembly. Whereas such an approach to management was successful in reaching an agreement regarding the two issues central to the water management plan, it pushed aside other issues, such as addressing the state of the dikes and was not able to take advantage of a larger local approval. This led to further challenges to the agreement concluded.

2. The definition of a water management plan may also be considered as a problem-solving process as expert knowledge is heavily relied upon. In such a situation, the process is put in the same category as a technical study and organized around the writing of a report. The Vilaine water management plan illustrates such a process and two reasons may explain such an orientation:

   1. To the extent that this water management plan covers an extensive area, it can be very difficult to call together all members of the local water commission for the purposes of discussion. Each thematic meeting was then prepared by a specific analysis conducted by the person in charge of the process.
   2. The individual in charge of the process possessed important skills and experience in the field of water management and was close to local elected authorities. Although the local water commission could have played the role of a deliberative assembly more than previously, the scope of its deliberations was strongly framed by the way issues were defined. In such an approach to management, it may be difficult to keep the motivation of other participants.

3 The planning process may turn to an exclusive collaborative process avoiding conflicting issues. Contrary to the previous examples, the emphasis here is placed on the quality of dialogue among the local water commission's stakeholders or on consensus building. The Ain Low Valley plan illustrates such a collaborative process: its definition was organized in order to give more latitude to the local water commission's debates and the advancement of its thoughts. The local water commission itself then defined the range of the issues to handle in the framework of the water management plan, and whereas each group of issues was examined by specialized subcommittees, the whole local water commission managed to progress at the same speed and approved one step of the planning process before turning to the next one. However, because of the concentration on a general collaborative process, the local water commission did not pay enough attention to the specific negotiation process concerning the management of the state electricity producer's dams, and the divergences among stakeholders only emerged at the end of the process, when the plan and its rules were submitted to the local water commission's vote. The plan was then approved by the local water commission without this stakeholder's voice, who further complained to the basin committee, considering it as a court of appeal.

These examples show that improving the planning process requires skills in mediation (Allain, 2005a). These should take the negotiated dimension of the process into account more clearly.

Negotiating in situations of public participation

The concertation planning system adopted in France clearly inspired the 2000 European Water Framework Directive. This Directive, however, also demands going further by introducing a public participation principle (article 14). In France, such a principle is applied in the framework of the master water management plan revision. The status of public participation in the planning process, however, remains unclear. To be in accordance with the Aarhus Convention, public participation should mean real involvement of the public in the decision-making process; nevertheless it is handled as a mere communication and consultation process. The public participation step, which is in progress, relies mainly on a short questionnaire sent to every inhabitant of a water agency's territory, along with a variety of planned public events. The large number of inhabitants and the vast areas of a water agency's territory make it difficult to conceive what the greater implication of the public involvement could be. Nonetheless, it is clear that the notion of public participation is not rooted in public practices, even if it is becoming more usual. It is particularly true among the French water practitioners – including public as well private stakeholders – who tend to consider themselves as initiates because of the technical nature of many water issues.

In that context, public participation experiences in the field of water management tend to stem from other fields or background changes, such as the development of participatory democracy, and may therefore clash with practices in action in water management procedures, such as the introduction of public debate (*débat public*) procedure in France shows. Created by the 1995 Environment Act and reinforced by the 2002 Democracy of Proximity Act, this procedure makes compulsory the organization of a public debate sequence lasting four months, early enough in the decision-making process for important infrastructure projects, such as motorways, railways, dams and airports to be debated. In such a perspective, a planner has to hold a debate with the public likely to be affected by the project about both its opportunity and its features, while citizen participation was previously restricted to the public inquiry procedure at the end of the process. A National Commission on Public Debate (*Commission Nationale du Débat Public* or CNDP), with the statute of independent administrative authority in 2002, is put in charge of the organization of these public debates. Among 37 public debates organized thus far, only two have dealt with issues of water management. The first one, which took place in 2003, concerned the building of the Charlas dam in the southwest. The second, in 2007, revolved around the transformation of the biggest sewerage station in France, located in the western suburbs of Paris. What did those public debates bring compared to concerted planning procedures?

The Charlas dam debate will be examined in greater detail:

1. An analysis of the arguments exchanged during the debate reveals strong objections to the decision-making process that previously took place in the framework of a Water Shortage Management Plan (Allain, 2005c, 2008). This process led to the dam project being presented as the most feasible solution. Opponents to the project, riparian stakeholders as well as environmental and anti-globalization activists, complained that debates about this plan were framed with taken-for-granted assumptions and out-of-date forecasts about water demands and by the arguments of powerful local stakeholders, such as the irrigation farmers.
2. The analysis shows that this debate framed negotiation about the political regulation of water management, on condition that local stakeholders accept going beyond a fight of positions for or against the project. Indeed, controversies between those in favour of the project (i.e. the planners), which relied on a technical solution likely to increase the supply of water to handle shortages, and those opposed designed a more open cognitive framework to think about the problems at stake. These controversies then led to questions about four assumptions underlying the proposed dam project, revealing room for manoeuvre:
   1. severity of water shortages;
   2. forecasts about the needs of water in agriculture;
   3. quantities of water likely to be supplied by upstream hydroelectric dams;

4 the regulation principle that relied on an interconnection between the Garonne River and the Gascogne area that is poor in water resources.

The controversy also revealed that all these aspects involved political issues requiring the re-examination of both the relationships among stakeholders at an adequate territorial level and national policies, such as that relating to electrical power. The public debate also showed that there was no obvious institution able to follow up on the thoughts raised, and which are still in process today.

Beyond these difficulties, the experiences of public participation show that even if concertation procedures contribute to making water management more explicitly political, they are themselves subject to power games among a limited number of institutionalized stakeholders. Furthermore, these procedures are not always appropriate as regards the scope of the issues at stake, which can go beyond the framework defined by the procedure, and, especially, beyond the watershed framework. In a field where technical aspects play an important role, public participation remains necessary in questioning the issues at stake more deeply and in revealing their political dimensions. The issue in the future will be to articulate both public participation and concerted procedures only involving institutionalized stakeholders.

Negotiating local water governance

Finally, I would like to suggest that concerted planning procedures cannot be separated from the more general issue of negotiating a system of river basin governance in the long term. Indeed, whilst a procedure such as the water management plan creates a permanent multipartite body, the local water commission, it does not provide for its continued functioning. The local water commission must therefore look for support, not only for the definition of the plan but also for its implementation, and that depends on local elected bodies. Power games among local elected bodies may occur and impede the planning process or the implementation of the plan. The efficiency of concertation planning processes then depends on the ability of a local elected body to assert its authority on the entire river basin territory. The case of the Drôme water management plan shows that such a condition cannot be taken for granted; whereas the definition of the water management plan was actually carried by a downstream elected body, the responsibility for the implementation of the plan was shared between downstream- and upstream-elected bodies (Allain, 2005b). The emergence of severe flood problems in recent years, however, made the local elected authorities aware of the need to repair many dikes in the river basin and led the General Council (*Conseil Général* at the department level) to take control over the river basin governance because of the huge amounts of money at stake and the need for coordination. Such an orientation may have effects on the water management itself to the extent that the General Council tends to prefer engineering works to the methods of 'soft' management promoted by the downstream-elected body until recently.

This new problem, however, also makes it clear that an agreement concluded in the framework of a water management plan is only a step and may be questioned further. In the Drôme watershed, concern for protection against flood risks may come into conflict with concerns about reloading the river bed due to severe incision problems, to the extent that the former implies controlling the water flows whilst the latter implies leaving the river to meander freely.

Conclusion

Social participation in French water management has been developing for more than 40 years, covering a wide range of procedures and practices. Yet participation is mainly expressed through concertation among institutionalized stakeholders and, among them, concertation planning procedures play an important role. An analysis of the implementation of water management plan procedures shows that social participation contributes to making water management more explicitly political and clarifies the political stakes underlying decisions of water management. Water management cannot be considered as a technical affair but concerns collective life and implies choices. Social participation, however, is confronted with new challenges that all require acknowledging and addressing as negotiated features of water management. Along those lines, it becomes clear that social participation must be considered as a tool likely to design and improve negotiation processes regarding the political regulation of water management, rather than an aim in itself.

Notes

1 Official government gazette (*Journal Officiel*), 16 October 1991 (p2912 and p2924), quoted by Gourdault-Montagne (1994).
2 The first aspect has been emphasized by Erhard Friedberg (1993), who argued that in any interaction stakeholders seek to influence each other in order to preserve their own leeway. The second one has been stressed by George Mead (1967), who put forward the suggestion that any interaction produces meaning that can bring the participants to revise their viewpoints.

References

Allain, S. (2000) 'Application de la loi sur l'eau et processus de négociation. Limiter l'irrigation sans nuire à la production agricole', *Gérer et Comprendre*, vol 60, June, pp20–30

Allain, S. (2001) 'Planification participative de bassin et gouvernement de l'eau', *Géocarrefour*, vol 76, no 3, pp199–209

Allain, S. (2002) *La Planification Participative de Bassin*, Rapport pour le Groupe Inter-Bassins SDAGE-SAGE. INRA, ENS Cachan, CNRS-GAPP, vol 1, 2A and 2B, www.sitesage.org

Allain, S. (2004a) 'Contrats de rivière et agriculture: quel pouvoir incitatif et quelle efficacité environnementale?', *Revue Européenne de Droit de l'Environnement*, vol 4, pp401–413

Allain, S. (2004b) 'La négociation comme concept analytique central d'une théorie de la régulation sociale', *Négociations*, vol 2, pp23–40

Allain, S. (2004c) 'Délibérations et action publique locale: une approche en terme d' Action Publique Négociée appliquée au domaine de l'eau', in B. Castagna, S. Gallais, P. Ricaud, and J. P. Roy (eds) *La Situation Délibérative dans le Débat Public, Vol. 2*, Tours, France, Presses Universitaires François Rabelais, pp11–35

Allain, S. (2005a) 'La médiation environnementale comme système de régulation politique. Application au gouvernement de l'eau', in J. Faget (ed) *Médiation et Action Publique. La Dynamique du Fluide*, Presses Universitaires de Bordeaux, Bourdeaux, France, pp135–150

Allain, S. (2005b) 'La gouvernance de bassin à l'épreuve de la régulation politique de la gestion physique des cours d'eau dans la Vallée de la Drôme', in S. Allain, A. Farinetti, N. Ferrand and A. Vincent (eds) *Gouvernance de Bassin et Gestion Physique des Cours d'Eau dans la Vallée de la Drôme 1-1*, final report of the programme *Territoire, Environnement et Nouveaux Modes de Gestion: La Gouvernance en Question*, CNRS-PEVS, Paris

Allain, S. (2005c) 'Décider de l'opportunité d'un barrage-réservoir ou construire une négociation territoriale explicite de la régulation politique du domaine de l'eau?' in S. Allain, J. Grujard and N. Raulet-Croset (eds) *Décisions et Délibérations dans les Projets de Barrage-Réservoir vis-à-vis de la Régulation Politique du Domaine de l'Eau*, final report of the programme *Concertation, Décision, Environnement*, Ministry of the Environment, Paris

Allain, S. (2008) 'Le débat public: outil pour une négociation territoriale?', in N. Blanc and S. Bonin (eds) *Grands Barrages et Habitants. Les Risques Sociaux du Developpement*, Editions QUAE, coll. Natures sociales, Paris, pp225–242

Forester, J. (1989) *Planning in the Face of Power*, University of California Press, Berkeley, CA

Friedberg, E. (1993) *Le Pouvoir et la Règle: Dynamiques de L'action Organisée*, Seuil, Paris

Gaudin, J.P. (1999) *Gouverner par Contrat. L'action Publique en Question*, Presses de Sciences, Po, Paris

Gourdault-Montagne P. (1994) *Le Droit de Riveraineté, Propriété, Usages, Protection des Cours d'Eau non Domaniaux*, TEC Lavoisier, Paris

Lascoumes P. (1991) 'Les contrats de branche et d'entreprise en matière de protection de l'environnement en France. Un exemple de droit négocié', in C. A. Morand (ed) *L'Etat Propulsif – Contribution à l'Etude des Instruments d'Action de l'Etat*, Publisud, Paris, pp221–235

Mead, G. (1934) *Mind, Self, and Society from the Standpoint of a Social Behaviorist*, 1967 reprint, University of Chicago Press, Chicago, IL

Strauss, A. (1978) *Negotiations: Varieties, Contexts, Processes and Social Order*, Jossey-Bass, San Francisco, CA

Walton, R. and McKersie, R. (1965). *A Behavioral Theory of Labor Negotiations. An Analysis of a Social Interaction System*, 1993 reprint, ILR Press 2nd edition, Ithaca, NY

Wildavsky, A. (1979) *The Art and Craft of Policy Analysis*, 1979 reprint, Macmillan Press

6

Social Participation in Mexican River Basin Organizations: The Resilience of Coalitions

Eric Mollard, Sergio Vargas
and Philippus Wester

> Dictatorship says 'Shut up!' Democracy says 'Yeah, whatever you say.'
> Michel Colucci, comedian (translation)

In line with current ideas on social participation, this chapter shows that a relationship exists between participation and democracy. However, our argument reverses the formula proposed by advocates of participation: we suggest that participation is not a stage that precedes democracy, but that democracy is a precondition for effective participation. To grasp this counter-intuitive argument, it is helpful to recall that developed countries with long-standing democratic traditions only recently – and cautiously – started promoting participation.

At this point, it is necessary to briefly define social participation and democracy. Social participation is considered here through official organizations where people's voices are taken into account for collective decision on a specific topic. Social participation is then defined as a more or less socially open negotiation with some transparency in public decision-making. However, this type of decision-making is only the first stage of a full process. The second stage, which is the enforcement of the collegial decision, is far from being systematically implemented although the decision seems more legitimated to the extent that it represents the voice of the people. Indeed, other people can challenge the representativeness of the committee membership making the decision and impede its enforcement. As a result, legitimacy and representativeness are only two factors among others that characterize a democratic regime. We will put emphasis on factors related to the level of regulation of social powers in a society, which is supposed to avoid the imposition of one voice against others. As stated by Norberto Bobbio (1996), democracy is a set of rules that establishes who is authorized to make the collective decisions and under which procedures. A rule decided by one, few, many or all would have to be obeyed by all. So, to avoid any imposed decision, we will see how important countervailing powers are, paying special attention to the public authorities and the importance of their regulation; we will consider democratic deficit, poor regulation and politics, everything having to do with the regulation of powers.

Mexico belongs to the Organisation for Economic Co-operation and Development (OECD), meaning that it is sufficiently developed to have working institutions and state machinery. It is an elective democracy where the multi-party system dates back to the early 1990s. At the same time, the country embarked on social participation, in particular in environmental issues (Foyer, 2003). In water management, new user associations successfully replaced the federal administration in large irrigation districts (Kloezen, 2002; Rap, 2004; Rap et al, 2004), and the 1992 water law created general participatory organizations for the management of river basins and aquifers (Wester et al, 2003).[1]

This chapter draws on the first independent nationwide assessment of the functioning and outcomes of river basin councils or *consejos de cuenca* (Vargas and Mollard, 2005). In the first part, we first describe the typical situation of a river basin that is crippled by conflicts regarding the apportionment and different uses of water. Synthesizing different studies and common opinions, we then try to specify the analytical model that led to a common interpretation on the functioning and outcomes of the *consejos*. We call it the standard model of participatory negotiation, where many observers consider participation in *consejos* as a façade behind which the power of the federal administration remains secure.[2] Indeed, this kind of participation appears to be incomplete, biased in favour of the administration and even useless. In the second part, we look into the shortcomings of this analytical standard model, particularly when user participation turns out to be actual during the cooperative decision-making process or during the enforcement of decisions. As a result, this new interpretative model challenges the standard conclusion as to the sole responsibility of government agencies in the poor outcomes of *consejos*.

For that purpose, we first elucidate the concept of coalition, leading us to a broader examination of the sociopolitical system in Mexico (Sabatier and Schlager, 2000). Beyond changes in a government regime or institutional reforms, it is indeed possible to identify more permanent links between local politicians (whatever the party), certain organized groups (farmers, for example) and the federal administration. The power of such a coalition constitutes a major obstacle to any environmental public action, as evidenced by the *consejos de cuencas*, river basin organizations.[3] More specifically, the process that lead to deadlock in the participatory process enabled us to identify the actual power holders: the federal administration, itself highly dependent on local politicians, who are in turn under the influence of organized groups.

After identifying the persistent asymmetry in powers, which enables organized actors to continue monopolizing governmental rents and public actions (that is, subsidies and development programmes) as well as to circumvent law in spite of institutional changes, we link the range of processes that hinder participatory actions with a democratic deficit as defined above. We discuss the notion of environmental democracies that suggests the need for cross regulation of countervailing powers, as exists in Western democracies (Crozier and Thoenig, 1975; Massardier, 2003). Mexico, however, is hampered by an incomplete system of checks and balances, as well as by the lack of independence of key actors (administrations, regional leaders, mayors). The enriched model then makes it possible to give an account of the Janus face of the administration: daily despotism in the face of incipient powers, as seen below in some *consejos de cuenca*, and institutional weakness elsewhere.

Interpretation of Participation: The Standard Model

After describing the physical and social characteristics of river basins in Mexico, we summarize the common interpretation of social participation in water management, which we call the standard model. We agree with the general standpoint – shared by researchers, observers and users – which describes the artificial character of social participation, often described as a façade behind which administrative intrigues continue unchanged. However, as this analytical model is unable to explain the failure of a few actual participation cases, we then analyse the methodological reductionism used to build this common interpretation as a transition to the second section, where we present a model enriched with political processes within the enlarged interplay of actors around water conflict management.

The Institutional Stakes of River Basin Organizations in Mexico

Water is unevenly distributed in space. Even if the rain falls uniformly, surface water flows or infiltrates and subsequently concentrates in springs and channels. Watersheds and river basins are areas of surface runoff in which water comes together and leaves through a single discharge point or alternatively concentrates in permanent or seasonal wetlands or a lake. In addition, the possibility of storing or diverting water towards privileged zones can convert natural heterogeneity into social inequality. Indeed, the history of water can be seen as one of inequality. Historically, shortages depended on the availability of technology and led to more or less permanent conflicts that varied with the regulatory modes of rights and powers (Wolf, 2003). Water also has a history of cooperation during which conflicts were overcome when large works were required. Cooperation provides an interdependent way of protecting people from being excluded from access to water by a third party. Very early on, water management required a higher entity than individuals or communities in the form of courts, possibly associated with religion, customs or with government agencies (Jaubert de Passa 1846).

There are many root causes of water disputes: unwarranted diversion of water upstream, lack of infrastructure maintenance, unintentional flooding or self-centred behaviour, difficulties in sharing water during dry years and more recently pollution, over-allocation or the allocation of new 'rights to reply' to urban, industrial and environmental water requirements. Population growth and the increased number of uses have multiplied the sources of conflict with, in recent times, a shift from local litigation to international controversies including regional meso-conflicts. This new scale of conflict was initially the consequence of giant hydraulic works and, more recently, the closure of river basins (Wester et al, 2008). A government can no longer resolve a conflict as before by tapping and distributing new water resources because all local resources are already being exploited.[4]

After the Spanish conquest of Mexico in the early 16th century, extremely large land holdings (haciendas) monopolized surface water. Depending on their financial means and the technology available, their owners diverted water, which led to conflict and litigation not only among themselves but also with the progressively dispossessed Indian communities. During the colonial era and after independence in the early 19th century, water was the concern of the municipal authorities, while the federal courts represented the ultimate recourse for water users, even though these courts were often biased or too expensive for the poorest users (Aboites, 1998). By the end of the 19th century, the federal administration had become increasingly responsible for water management and local litigation. From 1920 onwards, as a consequence of the revolution in 1910, as well as of the agrarian constitution, the federal administration put an end to despoliations of the poorest and was mandated to build large infrastructures for regional development. As its power increased, the federal administration replaced user associations and private contractors (Palerm, 2005). The nationalization of water at the end of the 19th century paved the way for development based on the assessment of hydrological river basins and provided for additional water uses that should not affect pre-existing rights. However, in spite of a centralized administrative framework, the failure of the federal administration to control illegal uses (for example, clandestine pumping from rivers, drilling and wastewater effluents), dependence on the one-party regime and local politicians, and corruption made it impossible to respect rational computations, administrative bans on water, or environmental needs (Güitrón et al, 2004). Current over-allocation of water rights, water shortages and conflicts are mainly the consequences of such past malpractices.

In the 1990s, Mexico did not have enough fiscal resources to maintain the state apparatus, in particular for agriculture. Participation was part of an economic package and participative water management was a response to state disengagement as well as a way to curb corruption, to re-legitimize public actions eroded by decades of underhanded dealings and to solve the increasing number of conflicts that arise when an authoritarian technocracy is unable to manage conflicts increasingly covered by the media. This is all the more true when the new scale of meso-conflicts results in their politicization (that is, politicians and public authorities are directly involved in the conflicts) and puts pressure on the administration, which formerly conducted its negotiations in secret and was only accountable to political authorities.

When municipalities, states and countries share one or more river basins, scaled-up conflicts are also confronted with the fact that the physical watersheds and territorial governments do not match. For example, whereas more than 90 per cent of the Colorado River basin is located in the US, use of its water is periodically responsible for conflicts with Mexico, the latter having developed water-demanding agriculture for export at the mouth of the river on the Gulf of California (Cortes, 2005; Maganda, 2005). Conversely, along the same border Mexico controls the upstream reach of the Rio Bravo (or Rio Grande) river basin, because of the Rio Conchos sub-basin that concentrates run off from the rainy Western Sierra. The power to retain water during dry years has regularly revived tensions with Texas in spite of an early international water treaty signed in 1944 (Bravo, 2005; Walsh, 2004). Both cases required the intervention of the two presidents while the problem comes from the management of a particular dam, farming reclamation releasing highly saline water or the diversion of water in dry periods by farmers upstream.

The relatively small Cuitzeo basin (Table 6.1) provides another example of conflicts caused by deforestation that resulted in erosion and silting of the reservoirs, pollution from a paper mill and sewage from the state capital of Michoacán, diversions for irrigation with polluted water and agricultural pollution itself. The fate of the second biggest lake in the country located downstream, as well as the fate of fishermen, depends on the water uses in the entire basin and their regulation (Marie et al, 2005). An additional difficulty for negotiators is defining the limits of this type of lake. Indeed, Lake Cuitzeo is considered to be shallow with an average depth of 1.4m and a maximum of 3m. Depending on whether the year is dry or wet, it can flood its banks or retract dramatically due to natural causes. Human activities accentuate this imbalance and we will examine how the river basin council failed to solve the widespread crisis that overtook the region.

Table 6.1 Surface area of river basins in Mexico

River basins	**Surface area (km^2)**
Cuitzeo	4200
Ayuquila-Armeria	9800
Lerma Chapala	54,000
Grijalva-Usumacinta	91,000
Colorado	632,000
Rio Grande/Bravo	920,000

Interests and limits of the standard model

The standard model synthesizes the common interpretation of the poor functioning of Mexican *consejos*. In considering this model, we take into account different studies as well as opinions gathered through surveys and interviews, which rest on a simplified description of governance in the *consejos* and result in an interpretation emphasizing the role of the administration. In this section, we show that many *consejos* follow this pattern, but not all. First, we show that the law did not design participation as a countervailing power to the federal water agency.

The 1992 federal water law created the *consejos de cuencas* as advisory organizations. The purpose was to improve not only the different facets of water management, but also to tackle particular issues decided by the public authorities. In general, these hot issues relate to an apportionment of water rights for new environmental, urban or industrial needs. The federal water administration established and, since their inception, has chaired them. The *consejos* have to be consulted even though the final decision remains the sole responsibility of the federal agency. Moreover, the administration frequently co-opts the representatives of each water use, who sit beside the representatives of the governors. The *consejos* have no financial autonomy, not even to refund the expenses engaged by the representatives when they attend a meeting, and even less to launch projects or research as an aid for decision-making. Only civil servants have their expenses refunded. Had the legislators wished to create a façade without reducing the power of the administration, this would have been the method they would have selected.

As shown in different *consejos*, the federal administration controls the entire proceedings of the *consejos de cuenca* and government officials decide the agenda of the meetings. They sometimes cancel meetings without previous warning, showing little respect for the representatives of civil society and reinforcing feelings of rejection. Meetings are usually infrequent, although they have not completely disappeared thanks to the renewal of representatives and the possibility of obtaining knowledge or funding from water management plans. Depending on the civil servants concerned, it can happen that some discussions deal with a schedule of investments.

Such routine authoritarianism, typical of an administration with no checks and balances, cannot usually be resisted by non-organized and often dependent actors (mayors, state administration, user representatives). Such situations affect approximately 50 per cent of the *consejos* in Mexico. Using approximate figures originating from our experience and the nationwide assessment of *consejos* mentioned above, we try to characterize the effectiveness of participation in a simplified way with measurable elements, as well as with the subjective concept of conflict (Table 6.2). The frequency of meetings, their attendance, the openness of membership and the particular role of the government staff depend on the life stage of a *consejo*, so that such figures can be misleading. We prefer to pay attention to the formal aspects only for Levels 1 and 2 and to give emphasis to the presence of conflicts for Levels 3 and 4.

Table 6.2 Scale of participation

Level	Features
1	preliminary or intermittent meetings
2	regular meetings over a period of several years but with no conflict between representatives, and no group decisions made
3	regular meetings, open conflict and difficulties in decision-making; or limited conflict when a group decision is the result of previously allocated funding
4	group decisions are actually enforced

Whereas *consejos* were created throughout Mexico by administrative decision, commissions and committees were created as the result of local initiatives (government or local society). A commission is an organization for a sub-basin and a committee is local. Several *consejos*, commissions or committees were sometimes created in response to a conflict, the implementation of a development programme or the allocation of a budget. After the end of the emergency or crisis from which the organization had originated and many meetings, the organization stopped working, as was the case of the Conchos River basin commission created at the time of the dispute with the United States. Other examples are the Apatlaco River basin organization in the state of Morelos, which was created to find a solution for the high levels of pollutants that were causing conflicts and the Cañada de Madero committee, which disappeared after social unrest due to the inability of the organization to deal with such problems.

In Mexico, we have not been able to find any example of Level 4, namely an enforced decision able to solve the problem for which the participation took place.[5] Fifty per cent corresponds to Levels 1 and 2 (roughly 20 per cent for Level 1 and 30 per cent for Level 2) for which participation is a mere façade controlled by the government agency. The remaining 50 per cent corresponds to Level 3. This estimate is optimistic due to the disappearance of a number of councils. Studies mentioned in Table 6.3 show that effective participation where antagonistic segments in the population have a voice almost inevitably leads to open conflict. As a result, a conflict becomes an indicator of Level 3 or actual participation.

Finally, our four-level categorization prompts discussion on the interactions between participation, conflict and the enforcement of any joint decision, as in the case of the Rio Bravo after the 2002 controversy or the cases of the Valley of Mexico City and the Balsas *consejos*. Instigated by the governor of Texas, the first dispute led the Mexican and US Presidents to sign an agreement to force peace by financing modernized irrigation in the upstream Mexican reach of Rio Bravo so that farmers would agree to give back part of the water they saved. The North American Development Bank asked for a participation clause under the control of the Border Environment Cooperation Commission (BECC). During the meetings, farmers did ask key

Table 6.3 Assessment of participation in some river basin organizations in Mexico

Consejo	Participation	Source
Ayuquila	2	André de la Porte, 2007
Costa de Chiapas	2	Vera, 2005
Grijalva-Usumacinta	2	Kauffer, 2005
Cuitzeo (*) (**)	2	Peña de Paz, 2005; Marie et al, 2005
Cañada de Madero (*)	2	López and Martínez, 2005
Lerma Chapala	3	Mollard and Vargas, 2005; Sandoval and Navarrete, 2005
Rio Bravo	3	Bravo, 2005
Papaloapan	3	Murillo and López, 2005
Colorado	3	Castro and Sánchez, 2005; Cortes, 2005

* No longer exists ** Project of a consejo

questions about the volume and the destination of saved water but the Federal Water Agency evaded the concerns. In this case, farmers' approval was determined by substantial funding, which avoided potential conflict on the future use of saved water. In spite of the positive opinion of the BECC on the formal participatory process led by the federal administration, we consider that the level of participation cannot be rated as Level 4 due to the absence of co-decisions, money being secured only with façade participation (Mollard and Vargas, 2006). The absence of conflict could even diminish the grade to Level 2. In the *consejos* of the Valley of Mexico City and Balsas, the conflicts on water were so intense that they had to be solved by political means outside the scope of the *consejos*. In both cases, participation has not been conflictive because hot issues were not discussed (Perló and González, 2005; Vargas, 2006).

The practices of the administration that are frequently cited confirm not only the participatory façade but also an unsuspected and perverse effect in that it not only deceives the public but also national and foreign observers. In the Grijalva-Usumacinta *consejo de cuenca*,[6] Edith Kauffer (2005) identified a political discontinuity with the construction of two separate master plans for each state in the sole Grijalva basin, as well as the systematic agreement given to administration-led projects. In the Chiapas *consejo*, which jointly represents the small coastal basins between the mountains and the ocean, the representatives of water users were wary of the administration and supposed that the purpose of the *consejo* was to legalize water uses and to apply a water tax (Vera, 2005). It is possible that an error was made by the government official in charge of the *consejo*, but it is also possible that the federal administration was testing the reaction of a minor *consejo* to this type of strategy; in either case, the representatives of civil society were not encouraged to work with the administration. Because of the many difficulties encountered, the frequency of the meetings dropped and fewer and fewer representatives attended. However, in this particular case, the administration-driven *consejos* succeeded in avoiding self-disbandment, which happened in Cuitzeo, as discussed in the next section.

The standard model correctly indicates that the *consejos de cuencas* are a failure resulting in lack of interest and discredit. The law did not want to or could not curb the power of the administration with any form of countervailing participation, probably due to the fact that any government agency is more prone to accept political instructions than uneasy and unpredictable citizens, who are considered by the political elite as being irrational, poorly informed or unskilled and having little knowledge about water culture. In its defence, one should not forget that for several decades, the one-party regime was crippled by corruption and personalized negotiations. Devolving power would have been a risk as many of the fractures in Mexican society, including in water management, could have deepened markedly. Indeed, the country has never had an open and legal way of resolving conflicts or institutionalizing social fractures except through corruption and clientelist agreements (see below) between local politicians, federal officials and territorial or corporate cacique-styled leaders (local political bosses). With the change to a multiparty system in recent years, the main drivers of the country have still not changed and it is unlikely that the *consejos* will change in the future.

Methodological limits of the standard model

The standard model suggests a lack of social participation for every *consejo de cuenca* and deduces that the administration is to blame. However, as we will see in the enriched model, an effective level of participation (50 per cent of the cases) has not alleviated the environmental crisis so far. Neither does the *consejo de cuenca* function better when the federal administration is excluded, as shown by the Cuitzeo *consejo*. The same is true of municipal management and some aquifer committees where participation failed even when the central administration was not represented (López et al, 2004). These two elements reveal the weaknesses of the standard model, which is unable to take into account the general failure of *consejos* and other participatory forums even if the federal water agency is missing.

The weakness of the standard model lies in the fact that careful examination stops at an *ex-post* description of a negotiation through records of meetings and interviews with direct stakeholders. It is thus impossible to identify who controls the key decision-making processes or to identify the social powers within the political governance. The solution recommended by the standard model, that is, more information and participation with less administration (Kauffer, 2005; Vera, 2005), is not well founded and is likely to be wrong.

As a result, the standard model of analysis overestimates the importance of speeches when practices would more accurately reveal the aims and the leeway of each actor, and when processes generated during the social interaction are the key elements in policy studies (Walley et al, 2007). The analysis of one actor's practices is not enough because speeches by other actors can reveal accusations directed at others. For example, in water conflicts it may not be easy to distinguish between the farmers who are widely assumed to waste irrigation water, the brokers who steal from the farmers, the politicians who do not work for the public good and the civil servants who are supposed to act contrary to the interests of the citizenry.

In addition, the concluding statement of the standard model 'more information and more participation' is in line with international doctrines such as good governance or integrated water resource management (Mollard, 2007a; Mollard and Vargas, 2005). Such convergence with ready-to-use maxims confirms the standard paradigm, but overlooks in-depth approaches that challenge action-oriented doctrines and systematically exclude the power dimension in negotiations.

To appreciate the political dimension of negotiations, that is the asymmetry of powers and over-determination imposed on the outcomes independently of the negotiating methods, it is necessary to identify social processes. It is then possible to look beyond appearances, for example, those of the supposed super power of the administration or those presented in actors' justifications.

The standard model of participation for *consejos de cuenca* is an apolitical model as long as it does not recognize contesting powers and their determining influence on the outcomes. This model is in line with research on management tools to promote and improve dialogues in accordance with formal international doctrines. Unaware of the social processes and the ability of organized actors to appropriate or hijack such doctrines, the risk is that organized actors acquire additional legitimacy by an appropriation of such apolitical doctrines. This kind of doctrine could then be counterproductive for public action and for solving environmental problems because such a scientific coalition between international doctrines, action-oriented disciplines and the standard model of governance builds a system of cross-legitimacy, which strengthens traditional coalitions between administration, political representatives and organized corporations, which are the very factors that lead to stalemate in negotiation processes.

Disciplinary fragmentation, superficial doctrines and the lack of a general theory capable of situating actors' practices and speeches within social and political systems are some of the many difficulties involved in going beyond the standard model. When territorial and institutional powers and their asymmetry (which continue after institutional and regime changes) remain undetected, this type of analysis is necessarily incomplete.

The Enriched Model of Social Participation

The standard model of interpretation of the *consejos* accurately shows that social participation is a façade behind which nothing has changed, confirming 'the more it changes, the more it remains the same' as Helen Ingram wrote in 1990 on water issues. However, the standard model is mistaken when it states that effective participation is the solution to environmental crises as we see it now. By incorporating the political dimension underlying the interplay between actors, the enriched model shows that participation implies a set of prerequisites that are seldom met, such as an operative, independent administration.

Political processes are varied and take place at different scales outside negotiations, including at the international level as seen for the doctrines mentioned above. To characterize some typical processes leading to stalemate in environmental negotiations, organizations in three river basins are examined that reflect a range of political dynamics: institutional innovation vis-à-vis the inadequacy of the official *consejo*, the political dependence of actors and inadequate representativeness and conflict politicization, which make it more difficult to bring antagonistic parties together.

Ayuquila-Armeria: diverting attention from genuine concerns

The commission of the Ayuquila-Armeria river basin belongs to the Middle Pacific Committee, which is co-chaired by the governors of Colima and Jalisco and the federal water administration (Silva, 2008). The commission discusses global diagnoses and management plans by avoiding important concerns, like the illegality of industrial and municipal effluents. Rather than enforcing the law, management plans are based on the multiplication of wastewater treatment plants. Indeed promising a better future is a way to stop social unrest among those who suffer from the poor quality of the river, while the first plants built still do not function at all or operate at a reduced capacity (Reynolds, 2002). Vacuous discussions that take place in the basin commission are not politically neutral because they enable the administration to avoid any confrontation with local authorities, in particular large municipalities that are unwilling to finance wastewater treatment, even though wastewater effluents are illegal. In other words, the administration decides on the agenda, focuses on diagnoses and management plans and recommends building new treatment plants to avoid the issue of simply applying the law and making existing treatment plants work.

Faced with deadlock by a local coalition that prevented an effective solution from being found, local organizations had to be created from scratch. This was the case in an inter-municipal initiative, which took over responsibility for the task allotted to the basin organization, as cited by the mayor of Tuxcacuesco:

> I have concentrated on the inter-municipal initiative because in the commission, I know that I can't obtain more concessions, I know I have no right to vote ... There are good intentions but there is a lot of conformity, nothing much can be done. (André de la Porte, 2007)

The initiative unites ten municipalities of the Lower Ayuquila and was supported by academics of the state university. This organization made it possible to improve the quality of the river, to create brigades to control forest fires and to promote separation of solid waste to reduce pollution by seepage.

The official commission functions poorly and concrete outcomes are rare in spite of top-level meetings. Members endorse the agendas decided on by federal representatives and each representative seeks to obtain subsidies without playing the role of a simple citizen, that is demanding law enforcement and sanctions against the municipalities responsible for pollution. Since the river basin committee is controlled by a coalition linking mayors, governors and the federal administration, it can prevent application of the law and the emergence of initiatives within the official commission. It is itself in a position of stalemate and this encourages civil society to innovate outside, as happened with the inter-municipal initiative. In this case, it is worth noting the position of the administration in the coalition siding with the mayors.

Cuitzeo: political dependency

The Cuitzeo basin is a closed basin without outlets. The downstream lake, which acts as a natural regulator, has inevitably become the indicator of the social management of water in the whole basin, an indicator that varies depending on whether the riparian residents suffer from floods or the fishermen from dramatic drying out and pollution.

The problems in the river basins are well known and so are their solutions: treating urban and industrial effluents and building small dams within the lake to ensure the durability of parts of the lake, all of which can be implemented at a reasonable cost. But simple solutions were too costly or not attractive from an electoral standpoint, prompting the governor to create a basin commission under his control. In 1997, he emphasized the benefits of the *consejo de cuenca* for the environment and for economic development, adding that the involvement of civil society is the key issue in finding solutions: 'for Michoacán the moment had come to make this new stage of undeniable democracy profitable' (Peña de Paz, 2005). It should be noted that this statement was made after regional unrest caused by the scientific discovery that the fish in the lake were inedible due to contamination.

Three main facts distinguish the dynamics of this *consejo*:

- the federal administration was excluded (it did not recognize the legality of the *consejo* but sent an observer);
- development and environment were linked;

- the representatives of the *consejo* were mayors but there were no representatives of water uses, who, according to the governor, would have been unable to address the development aspect.

The *consejo* was made up of 13 mayors and 20 federal and state civil servants. Thus, it was in line with the World Water Council (2004), which states that water has to be a policy issue and directly involve authorities. However, although the provision appears to be based on common sense, it does not take into account the social processes of politicization, whereby a powerful actor facing few countervailing powers can make any organization an instrument for his own interest. Initially, enthusiasm was reflected in the many meetings that were held, and in the discussions to find solutions. But the mayors' dependency on the governor (since subsidies come directly from the state or in the case of federal programmes due to state mediation) prevented them from dealing with truly significant issues and led to 'the traditional petitioner's requests ... for treatment plants, cleaning of canals, support for constructions, fish farms, fishing nets, etc.' (Peña de Paz, 2005). In other words, people drew up a list of projects at municipal level, but nothing that would solve the overall problems of the basin, and did not create an inter-municipal initiative, such as Ayuquila-Armeria. They even did not make a simple request to enforce the law on wastewater, which in itself would have sufficed to protect fishing activities.

Peña de Paz (2005) is right in underscoring the traditional character of the social relations, but what drives such permanence within participatory organizations? The bond between the governor and the mayors is a bond of dependence that prevents disputes and hence prevents any solution being found for real problems. The error the governor made in trying to turn participation to his advantage was to reveal how his state had been functioning through political bonds and clientelist negotiations with the aim of controlling and obtaining support from the electorate. Indeed the mayors' financial dependence allowed the governor to give preferential support to certain mayors, so that the new participatory approach flagrantly globalized municipal demands when no mayor showed any interest either in the environment or in an inter-municipal initiative. Indeed, the only group initiative was to dismiss themselves when, after many ineffective meetings, they realized there would not be enough money to share. This decision clearly demonstrated their independence from the federal administration because no other administration-run *consejo* has been able to disband itself. Consequently, the absence of the federal administration means that the latter cannot be necessarily blamed.

Lerma Chapala: the heightening of antagonisms

The *consejo* of the Lerma Chapala river basin represents an exceptional case of full participatory negotiation because of the struggle between two governors around the survival of the largest natural lake in Mexico. The lake had lost more than 90 per cent of its volume between 1980 and 2003 due to excessive diversion for irrigation and domestic water for four million people. The hydrological imbalance was accentuated by a rainfall deficit since 1980 (Wester et al, 2008).

The purpose of the *consejo de cuenca* was to restore the lake to a satisfactory level. The *consejo* met twice: in 1991, when the first surface water allocation agreement was passed but never enforced, and in 2003, when the lake was about to disappear. One of the successes of the process was a hydrological model (Güitrón et al, 2004) validated by all the negotiators, which rapidly resulted in the exclusion of the two most extreme requests: to maintain the lake at its maximum level and not to allow a drop in the lake before the dams had been filled with water.

The *consejo de cuenca* is organized around a monitoring and evaluation group (MEG) chaired by the federal water administration and composed of five governors and the representatives of six types of water use. As there are no general elections, but only the announcement of an assembly, few people attend. The representative of each use is co-opted in his state then elected by a restricted committee made up of the representatives of each state under the close scrutiny of the federal administration. Among different technical committees (water quality, for example), the group for management and distribution (GMD) is the most important due to the significance of rescuing the lake. As the *consejo* has no money of its own, each party has state-paid experts, who are either civil servants from the state administration or private consultants. The GMD evaluated different scenarios and their impact on the probable levels of the lake using a data-processing model provided by a Mexican research institute.

In 2003 and 2004, the GMD met regularly, sometimes every 15 days, and, at the end of 2004, all the governors signed an agreement. In spite of this apparently favourable conclusion, as mentioned above, the Lerma Chapala *consejo* only reached Level 3 for participation because of the lack of an enforceable agreement. Indeed, it is based on goodwill and revisable every year, meaning that in reality, no agreement had been reached at all.

On the one hand the negotiation was characterized by the absence of negative attitudes (everyone played his assigned role, including the administration's negotiators). It was a serious negotiation with a battle between specialists in hydrology within the GMD under the close scrutiny of the users' representative-based MEG. On the other hand, it revealed the processes responsible for deepening existing antagonisms. Any improvement of the dialogue would not have altered the final result given the power structure and the limited room for manoeuvre of the different actors. Such an external support would perhaps have modified the preliminary stages and provisionally reduced antagonisms, but it could not have influenced traditional, coalition-joined powers.

The politicians' leeway was limited by the need to avoid a mass demonstration, in particular by farmers. The peasant leaders' leeway was also limited as will be seen below. The farmers' resistance in the face of the risk of having their water rights reduced could have ultimately turned into violence with occupation of the dams or kidnapping of civil servants. This potential for violence hovered over the negotiation and made obtaining the farmers' agreement indispensable.

The power structure (obviously not in the hands of the administration), as well as the clientelist way of dealing with conflicts (as exposed in the Cuitzeo *consejo*), meant the outcome of the negotiation was foreseeable. However, two particular processes (politicization and the place of the leaders) rendered the negotiation process harder and making protection of the lake more improbable.

The politicization of the conflict, that is, the partisan involvement of the political authorities, illustrates the absence of checks and balances applicable to the governors, which had the effect of reinforcing antagonisms. The conflict opposed Guanajuato and the farmers, who are large-scale consumers of irrigation water in the central part of the basin, on one hand, and the state of Jalisco downstream where Lake Chapala and Guadalajara city are located, on the other hand. The controversy thus placed the governors of the two states in direct opposition.

The politicization started in Jalisco where the governor took a conveniently ecological attitude, although Jalisco spent much less on the environment than the other four states (INE, 2003). The governor specifically attacked farmers in Guanajuato although Guadalajara also pumps a large quantity of water from the lake (200 million m^3). The governor of Guanajuato appears to have felt trapped and his speeches antagonized both farmers and ecologists. Even though the two governors belonged to the same political party, neither the president nor the party head was able to calm down matters.

Politicization heightened existing antagonisms, particularly between farmers. Indeed, the water controversy became a conflict between authorities, and the farmers interpreted it as a moral justification for their arguments. They could ask for more, give up nothing and fight to the bitter end. They were not anti-lake, but defended their vital interests by arguing the natural variation in the level that had existed previously. At the GMD, governor-appointed experts criticized the hydrological database and the computational model and called for new knowledge and new models. In the Lerma-Chapala basin, the increase in difficulties is directly due to governors who acted without checks and balances. The governor of Guanajuato exemplified this all-or-nothing attitude when he initially managed the controversy neutrally and then became the main party in the dispute.

The second toughening process analysed in the enriched model is the choice of a leader within any social group. Although co-optation is a common practice that allows the administration to control social participation in the *consejos*, the federal water administration lost its opportunity as one leader had regional legitimacy vis-à-vis one governor and the farmers. The agriculture representative was a democratically elected president of the largest irrigation district, which covers more than 100,000ha. Although legitimate, a representative cannot really negotiate on behalf of the farmers; he can receive but not give away (for example, give back part of the water that is saved in exchange for the technical and financial support required to save it). As soon as he makes concessions on agricultural water rights, he can be disqualified even by a minority within the farmers' organization. He may then be replaced, either with a more demagogic, tough leader using the argument that the representative lacked legitimacy, or simply due to violence if the minority invades a dam. As the challenger nearly became the regional leader, the fear of losing control of the peasant unrest led many, including the authorities, to advise the legitimate leader to take the lead in the fractious movement and preserve the gains of the negotiation.

The model enriched with politics incorporates toughening processes, such as politicization in a controversy that encourages conflicts between authorities, or difficulties in leadership, as described above. As a result, this analytical model shows that participation will not be able to solve the environmental crises without modifying the power of the traditional coalition in spite of skilled staff and goodwill, as the different stages of the negotiation testify. Participation requires certain preconditions to be fulfilled and these seem to be lacking in Mexico.

Towards a Model of Environmental Democracy

In Mexico, the federal water agency has the legal power to make decisions on everything related to water, which it has misused on many occasions. But when confronted with a major conflict, the administration shows how dependent it is on elected representatives. This traditional coalition is based not only on opportunistic interests, but also on dependence. It is also true of farmers and mayors associated with the governor–administration duo. The assumption of the necessary independence of actors for an efficient public action is clearly revealed in the governance of *consejos*. In other words, insufficient regulation or lack of countervailing powers hinders the achievement of social participation. By regulation we mean, more specifically, cross regulation, a concept we will now discuss before drawing conclusions about its implications in Mexico and for environmental democracy.

As far back as the 1970s, some researchers questioned the exclusive use of organizational charts, formal hierarchies and institutions to analyse the governability in a company or a government agency (Crozier, 1977). In spite of institutional changes the power structure is sufficiently solid to resist, as it was during the introduction of regional jurisdictions in France or the end of the one-party regime in Mexico. To characterize this governance of powers, Crozier and Thoenig (1975) proposed the concept of cross regulation, which is based on the interdependence of the elective and bureaucratic channels from Parisian centralism to the mayors of small municipalities, especially before decentralization in the 1990s. Elected representatives and federal officials at each level needed one another and co-operation was essential to obtain a subsidy for a local project. The elected official relied on the administrative expert while the expert could get beyond the compartmentalization of government departments only through the elected officials.

Cross regulation at this period had shortcomings: secrecy, favouritism, top-down style and fear of public opinion. However, the concept reveals a smooth way for cross regulating powers by introducing a balance between central regulation and democracy. Cross regulation rested on two pillars that are absent in Mexico: independence and legitimacy. Independence for each actor produces collective interdependence, and legitimacy built up over time reinforces collective trust in institutions.

Our analysis of the *consejos de cuenca* underlined the strong asymmetry in powers, such as the coalition between the administration and the mayors in Ayuquila-Armeria, the dependence of mayors on the governor in Cuitzeo and the mere existence of governors in Lerma-Chapala. The mobilization of peasants represents disproportionate power vis-à-vis weak institutions at the price of violence if necessary, so governors have to take this seriously and prefer to share interests. Conversely, for farmers the coalition represents not only a form of interest sharing (in order to access government programmes) or the subjection of dependents but also a form of protection, which is accentuated when trust in institutions is missing. In Mexico, coalitions make the power highly asymmetric due to these different processes. The assessment of this imbalance leads us to examine some forms of regulation so that the independence of decision-makers could produce collective interdependence while avoiding a drift towards secrecy and favouritism.

The general process is the following: defective regulation of social powers generates asymmetry, which, in its turn, is accentuated by the mechanisms of coalition building. This is evidenced through the aggregation of dependents, the self-centred behaviour of power holders due to the absence of counter powers, and the need for protection when institutions are weak. Other processes influence the difficulties inherent in the exercise of participation. The identification of these processes through future research will help build a theoretical framework linking the environment and democracy. Perhaps such a political approach will put an end to normative doctrines – national or international – that are appropriated locally for their own interests by traditional coalitions. From a political standpoint, the role of international organizations and their doctrines has to be thoroughly studied too because this instrumentalization can be counterproductive to mitigating environmental crises in developing countries. To summarize, cross legitimacy between doctrines and traditional coalitions could be an evil to be rid of, whereas cross-regulations of traditional powers still has to be invented.

Notes

1 For the management of aquifers see Mollard et al, 2006.
2 Since 1989, the federal water agency has been the National Water Commission (CNA).
3 For the position in France see Mollard, 2007b.
4 With the exception of inter-basin transfers.
5 In a few cases, the *consejo* was able to find an enforced solution without conflict (industrial plants treating effluents), but was unable to manage further conflicts as occurred in Cañada de Madero.
6 This big *consejo* unites two large independent river basins and, purposely or not, makes the Chiapas Indians a minority.

References

Aboites, L. (1998) *El Agua de la nación. Una Historia Política de México (1888–1946)*, CIESAS, Mexico City

André de la Porte, C. (2007) 'Integrated water resources management: limits and potential in the municipality of El Grullo, Mexico', Master's thesis EPFL3735, Lausanne

Bobbio, N. (1996) *El Futuro de la Democracia*, FCE, Mexico

Bravo, G. (2005) 'Esquemas de participación comunitaria en la cuenca del Rio Grande/Bravo', in S. Vargas and E. Mollard (eds) *Problemas Socio-Ambientales y Experiencias Organizativas en las Cuencas de México*, IRD-IMTA, Jiutepec, Morelos, Mexico, pp356–367

Castro Ruiz, J. L. and Sánchez Munguía, V. (2005) 'La experiencia de un consejo de cuenca en un contexto binacional: el consejo de cuenca de Baja California', in S. Vargas and E. Mollard (eds) *Problemas Socio-Ambientales y Experiencias Organizativas en las Cuencas de México*, IRD-IMTA, Jiutepec, Morelos, Mexico, pp316–330

Cortes Lara, A. (2005) 'Hacia una gestión binacional de las aguas transfronterizas en la cuenca del rio Colorado', in S. Vargas and E. Mollard (eds) *Problemas Socio-Ambientales y Experiencias Organizativas en las Cuencas de México*, IRD-IMTA, Jiutepec, Morelos, Mexico, pp331–355

Crozier, M. (in collaboration with Erhard Friedberg) (1977) *L'Acteur et le Système*, Le Seuil, Paris

Crozier, M. and Thoenig, J. C. (1975) 'La régulation des systèmes organisés complexes', *Revue Française de Sociologie*, vol 16, no 1, pp3–32

Foyer, J. (2003) *Complexification des Conflits Sociaux au Mexique : l'Exemple du Conflit Socio-Environnemental autour de la Réserve de Montes Azules, Chiapa*, Institut des Hautes Etudes de l'Amérique Latine, Université de la Sorbonne Nouvelle III, Paris

Güitrón, A., Mollard, E. and Vargas, S. (2004) 'Models and negotiations in water management', *Proceedings from the Mexican Experience, IFAC Workshop on Modeling and Control for Participatory Planning and Managing Water Systems*, www.elet.polimi.it/IFAC_TC_Environment/Venice2004/poster/3v04mollard.pdf

INE (2003) 'Diagnóstico biofísico y socio-económico de la cuenca Lerma-Chapala', www.ine.gob.mx/dgoece/cuencas/download/res_ejecutivo.pdf, accessed December 2008.

Ingram, H. (1990) *Water Politics: Continuity and Change*, University of New Mexico Press, Albuquerque, NM

Jaubert de Passa, F. (1846) *Recherches sur les Arrosages chez les Peuples Anciens*, 1981 reprint, 4 vols, Éditions d'Aujourd'hui, Grenoble, France

Kauffer Michel, E. F. (2005) 'El consejo de cuenca de los ríos Usumacinta y Grijalva: los retos para concretar la participación y la perspectiva de cuencas', in S. Vargas and E. Mollard (eds) *Problemas Socio-Ambientales y Experiencias Organizativas en las Cuencas de México*, IRD-IMTA, Jiutepec, Morelos, Mexico, pp195–218

Kloezen, W. (2002) 'Accounting for water: institutional viability and impacts of market-oriented irrigation interventions in central Mexico', PhD dissertation, Wageningen University, Wageningen, The Netherlands

López, E., Marañón, B., Mollard, E., Murillo, D., Romero, R., Soares, D., Vargas, S. and Wester, P. (2004) 'Le gouvernement de l'eau au Mexique: légitimité perdue et régulation en transition', in A. M. Rivière-Honeger and T. Ruf (eds) *La Gestion Sociale de l'Eau: Concepts, Méthode et Application*, Territoires en Mutation 12, Université Paul-Valéry, pp223–243

López Ramírez, E. and Martínez Ruiz, J. (2005) 'Actores sociales y conflictos por el agua en la microcuenca Cañada de Madero', in S. Vargas and E. Mollard (eds) *Problemas Socio-Ambientales y Experiencias Organizativas en las Cuencas de México,* IRD-IMTA, Jiutepec, Morelos, Mexico, pp155–171

Maganda, C. (2005) 'Collateral damage: how the San Diego-Imperial Valley Water Agreement affects the Mexican side of the border', *Journal of Environment & Development*, vol 14, no 4, pp486–506

Marie, P., Mollard, E. and Vargas, S. (2005) 'Cuitzeo, una cuenca a escala humana. Conflictos, fracasos, porvenir', in S. Vargas and E. Mollard (eds) *Los Retos del Agua en la Cuenca Lerma-Chapala. Aportes para su Estudio y Discusión*, IMTA-IRD, Jiutepec, Morelos, Mexico, pp226–248

Massardier, G. (2003) *Politiques et Actions Publiques*, A. Colin, Paris

Mollard, E. (2007a) *Les Pratiques Internationales Exemplaires dans le Domaine de l'Eau. Qui Doit Apprendre?*, IRD, Montpellier, pp12

Mollard, E. (2007b) 'Jeux de pouvoir dans les négociations environnementales. Intérêts, portée et questionnement du cadre analytique de la gouvernance à partir du cas mexicain', in *La gouvernance: vers un cadre conceptuel*, IUED, Geneva (unpublished), pp12

Mollard, E. and Vargas S. (2005) '¿A quién preocupa la gestión integrada del agua? Entre indiferencia social y utopía peligros', II Congreso Iberoamericano Sobre Desarrollo y Medio Ambiente, 26 October, Puebla, México, www.iiec.unam.mx/CIDMA2005/interiores/Memorias_Cidma2005.pdf

Mollard, E. and Vargas, S. (2006) 'La participation sociale dans la gestion des ressources naturelles. Premier bilan pour l'eau au Mexique', Colloquium GECOREV Gestion concertée des ressources naturelles et de l'environnement, Université de Versailles-St-Quentin-en-Yvelines, 27 June

Mollard, E., Vargas, S. and Wester, P. (2006) *The Lerma-Chapala Basin, Mexico. Report for the Comprehensive Assessment of Water Management in Agriculture, Comparative Study on River Basin Development and Management*, IWMI, IRD-IMTA and Wageningen University, Wageningen, The Netherlands

Murillo Licea, D. and López Ramírez, E. (2005) 'Organización social y producción en la cuenca del río Papaloapan', in S. Vargas and E. Mollard (eds) *Problemas Socio-Ambientales y Experiencias Organizativas en las Cuencas de México,* IRD-IMTA, Jiutepec, Morelos, Mexico, pp245–275

Palerm, V. J. (2005) 'Políticas del estado en la administración y gobierno de sistemas de riego y redes hidráulicas', in J. M. Durán, M. Sánchez and A. Escobar (eds) *El Agua en la Historia de México,* Centro Universitario de Ciencias Sociales y Humanidades/Universidad de Guadalajara y El Colegio de Michoacán, Mexico, pp263–289

Peña de Paz, F. (2005) 'Espejismos en el lago de Cuitzeo. ¿Participación social en la gestión del agua?', in S. Vargas and E. Mollard (eds) *Problemas Socio-Ambientales y Experiencias Organizativas en las Cuencas de México,* IRD-IMTA, Jiutepec, Morelos, Mexico, pp103–127

Perló Cohen, M. and González Reynoso, A. (2005) *¿Guerra por el Agua en el Valle de México? Estudio sobre las Relaciones Hidráulicas entre el Distrito Federal y el Estado de México,* unpublished document from Universidad Nacional Autónoma de México-Coordinación de Humanidades/Programa Universitario de Estudios sobre la Ciudad, Friedrich Ebert Stiftung

Rap, E. (2004) 'The success of a policy model: irrigation management transfer in Mexico', PhD dissertation, Wageningen University, Wageningen, The Netherlands

Rap, E., Wester, P. and Pérez-Prado L. N. (2004) 'The politics of creating commitment: irrigation reforms and the reconstitution of the hydraulic bureaucracy in Mexico', in P. P. Mollinga and A. Bolding (eds) *The Politics of Irrigation Reform: Contested Policy Formulation and Implementation in Asia, Africa and Latin America*, Ashgate, Aldershot, pp57–94

Reynolds, K. A. (2002) 'Tratamiento de aguas residuales en Latinoamérica. Identificación del problema', www.agualatinoamerica.com/docs/PDF/DeLaLaveSepOct02.pdf

Sabatier, P. and Schlager, E. (2000) 'Les approches cognitives des politiques publiques: perspectives américaines', *Revue française de sciences politiques*, vol 50, no 2, pp209–234

Sandoval Minero, R. and Navarrete Ramírez, A. (2005) 'El reto de consolidar la participación social en la gestión integral del agua. El caso de la cuenca Lerma Chapala', in S. Vargas and E. Mollard (eds) *Problemas Socio-Ambientales y Experiencias Organizativas en las Cuencas de México,* IRD-IMTA, Jiutepec, Morelos, Mexico, pp52–63

Silva, P. (2008) 'Small-scale irrigation systems in an IWRM context: the Ayuquila-Armería basin commission experience', *International Journal of Water Resources Development*, vol 24, no 1, pp75–89

Vargas, S. (2006) 'Los conflictos y la gestión del agua en la cuenca del río Amacuzac: notas para la implementación de un proceso de abajo hacia arriba', in S. Vargas, D. Soares and B. Nohora (eds) *La Gestión del Agua en la Cuenca del Río Amacuzac: Diagnóstico, Reflexiones y Desafíos*, Instituto Mexicano de Tecnología del Agua, Universidad Autónoma del Estado de Morelos, Jiutepec, Morelos, Mexico, pp23–46

Vargas, S. and Mollard, E. (2005) 'Contradicción entre las expectativas ambientales de los agricultores y la defensa de sus intereses en la cuenca Lerma-Chapala' in S. Vargas and E. Mollard (eds) *Problemas Socio-Ambientales y Experiencias*

Organizativas en las Cuencas de México, IRD-IMTA, Jiutepec, Morelos, Mexico, pp64–82

Vera Cartas, J. (2005) 'Participación, consejos de cuenca y política hidráulica mexicana: el caso de la costa de Chiapas', in S. Vargas and E. Mollard (eds) *Problemas Socio-Ambientales y Experiencias Organizativas en las Cuencas de México*, IRD-IMTA, Jiutepec, Morelos, Mexico, pp276

Walley, J., Amir Khan, S., Karam, S., Witter, S. and Xiaolin, W. (2007) 'How to get research into practice: first get practice into research', *Bulletin of the World Health Organization*, vol 85, no 6, pp424–425

Walsh, C. (2004) 'Aguas Broncas: the regional political ecology of water conflict in the Mexico–U.S. Borderlands', *Journal of Political Ecology*, vol 11, pp43–58

Wester, P., Merrey, D. J. and de Lange, M. (2003) 'Boundaries of consent: stakeholder representation in river basin management in Mexico and South Africa', *World Development*, vol 31, no 5, pp797–812

Wester, P., Vargas, S., Mollard, E. and Silva-Ochoa, P. (2008) 'Negotiating surface water allocations to achieve a soft landing in the closed Lerma-Chapala basin, Mexico', *International Journal of Water Resources Development*, vol 24, no 2, pp275–288

Wolf, A. (2003) 'Conflict and cooperation: survey of the past and reflection for the future', in F. A. Hassan, M. Reuss, J. Trottier, C. Bernhardt, A. T. Wolf, J. Mohamed-Katerere and P. van der Zaag (eds) *History and Future of Shared Water Resources*, IHP Technical Documents, PCCP series no 6, UNESCO, Paris, http://webworld.unesco.org/water/wwap/pccp/cd/history_future_shared_water_resources/survey_water_conflicts_cooperation.pdf

World Water Council (2004) 'International Conference on Water and Politics', www.worldwatercouncil.org/fileadmin/wwc/Library/Publications_and_reports/Proceedings_Water_Politics/proceedings_waterpol_full_document.pdf

7

From a Participative Framework to Communities' Realities: The Challenges of Implementing Stakeholder Involvement in Quebec Watershed Management, Canada

Nicolas Milot and Laurent Lepage

Introduction

Ever since the beginning of the 1970s, the province of Quebec has fostered a social and political debate over water management. This debate intensified during the 1990s and led to the adoption in 2002 of the Quebec Water Policy (QWP). This policy has five intentions:

1 to reform water governance;
2 to implement an integrated water resource management;
3 to preserve water quality and ecosystems;
4 to continue the decontamination of wastewater;
5 to support the practice of water-related tourist activities (Quebec, 2002).

The implementation of the integrated watershed management (IWM) is a central aspect of the QWP. Based on a collaborative approach, IWM must generate a systematic protection of rivers and lakes, the acceleration of their clean up, the maintenance of all uses, the participation of the community and a greater effectiveness of the public programmes.

This chapter evaluates social participation related to the Quebec IWM approach. In Quebec, stakeholders' participation takes place in a particular context. In fact, the Quebec Government sets voluntary involvement of the stakeholders as the main avenue to improve water governance. However, this involvement presupposes some institutional adjustments that will come only with legislative recasting. Then, even if the IWM approach is officially acknowledged in the QWP, watershed management is already used in single-type problem contexts, particularly within the Ministry of Agriculture, Fisheries and Food and the Ministry of Natural Resources (Bibeault, 2003; Quebec, 2004a; 2007b). In the same way, important aspects of water management are under the responsibility of municipalities (treatment) and the Regional County Municipalities (land planning) (Quebec, 2006, 2007a). If social participation is put forward by the QWP, the effective integration of local debated decisions in the existing institutional system remains uncertain.

The evaluation of social participation is a difficult endeavor (Coglianese, 2002; Dorcey and McDaniels, 2001). In view of the novelty of the Quebec model, at this stage we can only proceed to an in situ examination of the implementation phase. This chapter is divided according to three major themes:

1 The evaluation of the participative framework, which can allow us to try to predict the advantages and limits of the collaborative procedure. This framework is analysed here according to two questions related to the implementation of collaborative approaches: the openness of the scientific and technical debate and the search for legitimacy of the collaborative processes (Callon et al, 2001; Korfmacher, 2001; Trachtenberg and Focht, 2005). To appreciate the challenges of local participation, we will recall some of the issues that were raised during the provincial debate that preceded the formulation of the QWP.
2 The implementation of the IWM implies a period of institutional adjustments. These adjustments are linked to the development of a new social dynamic that reflects the interactions among stakeholders (Milot and Lepage, 2007). In this sense, we consider social participation as an organizational fact that can be evaluated using concepts of the concrete systems of action analysis (Crozier and Friedberg, 1977; Mermet et al, 2005).
3 Finally, the experimentation of social participation in natural resources governance must go beyond procedural objectives and meet substantial outcomes. We shall open the discussion on these evaluative elements, specifying that it is essentially a prospective exercise and focus this part of the chapter around the concept of interdependence, which can be summarized as the biophysical and social dimensions that theoretically lead stakeholders to a situation of potential cooperation (Michaelidou et al, 2002; Ostrom, 1990).

Theory and Methodology

The participation framework: an analysis of expertise and legitimacy

In Quebec, the involvement of local actors in water management is new. In spite of certain local initiatives, the principal teachings result mainly from the contribution of citizens during the elaboration process of the QWP. Many authors based their analysis on a functional conceptualization of public action (Brun, 2007; Burton, 1997). According to this perception, social participation aims at filling some gaps, particularly in terms of transparency, local knowledge and representativeness of interests. However, the implementation of social participation is a more complex phenomenon. The framework developed by the provincial government has to deal with this complexity. In this chapter, we look at two potential problems that can be generated by a participation framework: the relative importance of technical debates and the legitimacy of the process.

Local watershed committees can be seen as a concrete system of actions where scientific and technical debates take place. In her analysis of the role of expertise in French watershed planning, Sophie Allain underlines the importance of considering this dimension (Allain, 2005). Although the collaborative process aims at reducing the traditional divides between the expert and the uninitiated and between the elected representative and the elector, Allain suggests that most of the time scholars have studied these issues at a macro level. Despite some micro-level analyses (scientific production within laboratory situation and the *Baie de St Brieuc* scallop issue), the relation between expertise and social participation at the local scale should be a subject of interest justifying further study (Callon et al, 2001; Latour, 1999).

Considering that water management begs technical and scientific knowledge this dimension is significant. In this regard we used the criteria suggested by Callon and his co-authors (Callon et al, 2001). First, we consider the possibility of open deliberative process – and the technical debate – for heterogeneous actors (the intensity of the debate, the procedural opening and the quality of the speeches). These characteristics are analysed for both the openness of the technical debate and the variety of participants taking part in the discussion. Another set of criteria pertains to the implementation of the procedure, for instance the access conditions, debate transparency and clarity of the organizing rules. We first applied these criteria to the participative framework that prevailed for the large deliberative exercise that led to the QWP. We then repeated the exercise but this time focusing on the framework proposed for the IWM collaborative approach. This opposition between a macro and a micro analysis – QWP definition process and IWM – is important considering the Quebec context. Historically, social participation related to water management has primarily been a provincial phenomenon. The involvement of local users remains an unknown factor.

The creation of watershed committees is associated with a search for legitimacy for both the participants and the organization itself. In general, the participation of social actors is legitimated by the nature of their contribution to the debate. Experts' participation is justified by their specialized knowledge, that of the elected officials by the fact that they formally represent the population and the participation of the administrative staff because they are directly involved in control of water use. Stakeholders' legitimacy in natural resources management is not as clear, the difference between **user** and **citizen** having to be elaborated. Certainly, the arguments rising from a functional perspective (transparency and local knowledge) remain true. However, participation in the IWM approach brings political decision-making and citizens' interests closer, whether individual or corporate. Inequity resulting from socio-economic differences among stakeholders is an issue (Murtagh, 2004). Of course, this risk of inequity must be taken into account. However, it would be erroneous to claim that the less powerful stakeholders cannot defend their interests in a collaborative process. Nevertheless, actors with less to negotiate largely depend on the participative framework, the latter reducing socio-economic inequalities and favouring the construction of certain legitimacy.

Several scholars have developed criteria to evaluate the framework structuring participation on that question of legitimacy (Cashore, 2002; Leach et al, 1999; Sullivan, 2001). We rely on two sets of criteria suggested by Trachtenberg and Focht (2005). These criteria evaluate social participation by taking into consideration legitimacy of process and of outcomes (Table 7.1). First, we consider the series of procedural criteria.

Table 7.1 Legitimacy criteria for collaborative approaches to watershed management

Procedural Legitimacy	Participants must appropriately represent a full range of nongovernmental stakeholders.
	Participants must fairly consider the concerns of the full range of nongovernmental stakeholders.
	Participants must genuinely consent to the policy decision.
Substantial Legitimacy	The outcome must be an improvement in the welfare of at least some nongovernmental stakeholders.
	The outcomes, and the means to attain them, must respect individual stakeholders' right.
	The outcomes must be a fair distribution of welfare among all nongovernmental stakeholders.

Source: Trachtenberg and Focht, 2005

Organizational analysis of social participation

The addition of new institutions devoted to water produces new social dynamics at the local level. These dynamics are diverse, depending on several factors and unavoidably show redefinitions of power relations between stakeholders.

In order to understand these social dynamics, we use the results of three organizational analyses produced with the collaboration of as many watershed committees. Our evaluation aims to decide whether social participation linked to IWM improves public water management or essentially produces new civic institutions. Based on the major concepts of the concrete systems of action analysis (Crozier and Friedberg, 1977; Lepage et al, 2003), we focus on actual participation: characterization of the concrete system of actors, strategies deployed by the stakeholders and limits of the participative process to produce changes in effective watershed management.

Reconnecting substantive and procedural aspects

This leads us to examine participation in connection with the substantial dimensions of IWM. Indeed, this approach is theoretically based on two sources of interdependence: the biophysical characteristics of the watershed unit and a particular vision of the community. First, we look at how social participation and collaborative processes takes into account characteristics of the river basin: upstream–downstream relations, multipurpose nature of the resource, multiple-scale issues and dynamics of the aquatic ecosystems (Grumbine, 1994; Manning, 1997; Slocombe, 1993). Second, we focus on two products of social participation particularly important in order to improve the coordination of stakeholders' activities: just distribution of outcomes among all stakeholders and introduction of interpersonal trust among them (Lubell, 2004).

Social Participation in Quebec Water Management

The slow construction of Quebec's water policy

In Quebec, a social and political debate over water governance started at the beginning of the 1970s. *Commission sur les problèmes juridiques liés à l'eau* (Commission on legal problems related to water) had already identified several inconsistencies in the public administration of water and had suggested some avenues to resolve these administrative quirks (Quebec, 1975). However, governments in subsequent years only proposed timid answers to these concerns.

In the middle of the 1990s, two different events brought back water management to the governmental agenda: the possibility that the City of Montreal's water services would be privatized and a proposed idea to export Quebec's freshwater during the summit on the economy and employment in 1996 (Gill, 1998). A strong political mobilization alerted public opinion to the water problems of the province (St-Pierre, 2005). That political dynamic widened with the creation, in March 1997, of Quebec's coalition for responsible water management, *Eau Secours!* The first demands of this coalition was to call for a public debate on water management, the adoption of an integrated water policy and the reinforcement of the role of the Ministry of the Environment relative to the management of water resources.

The government created a committee – chaired by the Prime Minister himself – which set out to analyse water according to four aspects:

- its protection;
- the production and distribution of drinking water;
- the production of hydroelectricity;
- the commercialization of freshwater.

These discussions were held behind closed doors. At the same time, the initiation of groundwater pumping led to new outcries relayed by the media. Prime Minister Lucien Bouchard felt obliged to promise that freshwater would not be exported or privatized without organized public consultation. This debate took form at the end of 1997 with the organization of the symposium on water management in Quebec (INRS-EAU, 1997). However, this event was strongly criticized by the spokesmen of the principal social and environmental groups. Although the symposium brought together experts, elected representatives, planners, farmers and bottlers, most ecologists boycotted the event because the discussions were to be based on a reference document that advocated privatization. Furthermore, the prohibitive entrance fee of US$250 diminished citizen participation. Following that event, the *Eau Secours!* environmental group circulated a petition calling for a real public consultation and also asking that this exercise be supervised by the *Bureau d'audiences publiques sur l'environnement* (BAPE, the office for public audiences on the environment). Environmental groups also demanded a moratorium on the exploitation of freshwater. After a year of hesitation, a moratorium was adopted in June 1998. Also, sensitive to public opinion, the government asked BAPE to conduct public consultation on water management in Quebec. This began on 15 March 1999 and was chaired by André Beauchamp. The Commission travelled around Quebec gathering the concerns of citizens. In the end 379 briefs had been submitted and 143 days of public hearings were necessary to collect the comments and the questions from all over the province. The commissioners paid particular attention to the participation of ordinary citizens (BAPE, 2000).

Three specific questions were initially proposed by the Commission:

1. Does Quebec need to export freshwater?
2. Does Quebec have to encourage the exploitation of groundwater?
3. Does Quebec have to consider the privatization of public water services (distribution and treatment of drinking water)? (BAPE, 2000)

However, the Commission was forced to open the public consultation and to adopt a comprehensive approach considering the broad range of topics expressed by participants:

- agricultural uses;
- exploitation of groundwater;
- public health;
- industrial and urban decontamination;
- public infrastructures;

- production of hydroelectricity;
- ratemaking of drinking water;
- impacts of forest activities;
- accessibility and protection of the aquatic ecosystems;
- integrity of the St Lawrence river ecosystem;
- integrated management of the water.

This last topic was discussed at length during the hearings as an approach that could solve many problems.

A collaborative approach for watershed management

After three years of studies conducted by the public administration and following the Beauchamp Commission Report, the Quebec Government adopted the QWP in 2002 (Quebec, 2002). The QWP made integrated management a building block for a new institutional framework. According to the QWP, integrated water management was divided into two sections: the integrated management of the St Lawrence River and management of other watersheds. The integrated management of the St Lawrence River was primarily defined in this policy as the continuation and the reinforcement of an existing federal–provincial programme set up in 1988, the St Lawrence Action Plan.

The Ministry of the Environment identified 33 river basins in the QWP in which implementation of watershed management had to be considered. This identification was based on the recognition of past local initiatives or the presence of major water problems. For each one of these basins, a committee was to be created with the principal mandate of producing a water plan. Each plan had to include a diagnosis of the environmental stakes, the identification of interesting ecosystems and wetlands, the ordering of the potential interventions and an action plan. Each water plan was to be elaborated using a collaborative approach. The Ministry of the Environment anticipated that these plans would be reviewed after six years (Quebec, 2004b).

Consequently, committees had to ensure that a collaborative procedure was established. Each had to set up a collaborative board that would respect the framework put forward by the central administration. Participation on these boards had to allow for an equal presence of municipal representatives, economic interests and community spokesmen (citizens, environmental and social groups). The stakeholders were to participate in the development of the water plan and also sit on the board of directors of those committees. Furthermore, the framework also allowed for broader forms of participation. This participation was to take the form of public hearings, held three times during the elaboration of the water plan.

The water plan's implementation, after its approval by a provincial interdepartmental board, was made official by the signing of a river contract between the committee and interested parties (Quebec, 2002). These contracts could be signed for a broad range of actions, and were to be gentlemen's agreements, negotiated on a voluntary approach and not legally binding. No legal proceedings could be instituted on the basis of these documents.

Finally, the QWP promised financial support from the Quebec Government. Since 2003, the Ministry of the Environment ensures an annual statutory funding of US$65,000 for each committee. This support represents the basic financing that can be enhanced by the community. No other mechanism was developed in order to link water management and resource users by a system of charges or an application of the user-pays or polluter-pays principles.

Three Watershed Committees as Case Studies

The organizational analyses were conducted with the collaboration of three watershed committees located on the basin of Du Nord River, St Francois River and St Anne River (Figure 7.1). The principal characteristics of these river basins are shown in Table 7.2.

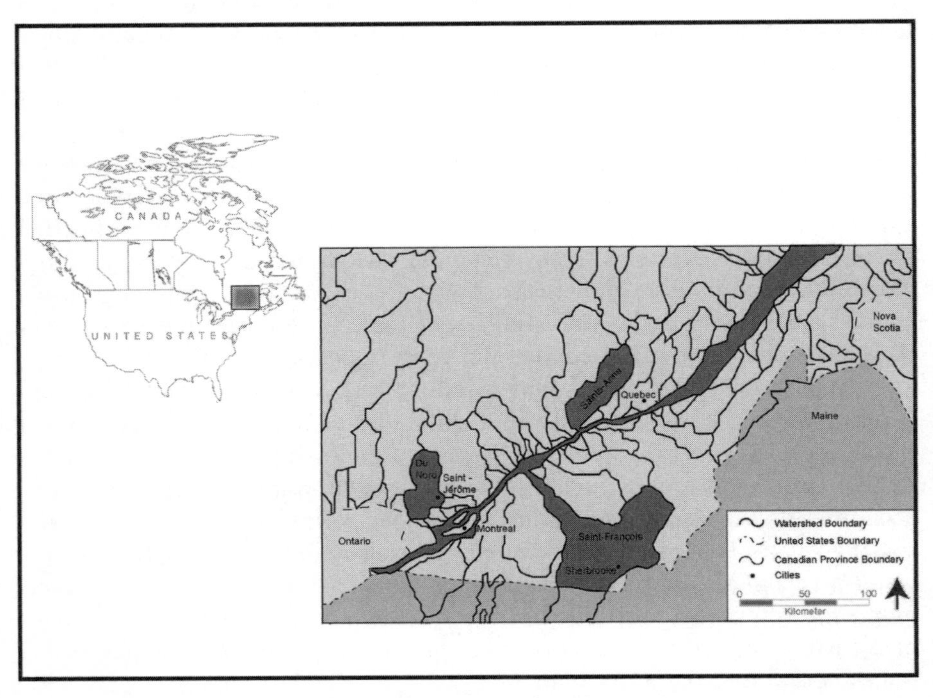

Figure 7.1 Case study areas

Table 7.2 Study cases specifics

		St Anne River	St Francois River	Du Nord River
Basin area (km^2)		2689	10230	2213
Basin population		16251	346513	154347 (+36381 seasonal residents)
Water quality*	Upper basin:	Good	Good to satisfactory	Satisfactory to questionable
	Middle basin:	Poor to very poor	Satisfactory to questionable	Questionable
	Lower basin:	Satisfactory	Questionable to bad	Questionable
Number of municipalities		21	102	37
Population of 3 biggest municipalities		9050	148170	64988
		2156	68845	35342
		1615	23827	11687
Land use activities**		Agriculture (12%) Forestry (82%)	Agriculture (23%) Forestry (66%)	Agriculture (10%) Forestry (68%)
		Urbanized areas (0.5%)	Urbanized areas (5%)	Urbanized areas (5%)
		Mining activity		Intensive tourism (ski, golf)

* Evaluation based on Quebec bacteriological and physico-chemical quality index.
** Land area percentage indicated when data are available.

An overview of the data points out the variety of contexts that these three committees have to cope with (ABRINORD, 2006; CAPSA, 2003; COGESAF, 2006). If the principal activities of these three areas are characteristic of the semi-urbanized and rural areas of southern Quebec (agriculture and forest activity), it is important to note the variation in population density that are respectively 6 people/km^2 for the St Anne River basin, 34 people/km^2 for the St Francois River basin and 70 people/km^2 for the Du Nord River basin (with a peak of 86 people/km^2 during the summer). These different population densities contrast with varied administrative realities, as the number of municipalities within each river basin ranges from 21 to 102. The St Francois River basin also has the particularity of being a transborder waterway, 14 per cent of its area being in the state of Vermont, US. Finally, there is contrast in the history of local watershed management. In the case of the St Anne River, a local watershed management initiative started at the end of the 1980s. The organization created then became the committee, in accordance with the QWP, in 2002. For the St Francois River, some initiatives of water management had already been experimented with, especially by lake residents' associations. The first stakeholders had, however, decided to create a new organization to implement IWM. No organization dedicated directly to water management was present in the case of the Du Nord River basin.

The data analysed in this chapter come from the review of the documents published by the Quebec Government and by the three watershed committees to structure the participation of stakeholders in provincial and local deliberations related to water management. We have also conducted 48 semi-directed interviews with stakeholders from these three case studies.

We present an evaluation of the frameworks related to the participation in the QWP elaboration process and to the watershed committees' activities. This choice is particularly relevant to emphasize the challenges linked to social participation in the hands-on management of a river basin. To look at these two types of participation we use here the criteria proposed by Callon and co-authors (2001) (Table 7.3).

Table 7.3 Evaluation of the participative frameworks related to Quebec water issues

	Criteria	Sub-criteria	QWP elaboration process (Beauchamp Commission)	Watershed management
Technical debate in the deliberative process	*Intensity*	Social participation is integrated in the early stages of exploration of the possible scenarios	Low	Medium to high
		Preoccupation in the composition of the group involved in the debate	High	Variable
	Procedural opening	Diversity and independence of the groups involved in the debate	Very high	Variable
		Representativeness of the groups spokesmen involved in the debate	High	Low to medium
	Effective quality	The relevancy of the discussions	High	Variable
		Continuity of the participation in the process	Low	Medium
Implementation	Equality in the admission to the debate		High	Medium
	Debates are traceable and transparent		High	High
	Clearness of the rules organizing the debate		High	Low

Source of criteria and subcriteria: Callon et al, 2001

The QWP elaboration process

The first part of this evaluation aims at understanding how the procedure allows the opening of technical debates. The analysis considers two distinct aspects: opening the debate itself to various forms of expertise and the opening to groups who could potentially get involved in the debate.

Social participation in the QWP's elaboration process is the result of a controversy that was carried by the media. The strength of the environmental coalitions, particularly in the Montreal area, which represents half of Quebec's population, encouraged participation in the provincial debate. In these circumstances, the early integration of social participation in the debate has to be understood as the result of civic mobilization, not as the result of a formal process. However, once it started its activities, the Beauchamp Commission showed an obvious preoccupation with assuring the diversity and representativeness of the groups that were to be consulted. This question was in effect raised during the Symposium of 1997. On the other hand, the two sub-criteria of quality, procedural opening and effective quality, give opposite results. The relevancy of the discussions that Callon and his co-authors defined as the capacity for the participants to state with acuteness and relevance their arguments and claims, was largely supported by the commissioner's attitude. As to the question of continuity, it is certainly a weakness that is typical of public hearings in a policy formation process. It would have been relevant to add to the commission's recommendations the importance of including social participation in any monitoring activity.

Finally, the criteria of implementation (equal access to the debate, transparency and traceability and clearness of the rules) can be considered as having been assured by the participation framework. The Commission's process was designed to assure the greatest participation and visits to every area of the Quebec territory. Debates were filmed and the briefs were published on the internet. Moreover, the process itself was clearly defined and known by participants.

Deliberative process of watershed committees

When we move from this macro debate on water policy to the micro reality of watershed committees, our evaluation gives contrasting results. These show the importance of the challenge to induce a technical debate on local scale participation. Thus, the framework has an important influence on what the policy really produces in the end. As for the process itself, which is based on stakeholder participation, social participation is integrated in the early stage of deliberation. On the other hand, this goodwill is framed by centrally defined objectives (production of the water plan and implementation of actions following a voluntary contractual approach). Also, stakeholders did not have much to say about the way that outcomes were to be executed.

The makeup of the collaborative board is designed to guarantee a good representation of the variety of local interests. However, the framework does not consider regional particularities, and some special concerns could easily be unrepresented. For instance, in some circumstances the economic interests and the local elected officials can be at the table, but there is no guarantee that social and environmental issues would necessarily be represented by a group or by citizens. This situation can often occur especially outside of urban areas, where the number of groups and their stability are variable.

Another consequence of the participation framework has to do with the proportionality between socio-environmental, municipal and community sectors. Some river basins contain several municipalities and the variety of municipal agendas imposes the presence of a great number of representatives. Thus, the collaborative boards must be very large to ensure the recommended proportionality. Several committees have more than 30 stakeholders directly involved in the deliberation. It is also necessary to question why many municipalities have de facto individual representatives, whereas other participants must be elected by their own electoral colleges. Finally, the quality of the interventions can vary considerably from one committee to another and depends on a number of factors. The moderator's capacity to lead the discussions between laymen and experts, users and non-users, elected representatives and citizens, is a crucial consideration, principally at the beginning of a watershed committee's life.

In examining the implementation criteria, the framework seems to ensure equal access to the debate, just as to the transparency and the traceability of the exercises. However, we have to point out the possibility of perverse effects linked to the formula chosen to guarantee the proportionality among the three sectors specified by the framework.

The clarity of the rules organizing the debates is a weakness. Obviously, the modalities of participation have not received the same attention as the question of proportionality on the boards. If the QWP and the framework refer many times to the importance of stakeholders' participation in the definition of common objectives, there is no precise indication as to how common objectives should be achieved. On the contrary, there are some contradictory signals; for example, both voting and non-voting members are mentioned, yet the idea of consensus is strongly promoted.

Procedural legitimacy of watershed committees' deliberative framework

The last observation is important considering our examination of the procedural legitimacy. If the fair representation of interests seems to be integrated in the development of a Quebec model of IWM, the criteria of equity between the concerns and the level of involvement of the stakeholders in negotiated decisions are not really considered in the framework. Although local actors can find arrangements for the first element in the formative years of each watershed committee, the results of their participation have to be integrated in all the water-related institutions, and no mechanism has been proposed yet. For instance the coordination of water plans and local land uses plans still have to be defined.

If many different modes of participation arise from different local dynamics, the variety of situations should be addressed in any future rewriting of the legal framework or through institutional adjustments. Without eliminating the possibility for each community to develop its own dynamic in accordance with its own water problems and socio-economic characteristics, the provincial coordination of the local IWM initiatives sometimes will have to step in. Also, reorganization of the institutional establishment, including those public administrators involved in water management, will have to anticipate these regional specificities.

Teachings from organizational analysis

The observations we present in this chapter are essentially qualitative. It is, however, possible to organize these observations to give a differentiated portrait of selected watershed committees.

First, the number of participants that have participated at least once in activities of watershed committees since the adoption of the QWP in 2002 is important (52, 61 and 82). On the other hand, if we look at the continuity of this participation, we note an important decrease. The percentage of stakeholders who have taken part at 50 per cent of the meetings from the start is 13, 18 and 18 per cent. These percentages are related to two principal factors: non-attendance of the stakeholders (justified or not) and frequent rotation of participants. This rotation is particularly strong in the case of municipal representatives, in spite of the fact that they have designated seats at the collaborative board. Several reasons can be suggested: the electoral reality that sometimes occasions the replacement of representatives; the waltz between planners, city councilmen and mayors as municipal representatives; and a low interest expressed by some participants. Finally, we note an average participation rate at any given committee meeting varying between 15 and 20 participants.

Stakeholders involved in the three watersheds have different reasons to participate in the committees. Some use the resource for economic advantages. Others have responsibility within public administrations related to water. Finally, some citizens are involved because of environmental conservation motives.

These observations raise questions. What are the reasons leading to such a strong variability in the stakeholders' participation? Who are the regular participants and what are their interests? Which principal issue structures the stakeholders' participation?

Multiple interests relative to social participation

The co-existence of multiple interests among stakeholders is the basis of collaborative approaches. However, if the majority of the participants interviewed expressed a great devotion towards their organizations and their river basin, it is interesting to note a variety of postures towards participation itself. This variety can be related to the sectors represented.

The farmers are actively involved in two of the three committees. In St Anne and St Francois River basin, they are among the most active participants. In the case of the Du Nord River, the downstream section – where we find the major part of farming activities – is characterized by weak participation among the agricultural sector. We cannot advance a clear explanation for this observation. However, the principal reason justifying the involvement of the other farmers can be considered. On top of the list, all the farmers wish to correct a negative perception of their activity, especially in line with water pollution: 'we recognize our responsibilities, but we find exaggerated the negative coverage from the media' (comment from an interview with farmers' spokesman). They want to be sure that they won't be identified as the only users responsible for water pollution and usually provide much information related to their activity in order to 'assure a correct description of farming activities'. Generally, other stakeholders have a positive perception of the farmers' participation. Many perceive a real effort from this sector to act on water issues. Moreover, the majority of the projects discussed within the committees are implemented in agricultural areas. It is finally important to note that the farmers met during this research study are all involved in their regional section of the Union des Producteurs Agricoles (UPA, the Union of Agricultural Producers), the main professional farmers' union in Quebec.

Municipal representatives, elected or not, express some hesitation relative to the creation of watershed committees. Municipalities are legally responsible for drinking water treatment services, whereas Regional County Municipalities deal with land use planning. The Regional County Municipalities are administrative entities that combine the municipalities of the same area, and the implementation of the IWM raised little enthusiasm from municipalities. The financial situation of most of watershed committees and the absence of a clear legal status for the water plan reinforce this attitude. Also, many municipalities already have serious problems enforcing existing standards and norms related to water quality. The cost of the treatment and decontamination infrastructure and costs related to monitoring activities and lack of expertise create situations where existing norms and policies are only partially applied, particularly in rural areas. Furthermore, since their only constant source of income is the property tax system, several city representatives fear a new system of local taxation to accommodate the reorganization of water management. The participation of the municipal representatives is therefore divided into two categories. On one side, some representatives are personally convinced of the usefulness of collaborative approaches and are actively involved. On the other, a majority of them are less committed or stay quiet during the deliberative activities. We observed, however, that all representatives have acknowledged 'the importance to be "there" to see what is done within the committee' (comment from an interview).

The logging industry is also an important economic interest within the three river basins. For activities held on public or private lands, we observe the same perceptions and interests. Foresters are worried about their impacts on water quality, but they have all said that the operational norms and new practices have changed a lot. They suggest that their work respects sustainable practices, 'even if there will always be some uncooperative colleagues!' If their participation is essential considering the importance of this activity, we have noted that there were few discussions about water problems in forest areas.

The residents form another group of stakeholders. These social actors are often already organized into local associations. The pre-existence of lake associations in the communities we studied is an important fact. These groups have historically dealt with water quality, sailing, aquatic sports and other water-related issues. In many cases, these actions have contributed to creating a strong feeling of 'belonging' to their immediate hydrologic basin, the lake. The implementation of IWM has been perceived by some residents as the multiplication of participation in water management or the arrival of a new competing arena. Nevertheless, these participants are often the principal spokesmen for environmental questions. They also represent important links with the rest of the local communities considering their high level of participation in many different organizations.

Finally, a small number of environmentalists are involved in committees. Some of their demands have forced the committees to produce early decisions, a phenomenon that we have noted in our three case studies. Indeed, each committee has been challenged at least once to 'object to a project which will produce negative impacts on watershed integrity'. In each case study, this claim has generated a debate between stakeholders referring to an ecological rationality (the committee must first and foremost defend watershed integrity) and those who defend a procedural rationality related to the mandate of the organization (the committee cannot position itself against one stakeholder – active or potential – in order to ensure the participation of all users of water). In each case, neutrality was the conclusion of the debate. When a problem cannot be collectively dealt with, it is often pushed down the agenda, at the risk of losing some players such as the environmentalists.

Multiple mechanisms and the preponderance of expertise

For more than 60 years, water management has been an issue that demanded public actions from provincial and local administrations. Policies and programmes have been developed through a constellation of organizations. Thus, the IWM approach proposed by the QWP is set in a complex system of institutions already involved in water managements. The results show that the watershed committees are not necessarily the only mechanism that links stakeholders to watershed protection. Several participants belonging to the logging industry and/or agriculture readily pointed out other avenues to deal with water problems (for example, the Prime-Vert Program and provincial norms for the logging industry). Several of these parallel initiatives are supported by budgets that even exceed the watershed committees' budgets. Furthermore, the first mandate of the committee – which is to produce the water plan – contributes to keep the deliberation at a technical level. Some stakeholders say they lose too much time in the discussions of the watershed committees and find other mechanisms better ways to get actions done on the river basin.

These observations must be considered on two levels. First, there is a potential risk to view the watershed committees' activities being confined to a scientific and technical discussion (eg water quality indicators, dilution capacity, erosion processes), rather than as political debate about solutions and actions. It is, however, too early to conclude that these discussions on expertise supersede or eliminate political debates. As an example, in 2006 and 2007 there were blue-green algae episodes, which led to technical and political debates that involved the Quebec Government, local administrations, watershed committees and other groups. Rapidly, stakeholders recognized that this particular problem cannot be considered only as a technical one. Causes and solutions were well known; the question was how to initiate actions to produce results. Second, the number and diversity of institutions guiding stakeholders' behaviours creates challenges in coordinating watershed management. This is probably one of the most important challenges that awaits the Quebec's Government reworking of water governance.

An organization that initially has to survive

The context in which the watershed committees were created is characterized by important constraints: financial, legal and relational. As the stakeholders involved in the committees are also board members of those organizations, many discussions focus on questions related to the maintenance and development of the organization rather than watershed issues. This situation produces frustration and has initially contributed to limiting stakeholder involvement. However, in the long term, this hard beginning could have a positive effect. Rules of the game having been dictated by the provincial government, watershed committees had to find inner strength to adapt to the general model. At the beginning, stakeholders discussed visibility, additional funding, organizational philosophy, operational structures and mandate redefinition. One interesting issue, which was observed in our three case studies, is the possibility for committees to go beyond deliberation to initiate actions.

The modest annual funding granted to the watershed committees forces the organizations to small-scale practical actions. Its mandate of being mainly an arena for deliberations makes it difficult to apply for additional financial support in specific problem-solving programmes from the provincial or federal agencies. Many committees confine themselves to environmental awareness programmes or landscaping activities. Others decide to keep their mandate as encouraging deliberation among stakeholders. It is difficult to pass a normative judgment on these strategies considering the variety of contexts. On one hand, committees fill the void of expertise by developing regular interventions. Also, many observers underline the importance for young watershed committees to carry out early actions or activities to build confidence and momentum for future activities (Barraqué, 1997; Margerum, 1999). On the other hand, there is a possibility of conflict in watershed committees' mandates. By initiating specific actions, those committees in fact become new stakeholders – judges in their own cases. They initiate actions, they increase their expertise and they finally develop new interests related to their new activities. The collaborative process must be adjusted to take this situation into consideration. We noted three organizational strategies that seem to answer this dilemma: the implementation of sub-basins committees where specific actions are debated, the division of the committees in two distinct sub-organizations (deliberation and action) or the carrying out of both tasks within the organization.

We do not believe that one response to this problem should prevail over the others. Local organizational dynamics are too different from one basin to the other to privilege just one solution. We believe that the strength of the Quebec model lies in the recognition of this indisputable variety of socio-ecosystems, calling for coordination rather than standardization of watershed management.

A strategic stake that justifies participation

All stakeholders expressed the desire to improve the watershed, yet the interviewees indicated that this is not the principal justification for their participation. Many pointed out the absence of an immediate crisis, which would loudly call for IWM. Some stakeholders said ironically they would need an event like the Walkerton, Ontario *Escherichia coli* tragedy of 2000 to provoke more interest in water management.

In this context, interviews reveal another strong preoccupation justifying stakeholders' participation: the strategic importance of being a member of the problem definition process. Consequently, the water plan is the first stake that structures relations between stakeholders. We think that this situation will have to change to guarantee the durability of community-based management. The IWM approach developed by the Quebec Government is not explicit about the function of the collaborative process during the steps that follow, implementation and monitoring. Cooperation between stakeholders could be limited to punctual problem definition events and affect the continuity of social participation in water management. If the objective is to create new civic institutions – which will lead to the formulation of rules by the community – rather than simply support the public institutions, the mandate of watershed committees will have to be revised to generate other types of cooperation among stakeholders.

Reconnecting with Substantial Aspects

The case studies permitted us to observe watershed committees in their initial steps – their structuring phase. Our observations must be placed in that context. For the final part of our evaluation, we suggest focusing on the concept of interdependence. Interdependence is a central component of natural resources governance (Ostrom, 1990). Contributions from ecological and social sciences clearly affirm the importance of this notion in understanding the processes characterizing socio-ecosystems. Water resource management is characterized by particular situations of territorial and functional interdependence related to questions of quality, quantity and ecological functions.

Ecological interdependence and stakeholders' participation

A first case of interdependence can be seen through the impacts of upstream activities on downstream stakeholders. This fact raised many questions during the elaboration of the IWM approach. Hence, each watershed committee has to take into consideration the proportional participation of upstream and downstream stakeholders. In our cases studies, however, we noted that the Du Nord River basin committee is confronted with difficulties in recruiting representatives from the downstream area. On the contrary, the upstream section, which counts several municipalities with citizens seeking contact with nature, is represented by many citizens and elected representatives who defend environmental concerns. It is obviously an important limit of the voluntary approach. Any future reform of the administrative framework of water governance will have to consider this limit, either by reinforcing the incentives to get involved in the process or by developing a coherent system of cooperation that integrates geographical representativeness.

A second ecological dimension is the multi-scalar nature of the watershed unit. The committees also produce strategies according to this factor of interdependence. By dividing their watershed into sub-basin territories, they create new institutional references geared towards local actions. The St Francois River basin committee, for example, started the implementation of local watershed committees in June 2007.

The multi-scalar dimension raises two potential difficulties. First, the development of a collaborative process in a sub-basin committee implies the redefinition of the participation objectives for the whole watershed management institution. In the case of the St Francois River Basin, it is too early to study this aspect and we can only discuss in a prospective manner. Our observations allow us to project that certain factors will make the creation of sub-basins problematic: stakeholders have limited hours to invest in the collaborative process, there are strong inhibitions to tackle very large watershed problems and stakeholders' preferences to be involved in a collaborative process closer to direct action. Second, sub-basin initiatives are not always an outcome of a watershed committee's deliberation. In some cases, the committee is not informed of it.

The multiplication of sub-basin management efforts will require strong coordination. This coordination will imply that all project instigators (ministries, para-governmental institutions, lakes associations and private corporations, for example) recognize the watershed committee as the focal point of watershed management. Again, it is important to note that the IWM approach is a new institutional reference for most of the stakeholders.

Water is also a multiple-use resource. This dimension seems to be more difficult to integrate into the collaborative process. Certainly, official documents recognize this aspect of the resource. However, this concern is not clearly translated into watershed committees' activities:

1. Water quality is the central issue for most of the stakeholders and this shows up in the deliberations. The blue-green algae events have probably contributed to enhance this preoccupation. The explosion of cases since the summer of 2005 (45 identified cases in 2005, 71 cases in 2006 and 181 cases in 2007) has pointed to phosphorus contamination as the principal water quality problem for Quebec lakes and rivers. This situation causes defensive strategies among stakeholders. Each one does not want to see its group being declared responsible for the problem.
2. The elaboration of the water plans is essentially a technical exercise that has neglected some important water uses. Subjective considerations like tranquility, landscape beauty or historical attachment are hardly integrated in frequent quantitative debates.
3. Social participation in water management is new for many communities and several efforts aim at facilitating consensus building. Intentionally or not, thorny subjects were evacuated in order to keep all stakeholders committed to the collaborative process. We believe that, as collective learning and trust build, these larger, potentially divisible issues will gradually be put on the agenda of the watershed committees.
4. The dynamic of natural processes is undoubtedly the most difficult aspect to bring in the deliberation of a watershed committee. Many stakeholders do not distinguish between static and dynamic balance of ecosystems, for example, when they discuss aquatic processes. As well, events like floods seem to be perceived as a question of physical control, rather than in ecosystem terms. The same trend is observed relative to the blue-green algae crisis. Political debates aim to identify solutions that will produce effects rapidly (eg banish phosphorus in soap or proceed with shore reforestation) without considering the fact that lake resilience will take time or, worse, that some lakes have already reached their eutrophication phase. These perceptions contrast with the idea of a better understanding and thus integration of natural processes into the IWM approach.

Social interdependence among stakeholders

Ecological dimensions justify the implementation of a watershed management approach because of the interdependence between the multiple uses of water. According to the QWP, 'IWM offers the most advantageous solution to sectorial management of water' (Quebec, 2002). Therefore, the quest for social interdependence can be seen as the principal motive to integrate collaborative processes within water management. This interdependence, which can be understood as social capital, is constructed by stakeholders' relations.

A first example is the **trust** developed by stakeholders towards other participants (Leach and Sabatier, 2005). This aspect is often presented as directly linked to successful collaborative approaches in environmental management. Trust facilitates the production of agreements, which are a sign of stakeholders' willingness to cooperate. Maintaining trust is a permanent concern for the employees of watershed committees.

Stakeholders' comments attest that this climate of trust has been constructed in our three case studies. It is, nevertheless, a work in progress, notably because trust is not necessarily a community attribute. We consider that the QWP has been elaborated on questionable premises about what are communities, as Agrawal and Gibson (2001) remark: 'many of these centrally imposed community programmes are based on a naive view of community'.

Finally, social interdependence should lead to a **fair redistribution** of the policy outcomes. The participation must lead to projects that will not always cater to the same stakeholders and create unjustified losses for the others. The Quebec collaborative approach that promotes consensual decisions seems to address this concern. The production of the water plan through consensual approaches assures that future actions will be collectively supported. However, a consensual approach does not necessarily mean fairness. Consensus building is a complex process and the outcomes can be explained by many different factors. The development stage of these case studies does not allow for further discussion at this point. Although the approach generally produces shared and fair decisions, it can also exclude some problems that could be too controversial. Finally, we still have to note that important stakeholders are not involved in the IWR process. Fairness is then defined within the watershed committee, not necessarily throughout the entire watershed community.

These observations about social interdependence reveal that the government's assumptions about communities are not given but built socially over time. In this situation, we can wonder if watershed committees have to be **neutral** organizations with the strict objective of being a deliberative arena for stakeholders, as it is defined in the policy, or if they should play an active role in promoting new, community-shared values. In fact, the Quebec IWM approach has been centrally defined without paying great attention to the communities' realities. Some communities reveal the capacity to go further with their realizations, just as others want to focus their activities on their supporting roles to public institutions.

Conclusion

Social participation in watershed management is recent in Quebec and our evaluation had to consider this process as still under construction. Consequently, it is not possible to conclude definitely on how the five initial intentions of the QWP were reached. However, our analysis leads to conclusions about the participation framework, the organizational dynamic resulting from social participation and the substantial elements of the IWM approach.

A series of interviews allowed us to foresee interesting characteristics of the participation process built in the Quebec IWM model. The participation framework encourages debate on the technical dimensions, notably by the collective elaboration of the water plan. Focusing on the organizational dynamics, we observed intense relations among stakeholders that in fact contribute to the problem definition process and to the elaboration of solutions. We also noted a variety of interests represented by the stakeholders. Finally, the upstream–downstream relations and the multi-scalar issues are most often integrated in the deliberations.

Nevertheless, our case studies gave us a glimpse of some of the limits to the Quebec model. First, our evaluation of the participation framework identified some unexpected impacts from the proportionality of stakeholder types in the watershed committees. The framework imposed by the provincial government leads to very large collaborative boards. The stability of these boards is thus difficult to maintain, as we observed in our three case studies. Also, the voluntary contractual approach defined by the Quebec Government is not uniformly applied in the communities. The lack of incentive leads to the absence of important users. Finally, the framework does not suggest a collaborative philosophy. Many different models can be developed, from a decision shared by all via a necessary consensus to a majority vote situation. If local dynamics must be respected in the development of collaborative values, the elaboration of provincial objectives will have to cope with the variety of applications and the different ways of undertaking IWM.

From an organizational perspective, we noted the predictable adjustment of the participative framework relative to local specificities. Our results show how social participation is rooted in different type of interests. However, these interests seem to be linked to one procedural motivation: the importance of taking part in the problem definition process. This observation is undoubtedly important and raises new questions. The presence of this procedural motivation, combined with the weakness of the financial support, ambiguous legal recognition of water plans and the multiplicity of mechanisms dedicated to water management, leads to the following question: what will become of social participation when the watershed committees have to go beyond problem definition to initiate actions (active management, monitoring, sensitization)? In the Quebec context, the fact that water quality issues have historically generated public actions will surely have an important impact. Stakeholders will have to distinguish between what should be done by the public administration and what has to be done by local communities.

References

ABRINORD (2006) 'Portrait et Diagnostic du Bassin versant de la Rivière Du Nord', Agence de bassin de la rivière du nord, St. Jerome, Quebec

Agrawal, A. and Gibson, C. C. (eds) (2001) *Communities and the Environment: Ethnicity, Gender and the State in Community-Based Conservation*, Rutgers University Press, New Brunswick, NJ

Allain, S. (2005) 'Expertise et planification participative de bassin. Une analyse en terme d'action publique négociée', in L. Dumoulin (ed), *Le Recours aux Experts: Raisons Et Usages Politiques,* Presses Universitaires de Grenoble, Grenoble, France, pp267–284

BAPE (2000) 'L'eau, ressource à protéger, à partager et à mettre en valeur', Quebec Government, Quebec

Barraqué, B. (1997) 'Subsidiarité et politiques de l'eau', in A. Faure (ed) *Territoires et Subsidiarité*, L'Harmattan, Paris, France, pp165–201

Bibeault, J. F. (2003) 'La gestion intégrée de l'eau: dynamique d'acteurs, de territoires et de techniques', *Cahiers de géographie du Québec*, vol 47, no 132, pp389–411

Brun, A. (2007) 'Gestion de l'eau au Québec: quand politique de l'eau et politique agricole se conjuguent à l'imparfait', *Quebec Studies*, vol 42, pp61–74

Burton, J. (1997) 'La participation du public à la gestion environnementale du fleuve Saint-Laurent: les zones d'interventions prioritaires (ZIP)', *Collection Environnement*, vol 22, pp147–161

Callon, M., Lascoumes, P. and Barthe, Y. (2001) *Agir dans un Monde Incertain,* Le Seuil, Paris

CAPSA (2003) 'Portrait hydrologique et multi-ressources du bassin versant de la rivière Sainte-Anne', Corporation d'aménagement et de protection de la rivière Sainte-Anne, St Raymond, Quebec

Cashore, B. (2002) 'Legitimacy and the privatization of environmental governance: how non-state market-driven (NSMD) governance systems gain rule-making authority', *Governance*, vol 15, no 4, pp503–529

COGESAF (2006) 'Analyse du Bassin versant de la Rivière Saint-François', Comité de gestion de la rivière Saint-François, Rock Forest, Quebec

Coglianese, C. (2002) *Is Satisfaction Success? Evaluating Public Participation in Regulatory Policymaking*, John F. Kennedy School of Government, Cambridge, MA

Crozier, M. and Friedberg, M. (1977) *L'Acteur et le Système,* Le Seuil, Paris

Dorcey, A. H. J. and McDaniels T. (2001) 'Great expectations, mixed results: trends in citizen involvement in Canadian environmental governance', in E. A. Parson (ed) *Governing the Environment*, Presses de l'Université de Montréal, Montréal, pp249–301

Gill, L. (1998) 'Le mouvement syndical et les enjeux économiques actuels', in Y. Bélanger and R. Comeau (eds) *La CSN. 75 Ans d'Action Syndicale et Sociale*, Les Presses de l'Université du Québec, Montreal, pp169–190

Grumbine, R. E. (1994) 'What is ecosystem management?', *Conservation Biology*, vol 8, no 1, pp27–38

INRS-EAU (1997) 'Symposium sur la gestion de l'eau au Québec: Volume 1', INRS – Eau, Quebec

Korfmacher, K. S. (2001) 'The politics of participation in watershed modeling', *Environmental Management*, vol 27, no 2, pp161–176

Latour, B. (1999) *Politiques de la Nature*, La Découverte, Paris

Leach, M., Mearns, R. and Scoones, I. (1999) 'Environmental entitlements: dynamics and institutions in community-based natural resources management', *World Development*, vol 27, no 2, pp225–247

Leach, W. D. and Sabatier, P. A. (2005) 'Are trust and social capital the keys to success? Watershed partnerships in California and Washington', in P. A. Sabatier (ed) *Swimming Upstream*, MIT Press, Cambridge, MA, pp233–258

Lepage, L., Gauthier M. and Champagne P. (2003) 'Le projet de restauration du fleuve Saint-Laurent: de l'approche technocratique à l'implication des communautés riveraines', *Sociologies Pratiques*, vol 7, pp63–89

Lubell, M. (2004) 'Collaborative watershed management: a view from the grassroots', *Policy Studies Journal*, vol 32, no 3, pp341–361

Manning, J. C. (1997) *Applied Principles of Hydrology*, Prentice Hall, Englewood Cliffs, NJ

Margerum, R. D. (1999) 'Integrated environmental management: the foundations for successful practice', *Environmental Management*, vol 24, no 2, pp151–166

Mermet, L., Billé, R., Leroy, M., Narcy, J. B. and Poux, X. (2005) 'L'analyse stratégique de la gestion environnementale: un cadre théorique pour penser l'efficacité en matière d'environnement', *Natures, Sciences et Sociétés*, vol 13, pp127–137

Michaelidou, M., Decker, D. J. and Lassoie, J. P. (2002) 'The interdependence of ecosystem and community viability: a theoretical framework to guide research and application', *Society and Natural Resources*, vol 15, pp599–616

Milot, N. and Lepage, L. (2007) 'The integrated management of the St Lawrence River: a social experiment in public participation', *Quebec Studies*, vol 42, pp17–30

Murtagh, B. (2004) 'Collaboration, equality and land-use planning', *Planning Theory & Practice*, vol 5, no 4, pp453–469

Ostrom, E. (1990) *Governing the Commons: The Evolution of Institutions for Collective Action*, Cambridge University Press, Cambridge, MA

Quebec (1975) 'Rapport de la Commission d'étude sur les problèmes juridiques de l'eau', Commission d'étude sur les Problèmes Juridiques de l'Eau, Quebec

Quebec (2002) *Water. Our Life. Our Future: Quebec Water Policy*, Quebec Government, Quebec

Quebec (2004a) *Étude sur la Qualité de l'Eau Potable dans Sept Bassins versants en surplus de Fumier et Impacts Potentiels sur La Santé,* Ministère de l'Agriculture, des Pêcheries et de l'Alimentation, Quebec

Quebec (2004b) *Gestion Intégrée de l'Eau par Bassin versant: Cadre de Référence pour les Organismes de Bassins versants Prioritaires*, Ministère de l'Environnement, Quebec

Quebec (2006) *La Loi sur les Compétences Municipales Commentée Article par Article*, Ministère des Affaires municipales et des Régions, Quebec

Quebec (2007a) *La Municipalité Régionale de Comté: Compétences et Responsabilités*, Ministère des Affaires municipales et des Régions, Quebec

Quebec (2007b) *Méthodologie d'Évaluation des Cas d'Érosion du Réseau Routier dans les Forêts Aménagées du Québec*, Ministère des Ressources naturelles et de la Faune, Quebec

Slocombe, D. S. (1993) 'Environmental planning, ecosystem science, and ecosystem approaches for integrated environment and development', *Environmental Management*, vol 17, no 3, pp289–303

St-Pierre, M. (2005) *La Lutte pour l'Adoption d'une Politique de l'Eau au Québec*, Centre de recherche sur les innovations sociales, Montreal

Sullivan, H. (2001) 'Modernisation, democratisation and community governance', *Local Government Studies*, vol 27, no 3, pp1–24

Trachtenberg Z. and Focht W. (2005) 'Legitimacy and watershed collaborations: the role of public participation', in P. A. Sabatier (ed) *Swimming Upstream*, MIT Press, Cambridge, MA, pp53–82

Part IV

Participation and Implementation of Water Management

8

Social Participation in the Irrigation Sector in Yunnan, China: Roles of the State, Water User Associations and Communities

Liang Chuan and Yue Chaoyun

Introduction

Social participation in the irrigation sector is particularly significant in Yunnan Province, China, which faces problems with low rate of utilization of the abundant water resources in the province. Massive construction of irrigation works have been undertaken to overcome the insufficiency of physical infrastructure. Although physical problems exist, the ineffective management of existing irrigation works is also due to social, political and institutional problems.

This chapter examines the role of the state in conjunction with the establishment of water user associations (WUAs) to consider whether there has been a move toward more democratic decision-making and irrigation management systems. We also evaluate WUAs and the participation of individual farmers at different levels. The focus is on the relation between state control, the governance of local government and social participation. It is asserted that the political will of Chinese governments for social participation at the grassroots level provides opportunities for social participation in the irrigation sector. However, social participation is undermined by the lack of transparency in the flow of information, problems with participation skills and lack of willingness of various levels of government to transfer power to the WUA. Institutional reform, labelled as decentralization, has transferred responsibility for the maintenance of irrigation works to the WUAs but without transferring the resources and power to fulfil this responsibility. By comparing villages with and without WUAs, the authors review the structure, decision-making mechanisms and operation of WUAs as well as the establishment process. The participation of the members within associations is also described. In the absence of political decentralization, social participation in the irrigation sector is characterized by the participation of the rural elite that has been consolidated by the establishment of WUAs. Rural elites, particularly village heads as the leaders of subunits of WUAs, are involved in decision-making within the associations, which initiate negotiation among the associations. However, participation at household level is limited. The process for the establishing WUAs is top-down and bureaucratic, and the representation of WUAs and internal decision-making mechanisms distorts wider participation of individual users in collective action.

The evaluation of WUAs and irrigation management was based on the views of individual farmers. The authors assert that the farmers' participation in irrigation management is primarily low level. The farmers mostly contribute to the maintenance of irrigation works as volunteer labourers as few farmers have the opportunity to participate in planning and decision-making about irrigation management. Their reaction to not being involved in water management takes the form of resistance in daily life. Social participation of individual farmers is constrained by their limited involvement in decision-making on the price of water and by the lack of transparency in decision-making on water resource governance.

The authors argue that the autonomy of WUAs is essential to promote meaningful social participation. Regulation from the civil affair administrative bureau stipulates that any association has to associate itself with a governmental agency as the supervising institute. This factor tends to distort autonomous engagement of the association and members, with the result that farmers' active social participation has been minimized.

Water Resources and Irrigation in Yunnan Province

Yunnan Province has abundant water resources given its many rivers and high annual precipitation (Figure 8.1). There are 908 rivers with an average watershed area of more than 100km^2. These rivers belong to the six river systems, namely the Yangtze River, Pearl River, Red River, Mekong River, Nujiang River and Irrawaddy River. The mean annual precipitation of the province is 1258mm, resulting in a total water volume of 222 billion cubic metres including 148 billion cubic metres of surface water (two-thirds) and 74 billion cubic metres of groundwater (one third). Yunnan Province is the third in China in terms of the capacity for self-producing water. The mean water volume passing through the province is approximately 194 billion cubic metres.

For more than 50 years, massive irrigation facilities have been continuously constructed in Yunnan in order to improve the rate of utilization of the water resource. The population has for decades been encouraged to participate in the construction of irrigation facilities, prevention of floods and

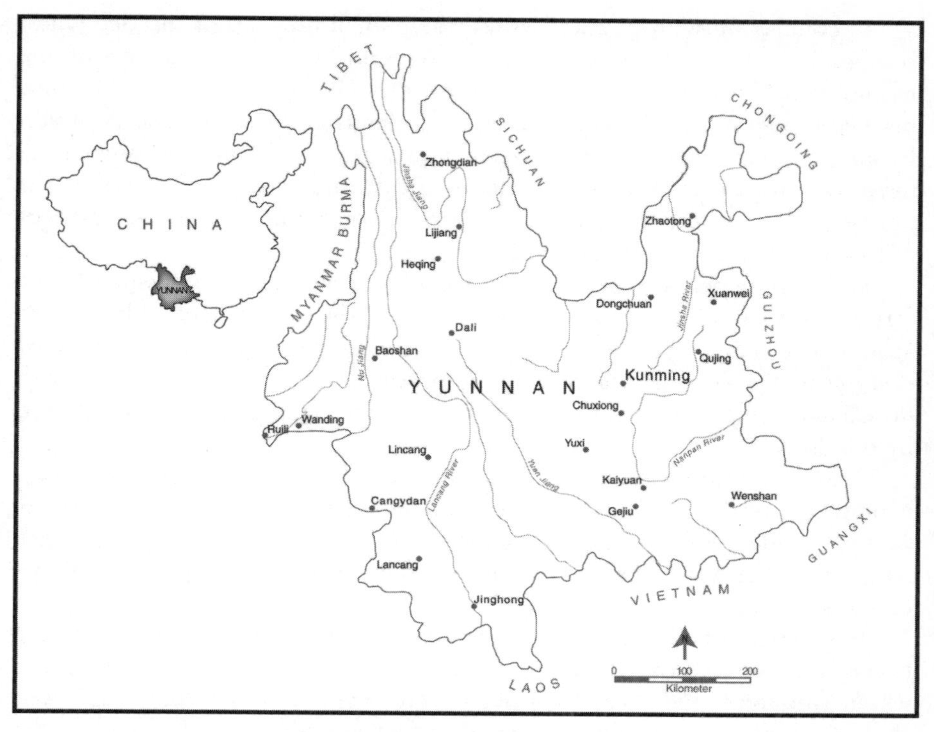

Figure 8.1 Map of Yunnan Province

conservation of water and soil. Irrigation facilities have played a vital role in promoting agricultural production in China. For Yunnan Province, it is estimated that the investment in the construction of hydraulic infrastructure and waterpower over the past 50 years is up to approximately US$8.49 billion (59.4 billion yuan). More than 300,000 hydrological works have been constructed including 5296 reservoirs of large, medium and small water storage capacity. The total water storage capacity of the reservoirs is approximately 9.6 billion cubic metres, and the hydrological works supply 14 billion cubic meters of water annually. With this supply, 14,568 million hectares of land are irrigated in the Province, and 347 million hectares are very efficiently irrigated (Yiqing, 1999).[1]

The irrigation sector faces the core problem of poor utilization of rich water resources. Only six per cent of water resources of Yunnan Province are utilized for irrigation, which is ten per cent lower than the average level of the whole country. As a result, water shortage is estimated at around 4.2 billion cubic metres in Yunnan. The low rate of utilization of water resources there is due to physical issues and the ineffective management of the existing facilities, which relates to social, political and institutional issues.

In connection with physical issues, the poor utilization of the rich water resource is affected by complicated topographical features because of the mountainous headwaters region, uneven rainfall in the different seasons, marked topographic differences between lowlands and uplands and pollution by industrial and domestic wastewater discharge. Ninety-four per cent of the territories are upland and mountainous areas, and most people live in the uplands. This is the biggest physical problem affecting the low rate of utilization of water resources in Yunnan.

Seasonal water scarcity occurs widely during winter and spring; the farmlands in question account for 30 per cent of the total cultivated land, with around five million people and 3.5 million domestic animals being affected every year. The summer and autumn rainy seasons are always associated with floods and soil erosion, and one fourth of the farmland in Yunnan is affected by floods every year.

Water pollution is a serious problem, and water quality is unsatisfactory. Monitoring of the 67 rivers over 9602km showed that 2093km of the rivers contain Grade 4 water or above, that is, 21.8 per cent of the total river length monitored. The water of six of the nine lakes is clean or lightly polluted, above Grade 4.[2]

In social and political terms, poor access to the rich water resources is due to insufficient investment in infrastructure, ineffective management of the water resources and backward water governance, which have all been challenged by the socialist market economy and the need for sustainable development. Lack of social participation in irrigation management is recognized to be the core issue responsible for ineffective management of the water resource in the irrigation sector. Since effective irrigation management is crucial for food security and the stability of society, social participation of irrigators in irrigation management is particularly meaningful.

Evolving Role of the State in Social Participation in the Irrigation Sector

The Chinese Government has enforced the democratic election of village leaders since the early 1990s. Irrigation is closely related to the interests of the farmers, and irrigation policies and management require their participation. In 2005, the government issued decrees to establish WUAs for agricultural irrigation management, and these require the social participation of the users in irrigation management. These associations are assumed to build democratic decision-making mechanisms and to take responsibility for the management of the irrigation systems under the guidance of the government. Social participation in the irrigation sector is supported by the policy in order to reform the irrigation management model.

The reform of the irrigation management model, officially described as irrigation institution reform, introduced the market system within water resource management. The irrigation service is defined as a semi-public service and irrigators have to pay for irrigation. Part of the water fee collected is used to pay the salaries of the temporary employees, since the number and function of staff of the management institute are fixed and funds allocated by the upper government do not cover expenditures by additional employees for the maintenance of irrigation works. Dialogue between water users and the government about the water prices have been initiated. Water fees collected from irrigation users are to be used for the maintenance of irrigation works; however, irrigation works are also destined for flood control, which is a public service. Public meetings about the prices of water for agriculture, industry and urban populations have been conducted. For instance, a public meeting was held in Luliang County in 2006 about the price of irrigation water. The Bureau for Water Affairs Management organized the public meeting, and several representatives from the administrative village committee took part in the public debate. The government collected the different ideas and suggestions for water prices for different uses, and made the decision taking into consideration the suggestions put forward. However, no legal requirement exists to include such suggestions in policies or decisions, and there are no public consultation procedures to ensure the voices of the weak are heard and respected.

Local government staff at the county and township levels are the direct organizers of WUAs,[3] and they have to be established by the irrigation administration within a set time. The local staff organize awareness-raising campaigns in the villages about participation in the WUAs and organize meetings of farmers' representatives to inform them about the setting up of the WUA, which will be housed in the same office as the township level water management station. A few farmers are included in the associations to maintain irrigation works and to collect user fees from their respective village, but the establishment of WUAs is not oriented for all-round social participation, and rather it is bureaucratic and top-down.

A series of institutional reforms was implemented in the water sector in China. This institutional reform by the local government comprised the following five tasks:

1 to define the categories of the service: public, semi-public or privatized;
2 to define the number and function of staff;
3 to define the financial subsidies required for the public service;
4 to define the water price for agricultural, industrial and urban uses respectively;
5 to set up WUAs to manage agricultural irrigation.

Irrigation administration bureaux (public institutes to manage public infrastructures) are responsible for managing the irrigation systems. The resources of the bureaux include public funds for staff salaries and operational costs, funds for flood control and the water fees collected from users, as well as income from leasing the water and infrastructure for multi-purpose use. Agricultural irrigation is defined as a semi-public service because of its dual function (irrigation and flood control). The farmers have to pay for the storage, distribution and transportation of water for irrigation.

WUAs are registered with the Civil Affairs Management Bureau and are currently managed by the Water Affairs Management Station (WAMS) of the township government. There are no procedures for social participation by users, and, in fact, the WUA is an extension of the local government in rural areas.

Social participation in the irrigation sector with local farmers as key stakeholders is affected by the roles of the government and the governance. Institutional reform aimed at involving local farmers in management through establishment of WUAs has reinforced the role of the government. Governance has changed regarding implementation of new policies for institutional reform, and the development of WUAs as well as the leading role they are assuming. Institutional reform that targeted the introduction of market institutions into the irrigation sector has started the process of incorporating and negotiating with the different key stakeholders. Nevertheless, the power for decision-making and financial management is still skewed toward the government relative to local farmers and the private sector.

Role of Water User Associations

As one of the components of institutional reform of water, the central government issued a decree to require local governments to set up WUAs. The Agriculture Water Users' Association of Banqiao Town, Luliang County is the case study presented here (Figure 8.2). Luliang is one of the counties of the Quqing irrigation module, which is the biggest of the 12 irrigation modules in Yunnan Province in the upstream portion of the Pearl River. Luliang County has well-constructed dams and a network of canals with an irrigation capability of over 27,000ha that ensure that 42 per cent of the fields are well irrigated.

The Water Affairs Management Bureau (WAMB) of Luliang County was responsible for setting up the WUA. The staff of WAMB organized an awareness-raising campaign about the WUA among the village leaders and local government leaders in the township government. The responsibilities and roles of WUAs were explained at the promotional meeting. A working group was set up at the town level under the leadership of the county governments, then the working group conducted investigations in each administrative village to obtain information about the number of the farming households. The working group then organized meetings at the administrative village level to raise awareness about the WUA.[4] The natural village heads, who are under the authority of the administrative village committee, as well as the leader of the branch Communist Party and farmers' representatives, attended the meetings. Representatives of the member households are elected by the households every five years, and one of every 15 farmer household heads is elected as their representative, holding the position for five years. The WUA used this

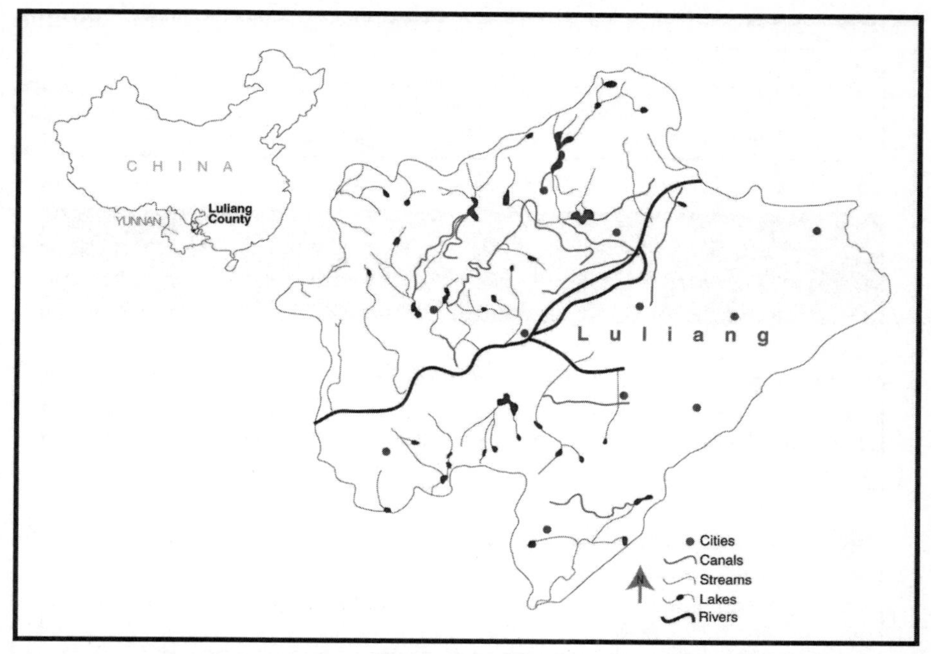

Figure 8.2 Irrigation works and the study site in Luliang County

institution for one level of the structure. The 20,000 heads of household are the legal members of the association. The representatives of the subunits are elected from among the candidates recommended by the township government or by the county government and are supposed to have the capacity to deal with conflicts, and to be both pragmatic and influential among the farmers.

The Banqiao WUA was registered with the Civil Affair Bureau of the county in August 2006, signalling the official establishment of the WUA. The WUA is affiliated with the township governments and there are four layers in the structure of the association:

1. At the top of the association is its legal representative, the governor of the Township Government, and the vice representative is the director of the WAMS, both being appointed by the government. The office of the WUA is housed in the same building as the WAMS.
2. At the next level down are 17 subunits representing 17 administrative villages.
3. The third layer has 1500 representative members elected from natural villages.
4. At the bottom level are 20,000 member households (heads of household) representing a total population of 86,000 (Figure 8.3).

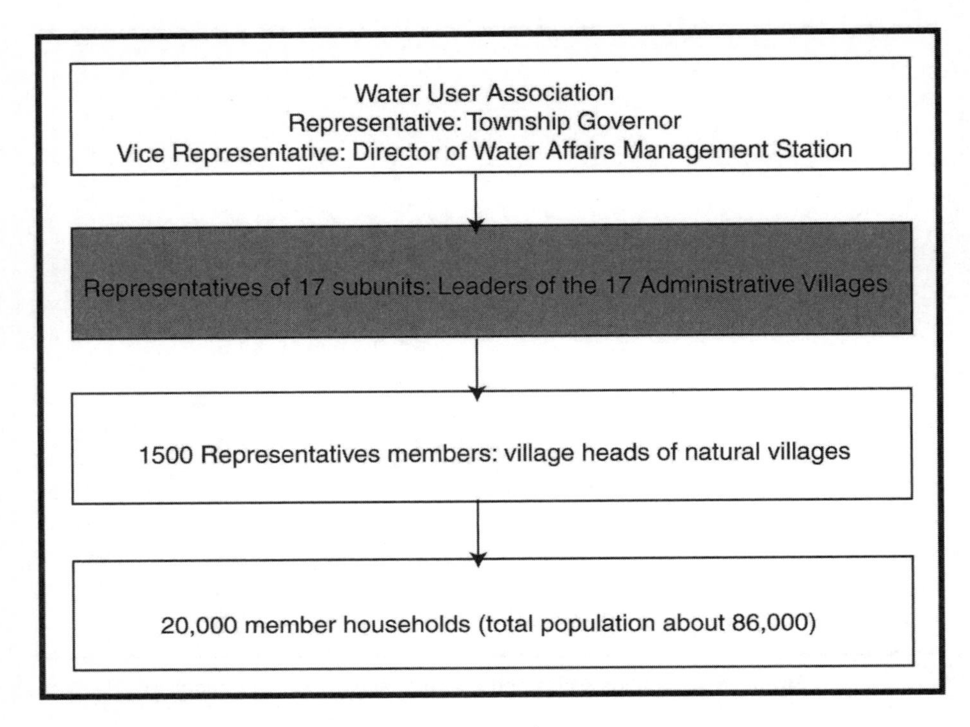

Figure 8.3 Structure of Banqiao WUA

Operation of the Banqiao Water User Association

The WAMB allocated special funds to the township government for the establishment and running of the association, which should be used exclusively on the operation of the association. In fact, the funds are under the control of the township governor who is also the appointed leader of the association. The funds allocated include money to pay the representatives of the subunits, and the WAMS provides offices, meeting rooms and other office facilities for the association. One per cent of the water fee collected from the farmer households is destined to pay for operating costs including the payment of labour. This costs US$29 or US$43 per year per administrative village. The rate is extremely low so that most of the village heads complain about the lack of operating funds for the users' association. That is why the actual rate they charge the users is higher than water use prices for irrigation provided by the state.

As WUA subunits, the administrative village leaders have the right to decide about irrigation. They decide the timing of irrigation and water access, whereas this was formerly decided by the township government.

The association organizes a general assembly three or four times a year before the start of the irrigation season to plan and implement the irrigation campaign. These mainly take place during winter and the following spring, during April or May. The two main irrigation seasons are in May and January, but in addition there are several small irrigation periods. All the subunit representatives attend the general assemblies of the association.

The association applies to the reservoir management station for water distribution. The principle behind water distribution is to irrigate downstream farmlands first and then upstream subunits. The leaders of the subunits of the association arrange the timing of irrigation and then the upstream and downstream subunits consult each other on timing. The order of irrigation among the different subunits means that, as subunits of WUA, the villages now have easier access to irrigation water than before. Coordination between water users and irrigation management bureaux has improved.

The construction of canals and ditches is planned and implemented by the local government without consulting the farmers, and the farmers are required to contribute their labour to build irrigation works. In fact, many farmers refuse to maintain the ditches and canals. The task of maintaining the irrigation facilities has been delegated to the subunits of WUA, and the workload of township government is thus reduced.

The subunits of the association are responsible for collecting the water fees from the natural villages, while the representative members collect the water fees from the member households. The association collects the water fees from the subunits and turns them over to the reservoir management bureaux. The water fee collection criterion is based on 7500 m^3/ha and US$43/ha. As an incentive, the subunits are allowed to collect a water fee of US$54/ha. Some subunits collect water fees at the rate of US$3–4 per capita. For some farmlands, water needs to be pumped for irrigation, and the owners have to pay additional US$0.70 for the cost of electricity. The water fee for irrigation is collected once a year; the association is required to collect the water fee before December each year.

A new irrigation water price of US$43/ha was implemented in March 2006 by the user association. According to the state policy, irrigation water is calculated at 6000m^3/ha by the central government, so the price defined by the state is US$34/ha, which represents a five-fold increase in price since 1985. The user association collects irrigation fees at a price that is US$9 more than the state-defined price. The leaders of WUA subunits are responsible for collecting the water fee. In practice, the irrigation water fee is collected by mu (a Chinese unit of area – 15 mu is the equivalent of 1 hectare) or per capita, US$43/ha or US$3–4/mu. The fact that the actual price is higher than the price defined by the state is one of the reasons that the farmers do not wish to pay for the use of irrigation water.

Yutang Administrative Village, a subunit of the WUA, is located in the central part of the canal system. It is considered by the staff of the Water Affair Management Station to be a good example because of the high percentage of payers in the village. According to the leader of this WUA, they were only able to collect 40 per cent of water use fees initially, and were only able to collect another 40 per cent after one year's persuasion. 20 per cent has still not been paid.

Participation of Individual Farmers

A survey with questionnaires was conducted in one subunit of the Banquiao WUA, located in the middle part of the river basin. Ten members were surveyed in three natural villages. A total of 30 respondents took part in the investigation, 29 of whom contributed questionnaires that could be used for analysis. Although the results have no statistical significance, they demonstrate the current irrigation situation and the preferences and attitudes of the villagers of this subunit of the WUA.

Responses suggest that farmers' access to irrigation water is uniform. Ninety-three per cent of the respondents irrigate fields by paying water fees and using the irrigation works since the cost of irrigation is lower than self-pumped river water. The farmers in this subunit have access to good irrigation facilities and the subunit of the association arranges irrigation in a timely fashion when needed.

The different topographic locations of the villages affect the preferences of the farmers with respect to the different ways of distributing water. These farmers are in the middle part of the river basin. The farmers in the village located in a flat area mostly prefer upstream to downstream distribution; indeed no respondent mentioned the opposite method. The farmers where there are differences in topography prefer flexible distribution since low-lying fields are easily flooded if the drains are not well built. A gate-man is assigned to manage water distribution to each village and when irrigation starts, the ditch gate has to be carefully closed to low-lying fields to avoid flooding.

The order of irrigation is decided by the leader of the association and the leaders of the subunits, and they set the irrigation of the downstream villages as their priority in arranging the order of irrigation. There are two reasons for this arrangement:

1 Wastage of water can be reduced as water can be utilized when it passes through the upstream area, since the main canals are not all cement-surfaced and leakage of water into the fields is good for soil conditioning.
2 Another is to take the rights of the downstream villages into account, as these are disadvantaged in their access to irrigation water.

The annual general assembly of subunit leaders from both upstream and downstream has provided the WUAs with a chance to discuss and decide upon the water schedule.

The investigation showed that 87 per cent of the respondents were willing to pay a water fee and that 13 per cent were not. This survey result corresponds to the report of the leader of the subunit. Among the farmers who are willing to pay their water fee, many delayed, and the representatives of the members had to urge them several times before it could be collected. The main reason for the delay was that the farmers had no cash in hand. The reasons given by the 13 per cent who failed to pay varied. Some old or poor farmers could not afford to pay; some refused because the system was not well managed, with either the drainage network not being well maintained or the water not being delivered.

Unlike the subunit located in the middle stream, the upstream village refused to pay water fees for years. The village located far upstream contributed land for the construction of a reservoir, which is why the villagers refuse to pay water fees. Their fields are often flooded in the rainy season due to poor drainage, and the villagers complain about the losses they have suffered. The association agreed to provide irrigation to this village free of charge.

The maintenance of the main irrigation canals is organized by the leader of the subunits. All members have to do this volunteer work before the irrigation season starts. The representatives of the members of each natural village work together with the leaders of the subunit to organize the volunteer work, and each family has to maintain the ditches leading to their farmlands.

The survey showed that 29 of the 30 respondents took part in maintaining irrigation works. Most (59 per cent) think that it was the users' obligation to maintain irrigation works. Some (34 per cent) think that it was in the public interest for each member to do the volunteer work required to maintain irrigation systems that are vital for agricultural production and a good harvest.

The construction of new irrigation works was mostly contracted out by the WAMB and the Water Affairs Management Stations. The local farmers were not involved in the construction of large irrigation works, and the money required comes from the public funds of the government, design being undertaken by technicians of the WAMB. Farmers were excluded from decisions concerning the construction of new irrigation works, and were not consulted. There was no participation in the design of large irrigation infrastructure supported by the government.

Farmers' participation at the grassroots level was differentiated by activities and attitudes. On the one hand, claiming access to water resources and volunteering work to maintain ditches involved individual-level participation. On the other hand, farmers' attitudes towards their obligation including water fee payment works against participation. Also, not being involved in the design of irrigation works is a problem. Individual participation by farmers is not very well organized and there is a lot room for improvement of participation at the grassroots level. Volunteer participation of individual farmers is, in fact, the basis for organized social participation at higher levels.

According to the survey, 58 per cent of the respondents thought that the number of conflicts about irrigation decreased after the creation of the users' association, 28 per cent of the respondents said that the number of conflicts was the same as before and 14 per cent of the respondents said they did not know. Seventy-nine per cent of the respondents reported that they had not been involved in conflicts in the preceding year, and six respondents (21 per cent) said they had been involved in quarrels with neighbours. Most of the villagers thought that the conflicts were mainly quarrels between individual villagers that were unavoidable; the land in low-lying areas is easily flooded for physical reasons, and quarrels are considered to be part of the farmers' daily life and can be settled on their own.

The results of the survey indicated that the WUA has helped by offering timely irrigation services and reducing the number of conflicts about irrigation. As for the question concerning the benefits of WUA, multiple choices and open answers were accepted. Forty-eight per cent of the respondents chose timely irrigation service and 34 per cent said that the WUA helped both to reduce the conflicts and to offer timely irrigation service. In total, 82 per cent of the respondents agreed that the association offered timely irrigation service. Ten per cent of the respondents thought that there had been no change after the creation of the users association and eight per cent said that they didn't know.

To understand the level of satisfaction of the farmers towards the management of the association, the question asked was 'how do you evaluate the management of the association and why?' Most of the respondents (83 per cent) were satisfied because it delivered irrigation in a timely fashion. They considered the WUA to be well organized. The association has the mandate to manage the distribution of water and the maintenance of irrigation works and farmers found water use to be convenient after its creation.

The farmers' expectations with respect to the WUA reflected their issues with it. The open question 'what do you expect from being involved in irrigation management?' targeted this point. The answers revealed that the transparency of the water management fee was an important issue. The respondents expressed willingness to be involved in management of irrigation by contributing advice and new ideas, and they also expected to receive a timely irrigation service.

The current irrigation situation and the preferences and attitudes of the villagers of this subunit of the WUA reflect the role of the association and the impacts on farmers' participation at the grassroots level. Indeed, the establishment of the WUA and its enforcement measures has promoted some effectiveness in irrigation management. Farmers' participation, however, was still limited in term of their roles in decision-making and planning. Interactions among users and participation processes also remain very basic despite the WUA being in place. Conflicts were not automatically resolved with the WUA, although certain matters were viewed as having changed for the better.

The Case of a Village without a Water User Association

In this section, the case in Menghai county of Xishuangbanna prefecture, where there is no WUA, is compared with Banqiao. Xishuangbanna Dai Autonomous Prefecture is located in the southwest of Yunnan Province, near the borders with Laos and Burma. It has a subtropical climate and generally has ample rainfall. However, a drought occurred in Jinghong and Menghai from December 2004 to March 2005, when only 7.3–9.8mm of rain fell.

Like Luliang county, Menghai county in Xishuangbanna is one of the 12 large-scale irrigation areas of Yunnan Province since it is also the main rice production area of Yunnan. Water and irrigation are vital for paddy cultivation and the period of drought and water scarcity that lasted from December 2004 to March 2005 seriously affected farmers' livelihoods.

Changtianba Hani village is located near Menghai Town, Xishuangbanna Dai Autonomous Prefecture. It comprises 30 households and has a population of 124. The average annual gross household income was approximately US$1100 in 2005. The main livelihood resources are tea, rice, sugarcane and animal husbandry (cattle, pigs and chickens). The village has slightly less than 17ha of paddy fields that require irrigation.

The township WAMS is the key player in the management of irrigation at the local level. According to the regulations, reservoirs with a capacity between 100,000m^3 and 1,000,000m^3 are managed by water management stations in the township government. The WAMS report to the WAMB at the county level. When several villages in the same basin use the same reservoir, the WAMS is responsible for the management. The administrative village committee and natural village heads assist in managing irrigation and are responsible for seeing the villagers construct irrigation facilities, as well as for implementing irrigation and maintaining the facilities at the village level. The village head has to report back to the WAMS about management (Figure 8.4).

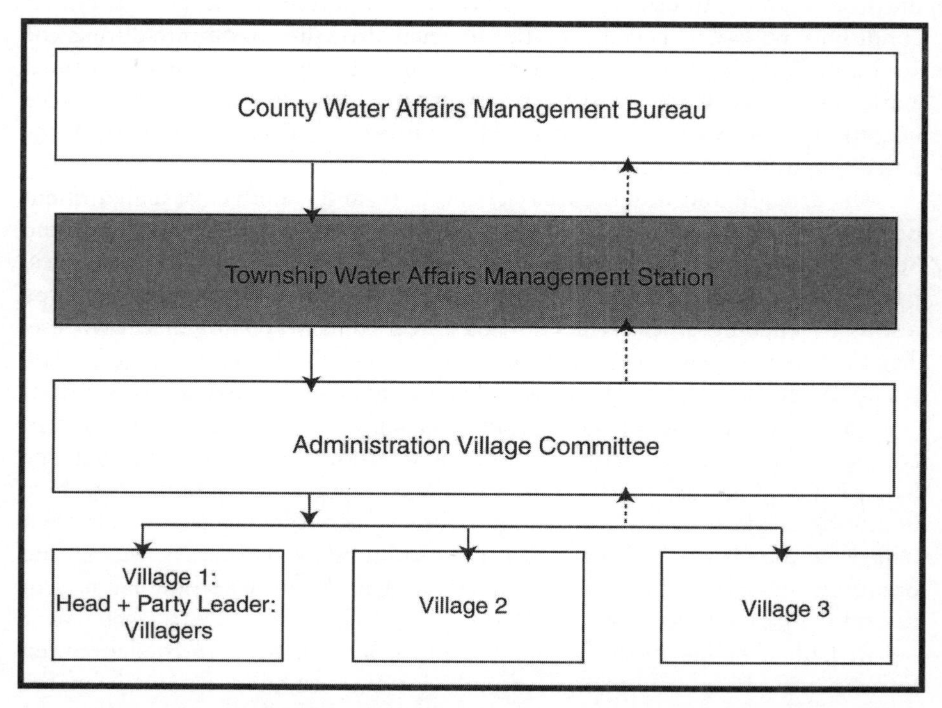

Figure 8.4 Administrative structure of irrigation management at the local level

During the period from 1999 to 2000, Changtianba Hani village constructed a small dam to ensure production in the paddy fields that belong to the village. Some of the funds came from a German project and some were contributed by the farm households. All the farm households in the village contributed a total of US$2855 through loans from the Township Credits Cooperatives. In addition, each farmer household contributed one-man-month of voluntary work to build the dam and the irrigation works. The WAMS affiliated with the township government was responsible for designing the reservoir. According to the village head, the technicians from the WAMS played a leading role in planning and designing the irrigation works and dam. All the construction materials were procured by the station without consulting the village head, and the villagers' participation was limited.

According to the regulations, small ponds and reservoirs with a capacity of less than 100,000m^3 should be managed by the villages themselves; however, the station leased the right to an individual businessman to breed fish and maintain the reservoir.

Since the sharing of rights and responsibilities between the WAMS and the village was not clearly defined, neither party took responsibility for management. The businessman with the leased right to use the water to raise fish also failed to maintain the structure. Thus, there was no one to take care of the dam and irrigation facilities. As a result, the gate was destroyed by thieves to catch the fish and water was discharged. The reservoir consequently did not contain sufficient water reserves for irrigation in the dry spring. As conditions worsened because of the unusual drought conditions during the spring of 2005, farmers were not able to irrigate their paddy fields on time. With no reserve water in their reservoir, they had to rely on rainfall for irrigation. Even with well constructed facilities, irrigation in the paddy fields failed due to bad management.

The drought and water scarcity resulting from the ineffective management of the dam led to conflicts between farmers who were competing for water. Some conflicts arose between farmers belonging to different villages, particularly between the Dai village and Hani village. Other farmers violated norms by opening other farmers' fields and stealing water for their own use. The farmers with water in their fields worried about losing their water, so they set up temporary huts in the fields to keep watch over the irrigation water.

The administrative village committee is supposed to arbitrate this kind of conflict between the villages, but it did not respond promptly. Although the irrigation conflicts are reported to the village administrative committee, there is always a delay in arbitrating the conflicts between farmers in different villages and there are no rules about how to punish violators. As the existing organization did not function effectively, individual farm households had to defend their own rights.

In China, rights over the small dams still have to be settled between farmers and the government. Currently farmers' rights are a focus of the government's agenda. In this context, Changtianba village leaders reported the ineffective management of the reservoir and the water scarcity for irrigation to the township government and the WAMS several times. At the end of 2005, the latter agreed to transfer the right to manage the reservoir to the village, and to acknowledge that the village had the sole right to manage it. With this new management right, the village leaders can now make decisions on how the facilities are operated and, in particular, they can decide to lease the right of water use and the leasing fee will become the common resource of the village.

Discussion

This section discusses the role of the government, the concept and the function of WUAs, social participation at the individual level and the interplay of stakeholders.

Since the state advocates social democracy and social participation, the latter has a good chance of being institutionalized in the Chinese irrigation sector. Chinese governments have shown their political desire to mobilize grassroots social participation. Social participation in irrigation was, in fact, one of the first legal actions undertaken by a Chinese government. The governments have committed themselves to public consultation in policy decisions that are related to the benefits of stakeholders or even in policies that affect the life of the people. However, the results of public consultation are affected by the lack of transparency in information, by the poor skills of participants in understanding the issues and making suggestions and finally by how the suggestions are integrated into policies.

Social participation in Chinese water governance is social engineering that actually has a cost, both financial and in terms of labour. Cost plays a fundamental role in the efficiency of social participation. There is the tradeoff between justice and the efficiency of social participation. While it has been efficient to use the administrative approach to mobilize and create WUAs, participation itself has been compromised because of an efficiency-centred approach. The establishment of WUAs has promoted the effectiveness of the irrigation management, but farmers' participation is still limited in terms of participation in decision-making and planning. Interaction among the users within the participation process has not been mobilized, although the WUAs are in place.

In the irrigation sector, representative participation was used to mobilize social participation. This kind of participation costs less than all-round participation. Accountable representatives are crucial for the meaningful participation of all the members. The election procedures for representatives and the implementation of the procedures can form the basis of the accountability of the representatives. Both of these procedures need to be improved, however. Information, education and awareness campaigns about the election for users association representatives should proceed beforehand, and every member should be mobilized to engage in the election.

On the other hand, the awareness of voters about their rights and obligations in electing representatives also affects the result of the elections. Due to the lack of commitment by voters and the non-implementation of election procedures, in many cases representatives are not elected but assigned.

Institutional reform in the irrigation sector has defined the mandates of the WAMB at different levels, and the nature of different water uses. The establishment of WUAs is designed to strengthen the management of irrigation. As a consequence of the reform, the task of maintaining irrigation works and collecting water fees has been transferred to the newly established users' associations. However, power and resources are still centralized in the upper level of the government. The accountability of the administration at different levels is not safeguarded since the allocation of public funds is usually not respected in terms of time and the amount of the funds.

Social participation in the irrigation sector is under the management of the government, as the WUA is an extension of the township government, and actually belongs to it. The management of irrigation is more effective than before; however, social participation is compromised due to the lack of accountability of representation and poor interaction processes among the members at the household level. The political will for social participation at the grassroots has opened the way for social participation in the irrigation sector. However, social participation at household level is often compromised by the lack of transparent information, the lack of participation processes, and finally the lack of representation and accountability of the government. In the case of the WUA in Luliang, farmers' participation was limited to volunteer labour to maintain irrigation ditches and paying user fees. Problems exist with farmers' lack of active participation, poor interaction processes and lack of transparency in the water fee management. Consequently, conflicts and resistance to paying water fees at the household level continue to be big issues.

The comparison of the two cases in Luliang and Xishuangbanna reveals the roles of WUAs. On the positive side, in the Luliang case study, the effectiveness of irrigation management was improved and users got a better irrigation service as well a reduction in the number of conflicts about irrigation between different villages. The subunit leaders of WUAs were more engaged in conflict solving and other irrigation management affairs thanks to the legal status of WUAs, and to the resources and decision-making mechanisms available, as well as to the increased interaction and communication between the different subunits of WUAs.

The power of the subunits of the users associations to negotiate with the water management stations about water access is strengthened compared with the power of the administrative village committee not involved in users' associations. When no WUA exists, irrigation management is still centred in the township WAMS.

Water distribution has become fairer since the WUAs laid down rules for the distribution of water based on the principle that downstream villages receive water distribution first. With this rule, downstream villages have access to water even during the dry season. Nevertheless, social participation in the WUA is limited, the process for establishing WUAs is top-down and bureaucratic, the election of the representatives of the association is fast and the participation of the farmers is not facilitated. The legal representative of the WUA is the local governor of the township government and the 17 administrative village committees are subunits of the WUA. The WUA structure overlaps with the township government and the administrative village committee. The farmers lack ownership of the WUA so this top-down approach becomes another form of extension from central to local government.

WUAs are local development associations, and, as Esman and Uphoff noted (1984), the local development association is a small-scale replica of the state, which is the formal source of authority at the village level. It is more responsive to state directives or higher authorities, and makes use of legal and administrative measures to manage natural resources in the communities.

Farmers' participation in irrigation management is primarily low level; they usually only participate in the maintenance of irrigation works as volunteer labourers and most do not have the opportunity to participate in planning and decision-making concerning irrigation management. Although they mostly respect the obligation to pay a water fee, they have no access to information about how the water fees are used. The majority pay for access to irrigation through canals, and some farmers have difficulty accessing irrigation due to problems with proximity to canals and ditches.

For individual farmers, the WUA represents an organization that manages irrigation. The creation of WUAs has promoted effective irrigation management as well as coordination between upstream and downstream users. However, farmers expect to participate in irrigation management by contributing advice and ideas and the representativeness of the WUAs needs to be broadened and become more inclusive.

Farmers' resistance is their reaction to not being involved in irrigation management. Although there are no obvious conflicts between users and managers, silent resistance occurs in daily life. Since farmers are not involved in decisions concerning the construction of irrigation works, the design and layout of canals, ditches and drainage does not fit their needs. They react to this problem by digging holes in the canal to access water. For some farmers, refusing to pay the water fee is a way to express resistance to the lack of transparency in the management of the fees and to the expropriation of their lands by the government in building reservoirs.

Conclusions

Social participation in the irrigation sector in Yunnan, China, is affected jointly by the roles of the state, the WUA and individual farmers. The state's role in promoting social participation involves regulation by the civil affair administration bureau concerning the conditions for a WUA to be approved as a legal body. On the other hand, the local government at township level is not willing to transfer power and resources to the users. Institutional reform, labelled as decentralization, has transferred responsibilities for the maintenance of irrigation works to the WUAs but without transferring the resources and power to fulfil these responsibilities. Consequently, the latter surrendered autonomy and ownership as well as the participation of farmers.

The WUA itself has promoted effectiveness of irrigation management but has not transferred the power structure. Recurrent social relationships among the participants, processes and patterns of relationships form the structure of an organization (Corwin, 1987). The structure of the water users association overlaps with the existing administrative village structure and the power of the local elite has been re-strengthened.

The effectiveness of irrigation management is not necessarily related to a system of participation by farmers that supports fairness and equity to participants. There is the tradeoff between irrigation management effectiveness and equity through participation. It is critical to identify a strategy to maximize the effectiveness of the management and equity through social participation.

Social participation promotes sustainability of irrigation management as well as equity. To promote meaningful social participation, comprehensive social engineering is needed:

1 The related legal framework needs to be improved including the regulation of registration procedures for associations.
2 The political will of the government at different levels is essential to realize political decentralization and to transfer power to grassroots organizations.
3 The internal structure and mechanism of the WUA should be built upon to improve members' interaction.

A diverse membership composition, differentiated social status among members, strong leadership and group solidarity are characteristics crucial to social participation (Esman and Uphoff, 1984). The composition of social status within an organization is related to power, attitudes towards public agencies and the individual member's ability to articulate his or her position. Therefore, external resources should be mobilized to facilitate process-orientated participation instead of quickly installing a structure. Education and awareness-raising about social participation should be conducted to promote the capacity of farmers as well as local government staff.

Notes

1 Effective irrigation area refers to the arable lands with basic irrigation works and water source.
2 Grade 5 is heavily polluted water; Grade 4 water is polluted; Grade 3 is lightly polluted; Grade 2 is clean; Grade 1 is very clean.
3 Township government is the lowest level of the Chinese government, above which is the county-level government, and then the prefecture-level and provincial governments.

4 An administrative village is based on several natural villages, while the natural village is the basic unit of the rural community as linked by clans.

References

Corwin, R. G. (1987) *The Organization–Society Nexus: A Critical Review of Models and Metaphors*, Contribution in Sociology, Greenwood Press, Santa Barbara, CA, p67

Esman, M. J. and Uphoff, N. T. (1984) *Local Organizations: Intermediaries in Rural Development*, Cornell University Press, Cornell, NY

Yiqing, T. (1999) *Yunnan Flood and Drought Disasters*, Yunnan Hydrological Department, Kunming, China

9

Participation in Water Resource and Services Governance in South Africa: Caught in the Acts

Zoë Wilson and Sylvain R. Perret

Introduction: New Legislation and Governing Principles

South Africa is a water scarce country, with less than 500mm of annual precipitation. The country also has a large rural population (40 per cent) that is logistically difficult to link to water and sanitation grids. Globally, the vast majority of people without access to clean water and improved sanitation are the rural poor. South Africa's service backlog is also disproportionately borne by the African population, both rural and urban. In 1994, approximately 45 per cent of the urban black population had access to piped water, while coverage for other urban groups was close to 100 per cent. The South African Cities Network explained it thus:

> It was the systemic function of the apartheid cities to ensure that white residents had all the social benefits of living in the city, and to deny black residents equal access to urban social goods and opportunities. The result is cities where very large proportions of the population are not included – materially or psychologically – in urban life. (South African Cities Network, 2004, pp77–78)

Since 1994, the South African government has undertaken massive reforms aimed at addressing poverty and inequalities inherited from the past regime. Inter alia, it has adopted ambitious new water legislation – the (National Water Act (NWA) of 1998 and the Water Services Act (WSA) of 1997 – that promotes equity, sustainability, representativity and efficiency through water management decentralization, new local and catchment-level organizations and institutions and improved access to services. Participation has been a guiding principle of reforms, including:

1 civil society consultation and feedback into the implementation of legislation and constitutional guarantees;
2 user involvement in the development and management of decentralized water resource entities, such as catchment management agencies (CMAs) and water user associations.

Integral to the concept of participation is the redress of imbalances of power in society. But redress is not the only official motivating factor behind participation. There is a growing consensus that engaging civil society in all aspects of resource management is likely to generate a healthy and empowered sense of collective ownership, which in turn contributes to the sustainability of the resource (Burt et al, 2006). The application of the principles underlying the new water legislation makes South Africa one of the few countries in which water is seen as an essential tool for achieving social justice and pro-poor economic growth (Van Koppen et al, 2003), and where decentralized management is seen as essential to integrated resource management more generally.

South Africa's commitment to reallocation through decentralization derives from both the principle of subsidiarity in new water management thinking as well as optimism about grassroots political processes. Significant progress has been achieved, especially in relatively well-capacitated municipalities. Yet, slow progress and contested gains in other areas, especially rural, have led some observers to conclude that decentralization in the water sector has not been the most effective means to redress past inequalities (Centre for Civil Society, 2003) or to manage the technological complexity associated with major country wide systemic upgrades and reform (Muller, 2007).

Thus, while South Africa has made considerable progress in terms of water governance reform in a relatively short time, and under huge inherited constraints, processes have been fraught with challenges, and in many instances yielded only partially satisfactory outcomes. Practical diagnoses, however, remain in their early stages. As the conclusion to a comprehensive assessment of participation in the establishment of CMAs in South Africa stressed recently, there is a need for 'deeper insights into the "hidden" dimensions of participation' (Burt et al, 2006, p6).

This chapter discusses the modes of participation that have emerged since the inception of the National Water Act of 1998 and National Water Resource Strategy of 2004, both of which are guided by the principles of participation, reallocation and decentralization. Using case studies, we identify four types of common mistakes leading to inadequate participatory initiatives:

- political challenges: conceptual traps and tacit beliefs that deny the politics of participation;
- methodological challenges: how the selection of participants and choice of forum influences the quality of participation;

- skill- and capacity-related challenges: how stakeholders' limited knowledge, access to information, or capacity disrupts the flow of meaningful participation;
- socio-historical challenges: how socio-cultural elements, inherited marginalization, and power asymmetries influence participation pathways and outcomes.

Below we draw upon a number of recent research initiatives and case studies involving literature review and project interviews, surveys and focus groups to identify and classify practical challenges encountered. The chapter draws on both water resource and water services case studies. In terms of water resource management, findings draw upon material from the Kat River valley (Fish-Tsitsikamma CMA), the Sand River catchment (Inkomati CMA), and the Olifants River CMA. Micro-level studies, especially from rural areas, detailing the functioning of water user associations and water committees are also considered. In terms of water services, a number of recent initiatives in a range of municipal settings are used, from the relatively well capacitated eThekwini Municipality (formerly City of Durban, Mvoti-Umzimkulu CMA), the missionary hospital community of Msleni in Northern KwaZulu-Natal (Thukela CMA), to high in-migration areas in the Western Cape (Berg CMA) (see Figure 9.1).

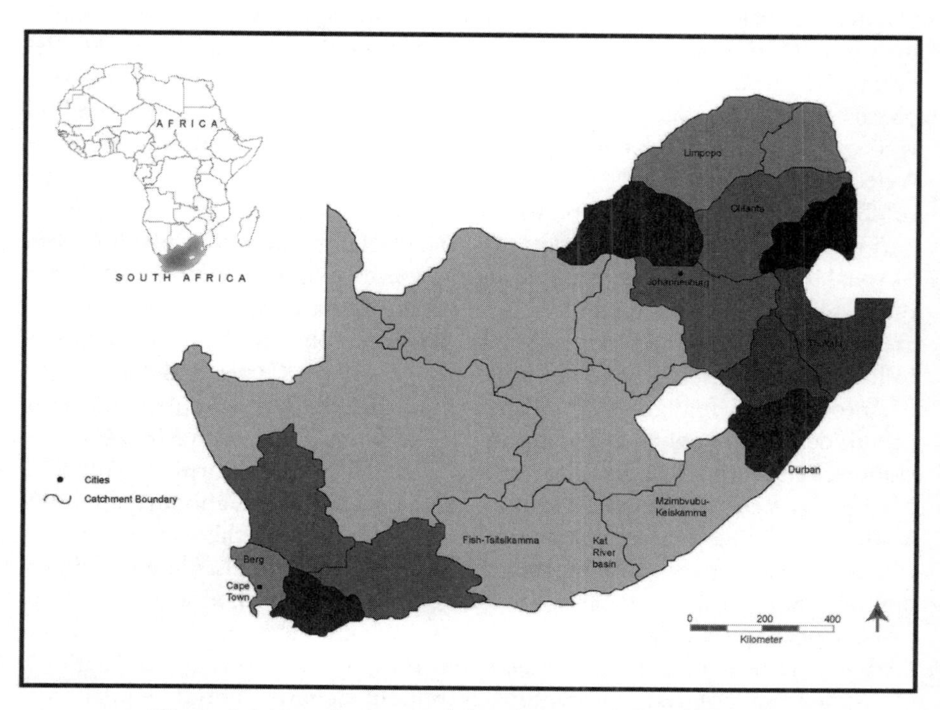

Figure 9.1 Map of select catchments/watersheds in South Africa

Analytical Typology of Challenges to Participation

Theoretical and political challenges: who speaks?

South Africa's democratic transition is the offspring of competing political traditions and underlying assumptions about the nature and role of the public. On the one hand, transition was heavily influenced by widespread commitment to socialist and communist ideals, as evidenced by the close – if often fraught – relationship between the African National Congress (ANC), the Congress of South Africa Trade Unions (COSATU) and the South African Communist Party (SACP), officially known as the Tripartite Alliance. From this vantage, glaring social inequalities are cast in race/class terms under the assumption that the large number of poor black historically disadvantaged individuals share common experiences and objectives, and that little conflict exists within the disadvantaged class. This approach competes with the classical liberal tradition, which analysts characterize as epitomized by the policy shift from the Reconstruction and Development Programme (RDP) to the Growth Employment And Redistribution strategy (GEAR) (Bond, 2006; Wenzel, 2007). From the liberal tradition is borrowed a notion of the ideal public sphere where basic institutions provide a relatively neutral playing field for free and equal individuals to cooperate over the distribution of resources (Wilson, 2006). The two approaches share a common pitfall, however. Both deny how contested power is at the local level, within the family, among neighbours, between in- and out-groups, etc. and thereby render the concept of representation relatively straightforwardly. In the classical liberal model, organized groups represent the outcomes of free and fair deliberative processes, and thus, like models from the left, which assume common interests/ voice, tend to render local and grassroots leadership as a simple emanation of collective will.

Both approaches, then, tend to oversimplify landscapes of political participation. In particular, it is common to overlook the political effects and payoffs of speaking for others – appropriating the voice of others (Alcoff, 1991). Yet, the historically disadvantaged do not necessarily share internally undifferentiated strategies and interests, as shown by numerous rural poverty analyses following Sen's seminal work in 1981 (Barber, 1996; Carter and May, 1999; McIntosh and Vaughan, 1996; Perret et al, 2005). For example, Tlou et al (2006) described how an initial group of 39 farmers in the 20ha Mngazi irrigation scheme (near Port St Johns, Eastern Cape, former Transkei, Mzimvubu-Keiskamma CMA) split into two separate yet neighbouring groups, while still sharing the same water resource and pump, on the basis of oppositional strategies over cropping system management but also latent political conflicts. South Africa's local politics are often fraught with violent and exclusionary practices cleaving on gender, class, racial, ethic and political grounds. Claims to representiveness are often made but less often stand up to scrutiny.

Similarly, popular or highly amplified debates about equity and participation sometimes speak to the most pressing concerns of only a small number of relatively well-networked South African groups. For example, as noted by Mike Muller, former Director-General of the Department of Water Affairs, internationally, water issues in South Africa have become synonymous with anti-privatization struggles. Yet 'privatization was not really an issue since only about 5 out of 284 municipalities had privatized their water services' (Muller, 2007, p43). Muller further argued that anti-privatization sentiment is not strong among local audiences, despite its widespread international appeal. This interpretation is supported by the results of the 2006 local government elections in which Operation Khayisa, the group campaigning against privatization and prepaid meters via the slogan *break the meter, enjoy the water*, gained just 0.31 per cent of the votes in Johannesburg, where it was most active. Similarly, recent research across various different kinds of communities in Durban revealed tensions between neighbours over illegal connections (Wilson et al, 2008), rather than popular support for the practice, as activist groups claimed. A significant proportion of respondents expressed concerns about the impact of illegal connections on the price of water, as well as the inconveniences of leaks and muddy areas.

Yet, participatory processes tend to reflect the assumption that consulting with vocal individuals, organized groups and/or traditional authorities is the appropriate means to unlock the creative, legitimizing and democratizing potential of participation. This assumes that local non-governmental organizations (NGOs), councillors and/or traditional leaders straightforwardly know and/or represent the interests of the community. Yet, for example, in a recent event at the University of KwaZulu Natal (UKZN), Dudu Kumelo described how her community was still in the process of trying to recover its fair share of compensation paid to the chief for community relocation related to the building of the Inanda Dam in 1988 (Kumelo, 2007). In stories told to her while trying to piece together how the land was expropriated, Kumelo recounted:[1]

> When these white people came to build the Inanda Dam, usually it was during the week only when mothers were at home. I remember my mother used to speak a little bit of English and when they asked for permission to sit under the tree that was in the family yard, she asked them what were they there for, and the answer was: 'Do not worry mama, the chief knows we are here.'

Similarly, around the country, water service authorities have expressed frustration at politically motivated vandalism where new water and sanitation infrastructure is laid without perceived due involvement and credit to local (formal or informal) authorities (MacLeod, pers. comm., 2005; Goldberg, pers. comm., 2006), even where this leaves the community as whole with little to no access to improved water or sanitation. Recently, for example, South Africa's national newspaper reported: 'Distribution of water, reliability of water supply, water storage **and vandalism** were problems still facing communities hit by a cholera outbreak at the end of 2000' [emphasis added] (Sapa, 2006).

Thus, local and poorly understood politics have been identified as a key barrier to more effective representation (McLennan, 2007; Muller, 2007). McLennan noted (2007, p7) for example:

> A different understanding of the process of delivery would require knowledge of the way in which local (social, economic or cultural) institutions, power dynamics and political processes work to form – and dictate the norms for – social relations and agency.

Our review of recent case study research supports this finding. To the extent, then, that parties interested in fostering better participation in service delivery adopt underlying theories of representation that render local and intra-group political dynamics invisible and deny the political pay offs of speaking for others, they will 'replicate or reinforce particular (unequal) distributional patterns despite attempts to shift these' (Cleaver, 1999; Fierlbeck, 1995; McLennan, 2007, p6; Post Uiterweer et al, 2006).

Methodological challenges

In the second, and related, category, we identify participatory methodologies that truncate or distort participatory or consultative processes by working with and through conventional channels, many of which, as noted in the previous section, are in fact highly contested, politicized and woven with inequities, such as gender and class. Here, we find participatory programmes designed to fulfil policy requirements, to legitimate established policy frameworks and/or bring key vocal players on board. Less emphasis is placed on scanning new landscapes of information or obtaining a diversity of ideas from the previously marginalized and disadvantaged. For example, Burt et al (2006, p14) report that the Mvoti-Umzimkulu CMA proposal (KwaZulu-Natal) was gazetted and then merely sent to libraries, district municipalities and traditional authorities. The response was poor with less than ten comments received. The proposal was then left at the offices of the Umgungundlovu Municipality, gathering three more comments there.

In a recent pilot study in Durban across highly different areas, surveys were conducted with non-politically active individuals who live and work in various areas and who have a professional responsibility for the health and well-being of community members, such as school teachers, health care workers, day care workers and pastors (Wilson et al, 2008). Results in some areas indicated that regular people were not fully satisfied with existing modes of consultation, and would be interested in more direct, unmediated, participation in decision-making. Results in Table 9.1 are segmented by area to provide representability to spatially differentiated groups.

Table 9.1 Perceptions of customer voice in Durban areas with conventional water services

People have a voice in water and sanitation services	**Agree**	**Disagree**	**Don't know**
uMlazi	74%	18%	8%
KwaMashu	42%	52%	7%
Newlands East	45%	42%	13%
Clermont	48%	31%	20%
Berea West/Westville	33%	38%	28%
Chesterville	33%	50%	17%
Hambanathi	48%	41%	9%

What is perhaps most striking is the general pattern across the municipality, with the exception of uMlazi, where satisfaction levels were notably higher. While we might have expected a great deal of variation between historically, racially and economically differentiated areas, this is not always the case. Rather, dissatisfaction with existing political processes held across demographics. For example, Berea West/Westville is a historically affluent area, while Chesterville is an adjacent new suburb comprising low-income Reconstruction and Development Programme (RDP) housing. Yet, we find similar levels of satisfaction with customer voice expressed. Note that we see the same congruence between responses in these two areas. More in-depth analysis of results is under way.

In a more conventional public outreach campaign conducted recently in Cape Town, however, only existing political channels were targeted. 'The project will work with existing community structures in each of the selected sites with the aim of integrating a consumer oversight role' (WRC, 2007, p45). Further, the project adopted a 'usual suspects' approach to building multi-stakeholder involvement:

> Establish a coalition of NGOs and CBOs [community-based organization] that are interested in service delivery oversight to play a supportive role for this pilot. Suggested bodies include Environmental Monitoring Group, Anti-Eviction Campaign, Khayelitsha Development Forum, Tsoha, SAMWU [South African Municipal Workers Union], Rates Payer's Associations, Contemporary Research Foundation, IDASA [Institute for Democracy in South Africa] and so forth. (p46)

The proposed configuration of organizations was meant to deliver participation, yet, as McLennan argues, whether such a selection of social institutions is:

> capable of ensuring real engagement is contested ... [with current approaches resulting] ... in a shift away from popular engagement to

> interest-group representation. **More significantly, only policy-literate, well-organized groups are able to access the formal pathways of consultation and negotiation over delivery.** [emphasis added] (McLennan, 2007, p16)

Similarly, a recent study of Grabouw in the fruit growing region of the Western Cape province (Berg CMA), involved over 60 in-depth interviews and focus groups and found:

> The municipality, the Groenland Irrigation Board, Department of Water Affairs and Forestry and farmers are capacitated in varying degrees as key decision makers on water use within Grabouw. However, the communities that are served by them are water users that are not (do not feel) part of these decision making structures. Furthermore, communities within Grabouw are neither homogenous nor organized (racial, rural–urban divides). (Peters and Wilson, 2006, p6)

Similar methodological challenges, unsurprisingly, are also found in rural areas. Recent history of the Thabina irrigation scheme (Olifants CMA), for example, illustrates the difficulties in navigating local politics to ensure that all stakeholders have adequate and effective representation. The scheme was earmarked for rehabilitation and transfer to users by the Limpopo provincial authorities, like many other smallholder irrigation schemes at the end of the 1990s (Perret, 2002). Part of the rehabilitation/transfer plan was the establishment of a water user association (WUA). As a pilot scheme for rehab, and as the first WUA officially established in South Africa, Thabina drew considerable support from provincial authorities in the form of farmers' technical and managerial training and infrastructure rehabilitation. However, the farmers' group of Thabina is characterized by internal conflicts and inequities, as well as a wide range of farming styles and strategies (Perret et al, 2003). Further complicating its effectiveness, the WUA had no jurisdiction over conflicts with illegal resource users located on the periphery of the scheme (Chancellor, 2006). Additionally, Veldwisch (2006) reported that only a small number of male commercial farmers had a real voice within the WUA management committee. Most of them are located in head-end positions along the main canal; some have private pumps and draw directly from the river, while tail-end farmers are typically deprived of water during dry seasons. Most of those deprived are female subsistence farmers. During the 2007 dry season, only a few powerful head-end farmers – members of the WUA committee – could irrigate, and issues around illegal water uses, lack of water for tail-enders and broken pumps remained pending. Denison and Manona (2007) noted that all other pilot schemes in Limpopo and Olifants CMAs (eg Boshkloof, Morgan, Dingleydale), which the authorities showcased in a will to quickly implement irrigation transfer and the establishment of WUAs, have faced similar and more severe failures and drawbacks.

The type of forum and the type of participant expected to make up the membership of that forum will exert a structurally determining force on the quality of participation. Rethinking participation with an eye to local politics may also, then, require redesigning processes of participant selection and self-selection, as well as the institutions of governance.

Skills or capacity-related challenges

In the third category are issues related to the skills and capacities of both civil society and implementers. This is expressed in information asymmetries, poor understanding of the policy environment and lack of management and communication skills. That is, participatory deficits are often the result of general knowledge gaps and capacity deficits rather than bad policy, per se. For example, results from a recent pilot study in Durban revealed that general knowledge about widely publicized policies remained low, making it difficult for people to comment effectively, even where wide stakeholder engagement is sought (Table 9.2).

Table 9.2 Understanding of free basic water policy in Durban areas with conventional water services

People are generally aware of and understand free basic water	**Agree**	**Disagree**	**Don't know**	**No answer**
uMlazi	41%	41%	19%	nil
KwaMashu	42%	48%	10%	nil
Newlands East	27%	56%	11%	5%
Clermont	40%	46%	6%	9%
Berea West/Westville	43%	43%	6%	nil
Chesterville	42%	33%	26%	nil
Hambanathi	45%	41%	7%	7%

As with Table 9.1, the general pattern of results holds across all areas, whether historically disadvantaged, relatively affluent or emerging middle income. This may indicate that the capacity – or even interest – of the general public in water management issues is low. In this case, building participatory governance structures in the water sector may face yet unacknowledged hurdles, possibly rooted in the history of water services as the exclusive purview of engineers and specialists.

Similarly, with respect to workshops in the Sand River catchment (Inkomati CMA), Burt et al (2006) report that the understanding of a rights framework for water allocation by all the people in the catchment was very poor:

> Everybody involved needed a clear understanding of what a rights framework meant – a commitment to the principles of a rights focus (non-discrimination, participation and universality), the defining of obligations, how to find out when and where violations of obligations were happening and how to remedy these situations.

What emerged was that the various stakeholder groups did not have adequate access to information, so they were not able to carry ideas forward into action with confidence.

Capacity levels in all areas of management appear to be a significant problem across many of South Africa's highly differentiated communities. Research conducted in Northern KwaZulu Natal (Thukela CMA) and the fruit growing region of the Western Cape (Berg CMA) reinforced the notion that general knowledge and capacity was a problem. In a recent study in Mseleni, a deep rural community in Northern KwaZulu Natal province, involving over 100 qualitative interviews and focus groups, researchers found:

> The cause of difficulties accessing water was laid upon: electricity cut-offs, poor engineering, poor social facilitation, lack of transport, lack of operations and maintenance, inappropriate technology, lack of finance, lack of capacity, lack of service delivery, an ineffective water committee and party politics (ANC versus IFP [Inkatha Freedom Party]). (Hazell and Wilson, 2006, p5)

Scholars such as Muller (2007, p39) have, as a consequence, questioned the logic of decentralization on the grounds that capacity deficits leading to local implementation barriers remain intractable:

> It is clear that decentralization, while supported from many perspectives, has not always achieved its intended objectives. Indeed, one author concludes that 'decentralization and participation can reinforce historical distributions of privilege'. (Burger, 2005, p483)

Similarly, the International Water and Sanitation Centre (IRC) reported recently:

> While short-term funding through the MIG (Municipal Infrastructure Grant) could be increased to expand service delivery in shorter periods of time, current allocations are going unspent, and the consensus from many organizations working in South Africa is that the municipalities often lack the capacity to implement programmes, and to monitor and evaluate performance. This lack of absorptive capacity ranges from an inability to implement programmes at a local level to institutional blockages and political interference in the budget appropriation and allocation process. There are multiple reports of funding sitting in interest-bearing municipal accounts rather than being spent on schedule. (IRC, 2006, p2)

Recently the South African Institute for Civil Engineers also noted that capacity deficits were severely limiting progress in water delivery: 'Eighty-three of the 284 municipalities in the country have no civil engineers, technologists and technicians' (Savides, 2007).

Thus both marginalized groups and implementers are not immune from lack of capacity and skills. Local and district municipalities are overloaded, understaffed and under-skilled (Post Uiterweer et al, 2006, p60). Roux et al (2006) further note that significant challenges remain in bridging longstanding disciplinary divides between researchers, policy makers and managers, as well as between the world views of experts and users. These divisions also disrupt the flow of information and slow progress. Thus, challenges associated with participation float in a larger context where cultural and cosmological divides and capacity deficits, especially in rural areas, profoundly distort and disrupt the institutional vision for the water sector, including its democratic aspirations.

Socio-historical challenges

In the final category, we find the legacy of apartheid militating against effective participation as poor basic education, a history of dependency and a general lack of confidence in politics and policies strongly limit the involvement of grassroot rural users and citizens. There are exceptions but in practice local communities, especially rural and peri-urban poor, still struggle against a range of inherited disadvantages that form a web of limitations to meaningful and constructive participation (Table 9.3).

Table 9.3 Perceptions of discrimination in Durban areas with conventional water services

Different groups of people get different levels of service based on discrimination	**Agree**	**Disagree**	**Don't know**	**No answer**
uMlazi	62%	31%	5%	3%
KwaMashu	41%	38%	18%	4%
New Lands East	42%	45%	11%	2%
Clermont	46%	37%	14%	
Berea West/Westville	52%	29%	19%	
Chesterville	58%	17%	25%	
Hambanathi	36%	48%	17%	

More equitable participation in the sector is particularly difficult to achieve where historical users remain relatively highly capacitated in the face of more equitable but fledgling water user associations and catchment management agencies. Lévite et al (2003) investigated the quality of participation in the development of the Olifants CMA (as the first CMA being officially established in 2006). Here, the ideal of participation has been difficult to achieve in practice. The authors highlighted local tensions and conflicts between a number of different and competing users, including mines, large-scale commercial farmers, municipalities and rural communities. Further, they described situations whereby existing discussion fora, involving only big users such as mines and municipalities, tend to steal the show and discuss water matters among themselves. Dialogue between sectors is very limited and only a minority of stakeholders are involved. In the Sekhukhune District Municipality (Olifants CMA, Limpopo Province), Post Uiterweer et al (2006) also reported that so-called stakeholder participation meetings were held in 2004, mostly on resource development, and instigated and lead by mines and large commercial farmers, while communities were not included. A similar case has been reported in the Steelpoort area (Olifants CMA) where Farolfi and Perret (2002) highlighted huge economic discrepancies between mining and smallholder agriculture in terms of water productivity in a context of acute scarcity, resulting in exclusive negotiations and arrangements between the National Department of Water Affairs and Forestry, local government and the mining sector. Yet, as Tables 9.1 and 9.2 also indicate, it is important to note that challenges may be less related to general demographic trends in the distribution of wealth and capacity, than to the sector more specifically. That is, even relatively well capacitated actors will struggle to participate in decentralized governance institutions if their knowledge of the sector is weak – which, among the general population, appears to be the case.

Under extant conditions, then, local people have limited access to information related to water management and water reform. In the Olifants CMA, Stimie et al (2001) and Perret (2002) found that rural communities are often not even aware of the reforms and the challenges at stake. Negotiations first take place between mines and central and provincial departments, which may only unevenly represent communities. Information is not only asymmetric, it is often nonexistent at the rural community level. Lévite et al (2003) further noted that meaningful participation of users in the management of water in the basin is constrained by lack of information on the state of the rivers and water use, as well as the historical absence of information and dialogue relating to the management of natural resources, which makes new information difficult to contextualize and link to embedded institutions and meaning systems. In many areas, a significant percentage of the population over the age of 20 had no formal education, and exist in an information vacuum characterized by uncertainty, where rumour abounds. Recently the national *Mail and Guardian* newspaper reported:

> During the field work carried out in 2005, researchers also discovered that some residents in Nqutshini, through which the Mhlatuze River flows [Thukela CMA], believed that the cholera outbreak was caused by

> whites fearful of black numerical superiority. Others believed that witchcraft had been responsible'. (Sapa, 2006)

Similarly, a recent study of perceptions of potable reuse also surfaced concerns that policy illiterate communities were easily manipulated by aspirant politicians making electoral promises of unlimited free water (Wilson and Pfaff, 2008). In this context, Hazell and Wilson (2006, p15), in their study in rural KwaZulu Natal (Mvoti-Umzimkulu and Thukela CMAs) found that:

> discourses which cast people as powerful agents in relation to water, contrast with the majority views (of participants), of the municipality and/or government as service provider, and people as passive recipients ... Some people felt powerless in relation to accessing piped water 'only white people know how to join pipes to their homes' ... (interview 12/4/06) ... 'there is nothing I can do, you can't change anything'.

Article 21 of South Africa's Constitution states that 'everyone has the right to take part in the government of his/her country directly or through freely chosen representatives'. Yet, the historical layers of institutions and challenges particular to the sector, many outlined above, remain important determinants of outcomes. In practice, many communities, and not only the disadvantaged, lack the capacity to engage with and construct forums for meaningful political action. Rural areas and thus the water resource management associations are most affected, especially socially marginalized, less connected rural households, for example, single women headed households (Merle et al, 2000; Perret et al, 2005). Perret (2004) highlighted that urban areas absorb the layers of best educated people in rural society, and male labour; women (often denied authority by traditional structures), children and the elderly are de facto key role players in rural areas. For such groups who already struggle to make a living and to have a voice within their very own community, participating in catchment management – often on a volunteer basis – would mean confronting innumerable hurdles.

Conclusions

South Africa's learning curve has often come at the expense of participatory processes that have played into or exacerbated partisan political conflicts, defined communities in ways that re-entrench local political asymmetries, bolstered an ethic of complaint that privileged political and politicized groups over regular citizens, eroded overall trust in government and failed to establish sustainable parameters for on going dialogue. The root causes of these effects, we find, fall into four broad and inter related categories: theoretical, methodological, skills or capacity-related and socio-historical.

Ultimately, participatory processes are often designed using an incomplete picture of related challenges. Case studies such as those presented here refute the operative theory of participation in the South African water sector. Local stakeholders do not necessarily share similar interests nor cooperate readily. They are caught in local politics, power games and potential or open conflicts. Often, existing forms of decentralization and participation play a role in causing power to adhere to the most vocal, historically advantaged or traditionally authoritative, thereby in actuality preventing the flow of contextualized knowledge, stalling implementation and diminishing the legitimacy and the effectiveness of initiatives.

Lessons drawn from our case studies concur in part or in whole with similar studies (Burt et al, 2006; McLennan, 2007; Muller, 2007; Roux et al, 2006). From the evidence presented in this chapter we distill four main observations:

1 Both participation and representivity are scarce resources over which actors compete and come into conflict. Power asymmetries and competing interests are a fundamental part of fair and inclusive participatory processes. A confluence of interests among diverse stakeholders – even those who appear at first glance to share the same objectives – should never be presumed to exist at the local level.
2 Any form of participation is not sufficient automatically to contribute to equitable and sustainable outcomes and, conversely, when ill-conceived may simply reproduce existing power asymmetries, or create new ones. Methods used for fostering participation often contain the seeds of their own failure. Over-reliance on traditional structures, vocal players, convenient participants, volunteerism, etc., can ultimately be determinants of how water management processes will (dis)function.
3 Newly established legal frameworks and structures, such as catchment management agencies and water user associations, do not in themselves ensure authentic participation, nor that legislative, ethical and democratic imperatives are fulfilled. Facilitation, capacity building and resources are needed. The sector is highly complex and not well understood among the general public, and this has significant implications for decentralization.
4 The history of apartheid brings its own unique challenges. South Africa has made a valiant attempt to rewire its institutions – but the informal structures and institutions of apartheid, as well as more general historical particularities of the water sector – continue to shape reform outcomes.

Finally, South Africa's history of injustice brings with it well earned scepticism. People are not necessarily willing participants, especially when consensus-driven processes do not even the local playing field, accommodate different perspectives, skills and knowledge, nor accord authentic influence over ultimate decisions. Further, water use is not necessarily the most important determinant of stakeholders' identity. People's motives for participation go beyond concerns about the resource itself, interwoven with the past and with the variegated moral geographies that inform broader hopes for the future.

Note

1 Kumelo also notes that women are traditionally excluded from decision-making at the chieftainship level (otherwise known as the Inkosi-level).

Acknowledgements

The authors would like to thank the Department for International Development and Newcastle University in the United Kingdom, the South African Department of Water Affairs and Forestry, eThekwini Water and Sanitation, University of KwaZulu-Natal Pollution Research Group, the Water Research Commission, the National Science Foundation and the Embassy of France in South Africa for generous support. We also thank the editors for guidance and invaluable efforts.

References

Alcoff, L. M. (1991) 'The problem of speaking for others', *Cultural Critique*, Winter, pp5–32

Barber, R. (1996) 'Current livelihoods in semi-arid areas of South Africa', in M. Lipton, F. Ellis and M. Lipton (eds) *Land, Labour and Livelihoods in Rural South Africa, Volume 2: Kwazulu Natal and Northern Province*, Indicator Press, Durban, South Africa, pp269–302

Bond P. (2006) 'Reconciliation and economic reaction: flaws in South Africa's elite transition', *Journal of International Affairs*, vol 60, no 1, pp141–156

Burger R. (2005) 'What we have learnt from post-1994 innovations in pro-poor service delivery in South Africa: A case study-based analysis', *Development Southern Africa*, vol 22, no 4, pp483–500

Burt, J., Du Toit, D. and Neves, D. (2006) *Participation in Water Resource Management: Book 1. Learning about Participation in IWRM: a South African Review*, WRC project report #TT 293/06, Water Research Commission, Pretoria, South Africa

Carter, M. R. and May, J. (1999) 'Poverty, livelihood and class in rural South Africa', *World Development*, vol 27, no 1, pp1–20

Centre for Civil Society (2003) 'The deepening divide: civil society and development in South Africa in 2001/2002, preface', *Development Update*, vol 4, no 4, pp3–10

Chancellor, F. (2006) 'Crafting water institutions for people and their businesses: exploring the possibilities in Limpopo', in S. Perret, S. Farolfi and R. Hassan (eds) *Water Governance for Sustainable Development. Approaches and Lessons from Developing and Transitional Countries*, CIRAD and Earthscan, London, pp127–146

Cleaver, F. (1999) 'Paradoxes of participation: questioning participatory approaches to development', *Journal of International Development*, vol 11, pp597–612

Denison, J. and Manona, S. (2007) *Principles, Approaches and Guidelines for the Participatory Revitalization of Smallholder Irrigation Schemes, Volumes 1 & 2*, WRC report # TT 308/07, WRC Publisher, Pretoria, South Africa

Farolfi, S. and Perret, S. (2002) 'Inter-sectoral competition for water allocation in rural South Africa: analysing a case study through a standard environmental economics approach', *XXXIX Convegno SIDEA*, Proceedings of the Annual Conference of the Italian Association of Agricultural Economics, Florence, Italy, 12–14 September 2002

Fierlbeck, K. (1995) 'Getting representation right for women in development: accountability, consent and the articulation of women's interests', *IDS Bulletin*, vol 26, pp23–30

Goldberg, D. (2006) Interview with special advisor to the Minister of Water Affairs and Forestry, 10 June

Hazell, E. and Wilson, Z. (2006) 'Mseleni case study', public access report prepared for Newcastle University and the Department for International Development, UK in reference to The DFID Engineering KaR Programme: Proposal W1-17, 2002–2007, Second Order Water Scarcity in Southern Africa

IRC (2006) 'Financing facilities for the water sector', Thematic Overview Paper #13 by Rachel Cardone and Catarina Fonseca, reviewed by James Winpenny, IRC – International Water and Sanitation Centre

Kumelo, D. (2007) 'The social and cultural benefits of Umgeni River that have been lost to the Inanda Dam', Presentation at Topic on 11 May, CCS/SDS Boardroom, room F208, Howard College Campus

Lévite, H., Faysse, N. and Ardorino, F. (2003) 'Issues pertaining to the implementation of user participation to solve water use conflicts, the case of the Steelpoort basin in South Africa', *20th ICID European Regional Conference: Consensus to Resolve Irrigation and Water Use Conflicts*, ICID & AFEID, 17–19 September, Montpellier, France

McIntosh, A. and Vaughan, A. (1996) 'Enhancing rural livelihoods in South Africa: myths and realities', in M. Lipton, F. Ellis and M. Lipton (eds) *Land, Labour and Livelihoods in Rural South Africa, Volume 2: Kwazulu Natal and Northern Province*, Indicator Press, Durban, South Africa, pp91–119

McLennan A. (2007) 'Unmasking delivery: revealing the politics', *Progress in Development Studies*, vol 7, no 1, pp5–20

MacLeod, N. (2005) Interview with Head, eThekwini Water and Sanitation, 14 June

Merle, S., Oudot, S. and Perret, S. (2000) *Technical and Socio-Economic Circumstances of Family Farming Systems in Small-Scale Irrigation Schemes of South Africa (Northern Province)*, PCSI report, CIRAD-Tera, #79/00

Muller, M. (2007) 'Parish pump politics: the politics of water supply in South Africa', *Progress in Development Studies*, vol 7, no 1, pp33–45

Perret, S. (2004) 'Matching policies on rural development and local governance in South Africa: recent history, principles and current challenges', in *Proceedings from the Workshop on Local Governance and Rural Development*, GTZ, University of Pretoria, 2 June 2004, Pretoria, South Africa

Perret, S., Anseeuw, W. and Mathebula, F. (2005) *Investigating Diversity and Dynamics of Livelihoods in Rural South Africa; Case Studies in Limpopo*, WKKF project on poverty and livelihoods in South Africa, final report, UP-Kellogg Foundation, Pretoria, South Africa

Perret, S., Lavigne, M., Stirer, N., Yokwe, S. and Dikgale, K. S. (2003) *The Thabina Irrigation Scheme in a Context of Rehabilitation and Management Transfer: Prospective Analysis and Local Empowerment*, Dwaf Project 2003-068, final report, Cirad-UP-IWMI, Pretoria, South Africa

Perret, S. R. (2002) 'Water policies and smallholding irrigation schemes in South Africa: a history and new institutional challenges', *Water Policy*, vol 4, no 3, pp283–300

Peters K. and Z. Wilson (2006) *Grabouw Case Study*, public access report prepared for Newcastle University and the Department for International Development, UK in reference to The DFID Engineering KaR Programme: Proposal W1-17, 2002–2007, Second Order Water Scarcity in Southern Africa.

Post Uiterweer, N. C., Zwarteveen, M. Z., Veldwisch, G. J. and van Koppen, B. M. C. (2006) 'Redressing inequities through domestic water supply: a "poor" example from Sekhukhune, South Africa', in S. Perret, Farolfi and R. Hassan (eds) *Water Governance for Sustainable Development. Approaches and Lessons from Developing and Transitional Countries*, CIRAD and Earthscan, London, pp55–74

Roux, D. J., Rogers, K. H., Biggs, H. C., Ashton, P. J. and Sergeant, A. (2006) 'Bridging the science–management divide: moving from unidirectional knowledge transfer to knowledge interfacing and sharing', *Ecology and Society*, vol 11, no 1, pp4

Sapa (2006) 'Report: water problems remain in rural areas', *Mail and Guardian*, 12 July, www.mg.co.za/articlePage.aspx?articleid=277212&area=/breaking_news/breaking_news__national/, accessed 12 June 2007

Savides, M. (2007) 'Engineering skills in short supply', *The Mercury*, 6 June, edition 1, www.themercury.co.za/index.php?fArticleId=3868957, accessed 26 June 2007

Sen, A. (1981) *Poverty and Famines: An Essay on Entitlement and Deprivation*, Oxford University Press, New York, NY

South African Cities Network. (2004) *State of the Cities Report*, SACN, South Africa

Stimie, C., Richters, E., Thompson, H. and Perret, S. (2001) 'Hydro-institutional mapping in the Steelport River Basin, South Africa', IWMI Working Paper 17, Pretoria, South Africa

Tlou, T., Williams, C. J., Perret, S., Mosaka, D. and Mullins, D. (2006) *Investigation on Farm Tenure Systems in Small-Scale Irrigation*, WRC project report #1353/1/06, Water Research Commission, Pretoria, South Africa

Van Koppen, B., Jha, N. and Merrey, D. J. (2003) 'Redressing racial inequities through the National Water Act in South Africa: the prospect of revisiting old contradictions?', unpublished research report report, International Water Management Institute, Battaramulla, Sri Lanka

Veldwisch, G. J. (2006) 'Local governance issues after irrigation management transfer: a case study from Limpopo Province, South Africa', in S. Perret, S. Farolfi and R. Hassan (eds) *Water Governance for Sustainable Development: Approaches and Lessons From Developing and Transitional Countries,* CIRAD and Earthscan, London, pp75–91

Wenzel (2007) 'Public-sector transformation in South Africa: getting the basics right', *Progress in Development Studies*, vol 7, no 1, pp47–64

Wilson, Z. (2006) *The United Nations and Democracy in Africa: Labyrinths of Legitimacy,* Routledge, New York, NY

Wilson, Z and Pfaff, B. (2008) 'Religious, philosophical and environmentalist perspectives on potable wastewater reuse in Durban', *Desalination*, vol 228, pp1–9

Wilson, Z, Malakoana, M. and Gounden, T. (2008) 'Trusting consumers: involving communities in water service decision-making in Durban, South Africa', *Water SA*, vol 34, no 2, pp141–146

WRC (2007) 'K5/1616: The state of community consultation in the provision of water services', final draft report, WRC, Pretoria, South Africa

10

The Role of Locally Managed Water Aid: Effective Partnerships in Sri Lanka

Margaret Shanafield and Palitha Jayaweera

Introduction

Domestic water supply projects across the developing world are driven by two equally important but often conflicting motivations. There is an immediate need for action towards providing sanitary drinking and domestic water supplies in the world's poorest communities. However, the methods used in implementing these aid projects need to impart not just a temporary fix, but a lasting solution.

The immediate need for access to improved domestic water supplies is addressed by the Millennium Development Goals (MDGs) as developed at the UN Millennium Summit in September 2000, in which participating countries agreed to halve their populations without access to safe drinking water by 2015. Participants from many sectors, including national and local governments, large banks, non-governmental organizations (NGOs) and community organizations all play a part in this process. A participatory approach to project implementation is most often used, though project results must be subjected to the common critiques regarding actualized levels of community initiative and empowerment. In addition, the project lifespan (and therefore involvement of any foreign partners) is usually short, while long-term maintenance is, in many countries, mandated to the local governments. Therefore, lasting impact from these projects depends partly upon the integration of new facilities into existing operation and maintenance schedules.

Within the context of international water development projects, this study follows the Water and Environmental Sanitation (WES) project implemented in 36 rural communities of Sri Lanka between 2005 and 2007. WES was implemented by community-based organizations in each village with the technical inputs and guidance of Community Self Improvement (COSI) Foundation for Technical Cooperation and funding from Plan Sri Lanka. COSI is a small, Sri Lankan-run NGO started by co-author Palitha Jayaweera and his colleagues in 1994, and is run as a not-for-profit organization governed by a board of directors. Mr Jayaweera is currently programme director and a board member of COSI. Margaret Shanafield joined COSI for several months in 2004 as a volunteer and independent consultant. This chapter meshes the authors' Asian and American perspectives, and strives to present the WES project as it relates to both Sri Lankan culture and current views on development methodologies.

The authors evaluate the success of the WES project with respect to the goals of water provision in rural areas and the impact of the chosen methods in ensuring long-term solutions to community water supply. Small, local organizations such as COSI provide several advantages to development projects, and may be an important link between the many levels of participants working towards achieving the water and sanitation MDGs.

Background: Water and Sanitation Development Needs in Sri Lanka

Sri Lanka is one of 58 countries that have developed a national target for drinking water and sanitation access. The country-specific target set by the government of Sri Lanka sets higher goals than the basic MDGs of halving the population without access to improved facilities. Access to an improved water source should increase from 82 per cent in 2001 to 85 per cent by 2010 and 100 per cent by 2015, while access to proper sanitation should increase to 93 per cent during the same time period (United Nations Development Programme, 2006).

Though the MDGs provide an ambitious plan, and many organizations are working towards reaching them, overall improvement is difficult to measure. World Bank estimates put access to clean drinking water at only 79 per cent in 2005 (World Bank, 2006). This statistic may mean a slight improvement over the 77 per cent reported by World Health Organization (WHO) and United Nations Children's Fund (UNICEF) for 2000 (UNICEF, 2001), or a decrease in access from the 2001 national data of 82 per cent, depending on which statistic is more accurate. Indeed, monitoring the effectiveness of programmes designed to address MDGs may be difficult; nonetheless, they have become the most widely accepted yardstick for evaluating development efforts and government objectives (World Bank, 2005).

The population of Sri Lanka is roughly 20 million people, the majority of which live in rural communities. Between 1990 and 2002, the percentage of urban and rural dwellers in Sri Lanka remained constant at 21 and 79 per cent, respectively (WHO/UNICEF Joint Monitoring Programme for Water Supply and Sanitation, 2004). Analysis of national data from 2004 estimated that 48 per cent of the rural population did not have access to safe drinking water. Therefore, successful attainment of the MDG drinking water and sanitation access targets requires concentration on the rural areas of the country. Whereas urban dwellers are largely served by treated, municipal piped water, only 15 per cent of the rural population has access to piped water (Amarasinghe et al, 1999). Bad water quality and poverty also contribute to drinking water access problems. Because the vast majority of the Sri Lankan population depends on untreated groundwater for domestic water supplies, fluoride and other ion concentrations pose an acute problem. In certain areas of the Dry Zone (particularly in the North Central Province) (Figure 10.1), groundwater contains unsafe concentrations of fluoride, leading to a high prevalence of skeletal fluorosis. In contrast, groundwater in the Wet Zone contains only minute concentrations of fluoride, resulting in dental caries (Dissanayake, 1991).

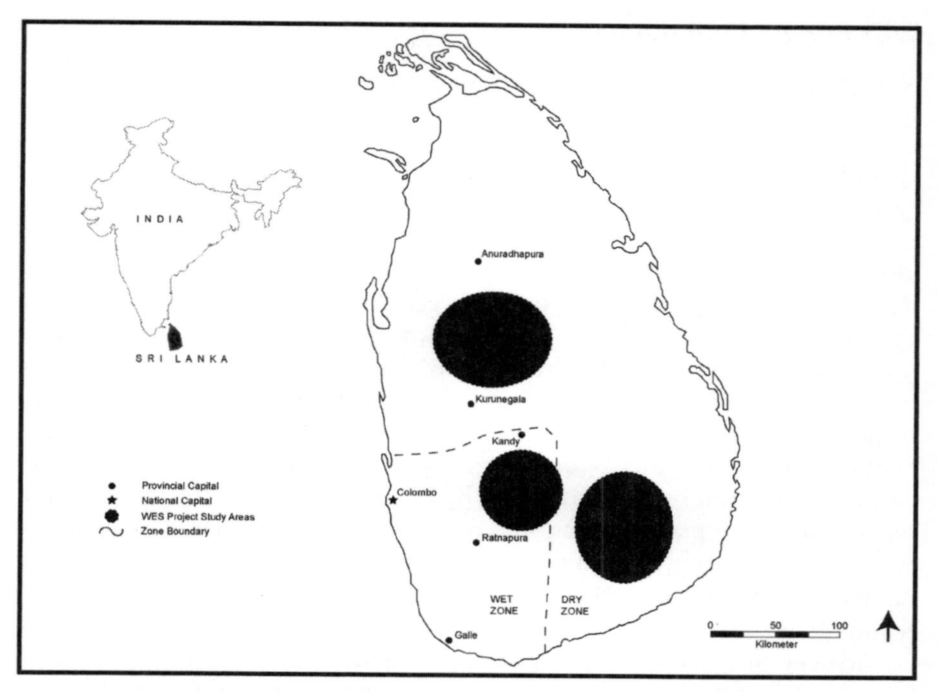

Figure 10.1 Map of Sri Lanka showing wet and dry zones, and WES project areas in the Northwest, Southeast and Plantation regions

Development of rural water and sanitation facilities in Sri Lanka is delegated by the Sri Lankan constitution to the local governments, under guidance from the National Water Supply and Drainage Board. The National Policy for Rural Water Supply and Sanitation provides the core of the government's rural water supply strategy. However, in the past one of the hurdles toward improving rural water supply has been that the rural government agencies lack the resources to be effective. The government therefore encourages international involvement in this process, especially in the form of technical expertise and financial investments.

One large, international NGO that has been working in Sri Lanka is Plan Sri Lanka. Although Plan began working in the country as early as 1980, it did not become involved in water and sanitation projects until recently. In May 2004 Plan Sri Lanka entered into a partnership with COSI for the WES project to supply 36 rural communities with water and sanitation facilities. Financial support from international agencies such as Plan allows COSI to use its technical and cultural expertise for water and sanitation implementation in Sri Lanka and neighbouring countries.

WES Project Methodology

Villages were selected for participation in the WES project based on priority lists developed by Plan during prior work; in many cases, the villages themselves requested to become involved in the project. The villages were spread among six districts in the Northwest (NW), Plantation (PL) and Southeast (SE) programme units (Figure 10.1). Plan and COSI adopted a participatory method described as a 'demand-driven, people-centred' approach for the WES project. The success of this method depended on extensive, two-way communication between the implementing group and the community, with community involvement in all stages of development. Specifically, the project was designed so that with technical guidance, the community would choose the type of water supply and sanitation that best fit members' needs, play a significant role in the planning and design of those facilities and contribute a certain level of capital and labour. The village contributions were intended both to increase village engagement and ownership in the new facilities, and to ensure that appropriate technologies were selected. This prevented project efforts from being seen as handouts and as leading to the construction of facilities that were expensive and difficult to maintain.

To achieve the goals of community involvement and communication, many levels of support structures were present. Community facilitators and local NGOs, called operating partners, were hired by COSI to manage daily operations at the community level. The community facilitator worked in the village full-time. In some areas, such as in the Plantation region, villages consisted of dense clusters of long houses primarily populated by tea plantation workers, while in other areas the villages were more geographically spread out and communities were politically defined. The primary duty of the facilitator was to identify existing community-based organizations (CBOs), and determine which, if any, were appropriate to work with the WES project. Many varieties of CBOs exist in these rural communities. Some have been set up by other development projects, while others grew from within groups of neighbours to fill community needs. For example, the farmers of a certain community may form a group to work together so that everyone cooperates to get the fields planted or harvested. Historically, village committees also managed local reservoirs and irrigation systems.

Determining the proper CBO to work with the project was often a difficult task, since existing groups rarely included – or represented – all members of the community. Therefore, the facilitator, in conjunction with the operating partners and COSI staff, worked with the CBO to incorporate new members (if the CBO was willing), or build community support for a new organization specific to the WES project. In addition to selecting the CBO, the facilitator worked with the community to gather demographic information and communicated any issues to COSI staff. Monthly training sessions for the facilitators and operating partners were conducted by COSI to guide the process.

Once demographic information was collected by the community facilitator, small neighbouring groups within the community met with the facilitator or operating partner to discuss their needs. The organization selected for the WES project then used input from the smaller groups to build community action plans to improve their quality of life in five areas:

1 community strengthening;
2 water supply;
3 sanitation;
4 hygiene education;
5 environmental improvement.

This broad focus differed significantly from early development projects in Sri Lanka, which focused only on water development.

The five action plans formed the development strategy for each village, and specified the appropriate and affordable type of water supply, the number and type of latrines needed, and the level of education needed to ensure that water and sanitation facilities were hygienically maintained. Several specific, participatory activities were used to aid in the process of community action plan formation. It is said that a picture tells a thousand words so during community group meetings, members worked together to draw pictures of what they considered to be a well-off household, an average household and a poor household in their village. Although the pictures varied greatly between villages, depending on the economic well-being of the project area, meaningful similarities existed among the well-off households of all villages; all had clean, well-organized yards and a water source and latrine either in the yard or connected to the house. COSI staff and facilitators discussed these values, the importance of water source and environmental sanitation to village welfare with the community. Groups of villagers then made maps of their neighbourhoods, including the economic well-being level described by their pictures and the existing water sources. The maps helped community members and facilitators to better understand the existing water and sanitation situation and to begin discussions of options for new facilities in light of available resources.

Another participatory activity that helped facilitate community discussions on the economic and social importance of clean water and environmental sanitation was to build a problem tree. Facilitators met with community members to identify their key problems, and map out the causes (or roots) of the problem and the effects that grow out of the problem. One commonly identified problem in many villages was sickness, often stemming from lack of access to a sufficient, clean water source or from unsafe disposal of human and household wastes. Sickness in children leads to absence from schooling and stress for concerned family members. Among adults, sickness leads to inability to work (and therefore lower economic well-being) and possible social tensions. For example, at one meeting a villager admitted half-jokingly 'if I get sick, my neighbour may start looking at my wife, thinking she might become available'.

The WES project lasted for two years in each village. The first ten months were devoted to project development and the formulation of the action plans. During the final three months of project development, the pre-construction phase, villages committed to financing 20 per cent of implementation costs. Depending on the type of water and sanitation facilities the village chose, this financial contribution was fulfilled either through contribution of labour or required monetary input. During the second year of the project, community members worked with their operating partners on the actual construction of the facilities, while COSI staff and facilitators helped to develop hygienic household and environmental sanitation habits.

One additional aspect of the WES project that stemmed from Plan's background sponsoring children was the parallel focus on educating children in a school setting. The rationale for this focus was that within the community, children are most impacted by the level of sanitation and drinking water facilities in their villages. Since children are more susceptible to diarrhoeal and waterborne diseases, their successful development is directly dependent upon the availability of clean drinking water and proper hygiene practices. Hygiene and environmental sanitation education is best aimed at children, who can then propagate healthy habits within their families. Therefore, special attention was given to hands-on educational activities to teach children good hygiene practices and help them understand the concepts of environmental sanitation at home. Priority was also given to ensuring that schools have safe drinking water sources.

Identifying Community Need: Existing Water and Sanitation Facilities

Initial demographic data collected in the villages showed a wide range of economic and social well-being (Figure 10.2). A total of 9848 households are included in the project, with 3995 in the Northwest, 1575 in the Plantation and 4278 in the Southeast project villages. Villagers were surveyed to determine existing access to water sources and sanitation facilities, as well as their hygiene habits. Of these surveys, response was most successful for water related questions, with a response rate of approximately 70 per cent of households in the Northwest, over 95 per cent in the Plantation area and 90–95 per cent in the Southeast. Quality and quantity of water sources was best in the Northwest, though only slightly over half the households in the project area reported they had access to clean water, and less than 40 per cent had sufficient water all year (Figures 10.2a and 10.2b). In contrast, proximity to water sources was best in the Plantation area, where almost 80 per cent of the households had a water source either inside the home or within 250m of their yard (Figure 10.2c). Water sources also varied between the three regions. In the Northwest region, 70 per cent of households shared a common well as their main water source. A common well usually refers to a hand-dug well located in a public place and shared by several families. Twenty per cent used a tube well (borehole well) and ten per cent relied on surface water sources (i.e. springs, rivers or tanks). In contrast, 86 per cent of households in the Plantation region had a localized groundwater source (usually a handpump) and the remaining households had piped water. In the Southeast region, a diverse set of water sources exists, with common wells, surface water and tubewells as the main sources. Based on their location partly in the wet zone, and the prevalence of groundwater and piped water sources, it was surprising that the plantation villages reported relatively bad water quality and quantity. This may have reflected unsafe environmental sanitation practices (not keeping the area around a groundwater source clean of wastes), and highlights the importance of sanitation facility development and hygiene education in conjunction with rural water development.

The initial demographic surveys also identified a strong need for sanitation facilities and hygiene education. Hygiene and sanitation practices are strongly linked with malnutrition rates in children (World Bank, 2005). Several previous projects have also highlighted the greatly increased health benefits when hygiene and sanitation improvements are made in conjunction with improved water supply, as compared to water improvements alone (Bajard et al, 1981; Rijsberman, 2004; VanDerslice and Briscoe, 1995). In all villages surveyed, a high percentage of households reported they openly disposed of wastewater, and did not have a system of stormwater drainage for yardspace. The percentage of households lacking any kind of toilet ranged between 18 per cent in the Southeast region villages and over 38 per cent in the Plantation region. Additionally, although response to hygiene questions was lower than expected, low rates of hand washing after defecation and low rates of boiling of drinking water were reported.

Critical Evaluation: Meeting Immediate Needs and Beyond

One of the hurdles in meeting the MDGs is that not enough projects have been initiated to address global needs. Even if the MDG goal of halving the population without access to clean drinking water was met, 650 million people across the globe would still be unserved. At the present rate, it is likely that a

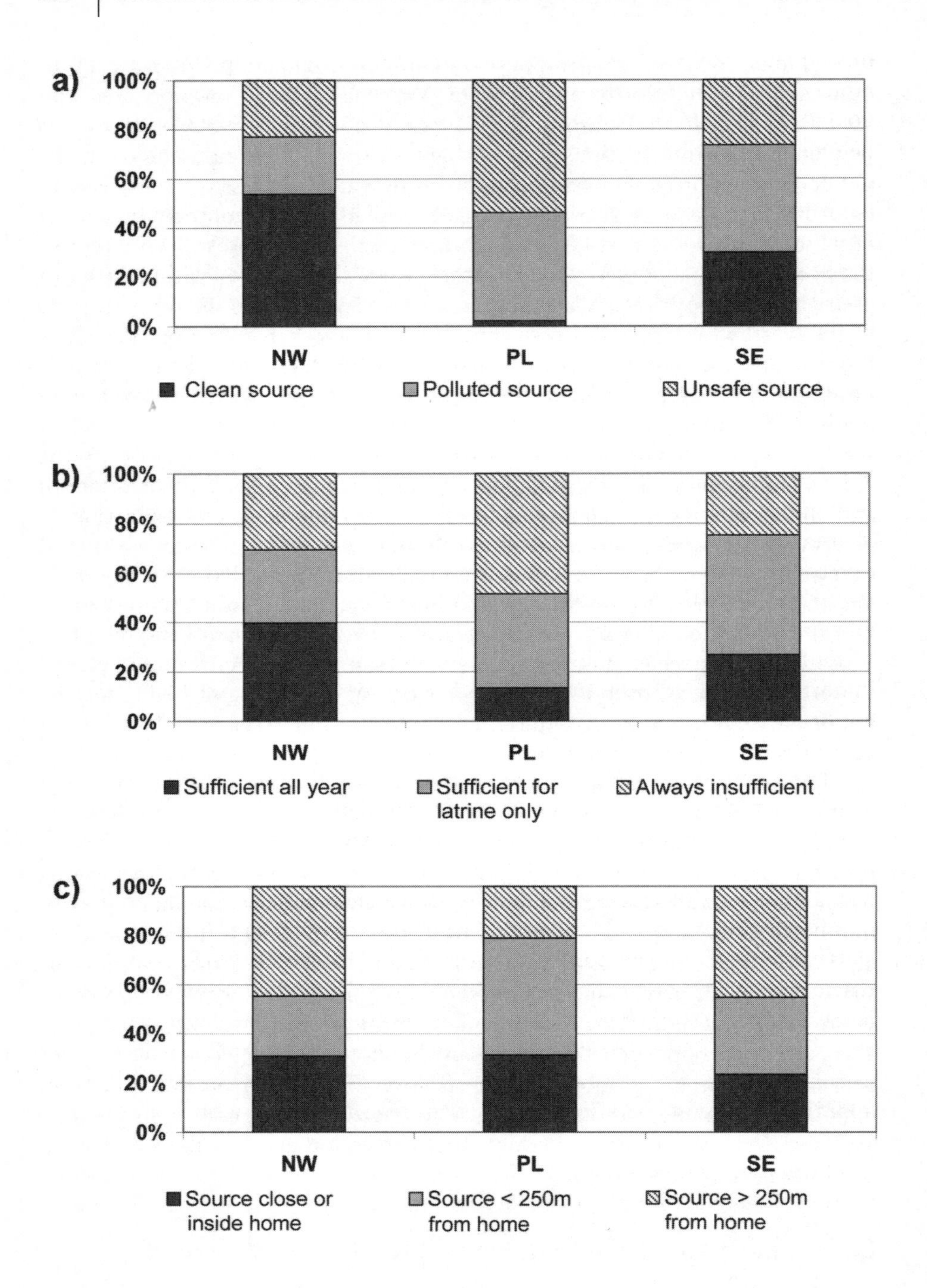

Figure 10.2 Village water quality (A), quantity (B) and accessibility (C) in the WES regions at the beginning of the project

much higher population of people will still lack access to clean drinking water, largely due to inadequate commitment of funding and actions towards this goal (Gleick, 2004). International conferences on the subject result in solutions belonging to two very different camps; either go big (privatize and build large dams) or focus on the communities. While many of the most public figures advocate for increased privatization of water services or the construction of large dams to increase reservoir capacity, experiences from countries around the world suggest that the most efficient way to tackle water and sanitation access issues is by starting with localized, community-oriented water schemes at the household level (WSSCC, 2000) or with small communities of 20–50 families (Sooryamoorthy and Palackal, 2003). Therefore, the achievements of initiatives that serve many communities, such as the WES project, play an important role in filling water and sanitation access needs.

How has the WES project contributed to the MDG goals? As of 2007, the WES project had concluded in all 36 areas, though due to some hurdles in project implementation, various activities were still ongoing in some villages. A total of 8657 families in three regions participated in project activities. Several hundred new water supply systems were constructed by villages with the help of COSI expertise, including 134 (private) family wells, 15 (common) village wells, 19 gravity water schemes (i.e. from reservoirs), 13 domestic rain water harvesting tanks, and 7 tube wells with handpumps. Perhaps even more important were the sanitation facilities; 3935 new water-sealed latrines were constructed and 374 water-sealed latrines were rehabilitated as part of WES. Domestic-level waste management systems were provided for all households and participation in the hygiene promotion activities was compulsory.

Community involvement

For a thorough evaluation of the project, we must also look beyond the numbers, at the success of the development strategies employed in this project. This task requires an evaluation of the level of success in achieving the participatory development ideals elaborated in-depth by, among others, Robert Chambers in *Whose Reality Counts? Putting the First Last* (1997). Chambers theorizes that aid workers should 'facilitate, sit down, listen and learn' and then 'share methods which local people can use for their own appraisal, analysis, planning action, monitoring and evaluation' (Chambers, 1997, p103).

In practice the participatory approach can easily fall short of its goal of putting the people's wishes first. Limits on time and resources often conflict with efforts toward achieving participatory ideals in real-world situations. While the best solution would be for individual communities to organize and put forward their own water management schemes, this happens only very rarely. Additionally, the local and national governments usually tasked with serving their rural populations with appropriate facilities rarely have the means, experience or funding to adequately meet the needs of the people.

Therefore, the struggle to serve the poorest communities with clean drinking water is often taken up by national and international aid groups. These NGOs are accountable to their donors, and quantifiable results are normally expected after only a short project lifespan. This leaves little time available for the community mobilization phase in which participatory approaches can be applied and community members consulted.

Despite these restrictions, examples of successful community involvement are not uncommon (Bessette, 2006; Pickford et al, 1998). However, several authors have vehemently pointed out the common pitfalls of current participatory approaches and even go so far as to suggest that the participatory approach should perhaps be abandoned altogether. Many criticize past participatory projects for disregarding existing community decision-making processes and strengthening the already powerful at the cost of the poorest. How does the WES project stand up to these accusations?

Neither of these cases was seemingly a problem in the WES project. However, we argue that neither the idealized local knowledge, nor simply avoiding the criticisms of it, leads to the selection of appropriate technologies. The ideals described above put the facilitator into an arguably passive role; one in which he or she merely provides the community with information on appropriate technologies and does not seek to influence community opinion. Villagers have local knowledge and know what will work best for their needs. The authors compiled by Cooke and Kothari (2001) refute the assertion that participatory approaches lead to an accurate description of local need. To prove this point, several projects are discussed that demonstrate the possibility that development workers may pose the question of need in such a way that community opinion is manipulated to produce a desired outcome. Further, they show that attempts at soliciting community involvement in decision-making may be skewed by villagers voicing preferences based on what they anticipate the project is willing to provide, rather than on what best fits their lifestyle and customs.

Far from being a passive observer and mere source of information, we argue that the development worker should have the knowledge and experience to properly guide the community towards a sustainable selection of facilities. These facilities must be easily maintained, since it is unlikely that communities will receive significant operations and maintenance support from the local government in the future. The majority of project costs are covered by the international funding agency, whose specifications must also be considered. To satisfy the international funding agency, facilities must be simple to construct, so that the project is finished within a reasonable timeline. However, they must comply with local culture and traditions, so that they are not abandoned at project completion. As fellow Sri Lankans, COSI staff understand village traditions and Sri Lankan domestic practices, which speeds the community mobilization process. Moreover, as experienced development workers, they understand what technologies work, and which ones are unsuitable due to individual restrictions in specific villages. The combination of these two factors put COSI in a unique position to guide communities towards the proper selection of resources; more of a two-way dialogue and consistent with many of the examples of participatory projects deemed successful.

A clear example of this two-way dialogue is seen in one of the villages served by the WES project. During initial CBO meetings, villagers requested a piped water supply. Piped water and flush toilets are often requested as the preferred drinking water and sanitation facilities, since this is what is seen on television. However, evaluation of existing resources showed that the village was several miles from an electrical hookup, though close to a medium-sized reservoir. After the costs and benefits of several types of facilities were explained, the village agreed upon a gravity feed system from the reservoir. In this case, the facilitators did not just rely on the local knowledge as proposed by Chambers. Neither is this a case of the manipulation denounced by Cooke and Kothari. Instead, it demonstrates the need for a middle ground, in which facilitators listen to community members, but also use their own experience to help guide the community towards the most viable option.

The issue of 'strengthening the already powerful at the cost of the poorest' (members of the community) also warrants discussion. On one hand, research shows that existing CBOs, rather than newly formed groups, provide the strongest base for participation in development projects. In communities with active CBOs and associations, households are more accustomed to working together and greater design participation is likely (Isham and Kahkonen, 2002). Strong CBOs also reveal social ties that prohibit the misuse of facilities, and strengthen the ability for monitoring and maintenance (Sooryamoorthy and Palackal, 2003). New groups, in contrast, often lack accountability, and are not based on important hierarchies within the society. Therefore, the authors argue that as long as community leaders do not neglect community members or misuse their power for their personal benefit, strengthening the existing CBOs, which are most often led by the wealthiest community members, does not detract from the merit of a development project. Further, in some cases it may even be more beneficial, as wealthier community members become willing to cross-subsidize local water schemes and afford a better quality product for everyone (Sooryamoorthy and Palackal, 2003, p5). The formation of new groups to avoid strengthening the already powerful might, on the contrary, only serve to disturb the existing decision-making processes and detract from project success.

Although adopting an already-existing CBO for development projects is preferable, care must be taken to ensure that the organization represents, or is willing to expand to represent, even the poorest families in the community. In the WES project, identifying the proper CBO in each village proved one of the greatest difficulties. Existing community groups had been set up by previous projects; however, these groups rarely represented all of the families to be included in the water schemes. Nevertheless, these groups were now used to being in control of community decisions, and did not welcome the idea of having new leaders. Therefore, one of the challenges for community facilitators was to educate these leaders as to the goals of the WES project, and ensure that all community members felt motivated to participate in the group. This issue compounded the community mobilization process in some villages, where it was already difficult to get the farmers to leave their work in the fields to attend meetings and to commit resources to what they viewed should be a free resource.

Areas of disconnect

Several additional issues added difficulty to CBO selection in the villages. At the beginning of the project, initial momentum for community involvement was slowed in some villages where the facilitator did not have sufficient knowledge about the WES process. Many facilitators were unable to keep the community aware of activities connected to the specific phase of the project. These problems dissolved as facilitators became more knowledgeable through the course of the monthly trainings. More enthusiastic replacements were found for facilitators who did not show initiative and improvement after the initial months of training. Strong guidance from the facilitators working on a daily basis in the villages was crucial to strengthen CBOs and communicate community involvement and progress.

Another difficulty was the community organizations' limited exposure to community development interventions. For example, in some cases community leaders of CBOs had not previously participated in a development project and did not clearly identify the objectives of their organizations, since they did not understand their roles within the project. Other CBOs did not have clear knowledge and awareness of project activities. This problem led to low turnout at meetings; if the project goals were not promoted by community leaders, local residents did not prioritize meetings over their other responsibilities or work.

At the household level, lack of understanding of the importance of clean drinking water and sanitation may partly explain the lack of communication on the topics of sanitation and hygiene. In communities where the CBOs did not take initiative, basic demographic information and household self-assessments were not properly explained, or not collected in a timely manner. Lack of thorough, accurate information made it difficult to discuss community needs and develop community action plans. Moreover, documentation of attendance at meetings and the topics covered during the meetings was often inadequate, making it difficult to assess progress and to report on the status of community knowledge or needs to COSI.

The recent growth in the Sri Lankan garment industry also played a role in CBO dynamics. As community members go outside the village to work in garment factories and other industries, they become less in tune with village issues. This has led to the issue of sleeping villages; communities where the villagers no longer work in the area but commute to factories for work, leading to social and cultural problems within the community as traditional ties are lost. The WES project had difficulties with CBOs that had leaders who worked outside the village. These leaders interacted less with the community and were thus less able to promote and organize village support for the project.

At the regional level, lack of existing monitoring mechanisms hampered measurement of programme implementation progress. This resulted in delays in identifying the proper CBOs or selection of CBOs that did not represent the entire community. On the other hand, some activities were difficult to implement because the number of families within the group was too large, making the necessary level of discussion more challenging.

Difficulties in CBO selection even had repercussions at the international level regarding the relationship between COSI and Plan. According to the timeline set up for the project by Plan, the project was to last two years in each community. This left approximately one year of community mobilization for the community to participate in the planning and organization of their local water schemes. In some communities, this deadline was easily met; COSI staff could easily identify villages in which the facilitator was very active and had gained community support and been able to strengthen community groups. Where this was not the case, problems arose. In communities where the community facilitator was not able to rally the community or to solve power struggles in CBOs, delays occurred and sometimes resulted in poor community involvement in water scheme planning. In some communities, the facilitators needed to be replaced part way through the project, resulting in similar delays. Due to a certain amount of rigidity in the project timeline to meet agreements with the funding agency, it was not possible to significantly delay implementation of water schemes. These difficulties in the CBO selection process again highlight the conflict between prioritizing a successful community dialogue and meeting international funders' regulations.

Finding long-term solutions, re-establishing maintenance systems

Sri Lanka is often described in historical accounts as a hydraulic civilization; one in which the ruling power kept control of the population through strict regulation of vital water resources (Wittfogel, 1957). Although this term has deeper implications, it is used here to demonstrate that the importance placed on maintaining water resource facilities did lead to the successful operation and management of tanks, canals and other structures for over 1000 years.

As early as the third century BC, most parts of Sri Lanka were inhabited (Mendis, 1986). Due both to its agricultural economy and the seasonal nature of rainfall in much of the country, small village tank systems were crucial to the earliest settlements on the island (Shand, 2002). Much of the arable land surface in Sri Lanka is located in the dry zone (Figure 10.1), an area within the lowest peneplain of the island, and receives a mean annual rainfall of 1250mm only during the period of October to January, necessitating the construction of surface storage systems for irrigated agriculture (Panabokke et al, 2002).

State involvement in water development began in the first century AD, when a succession of kings ordered the construction of a series of large tanks and canals to meet the pressures of a growing population. These large tanks and canals were the property of the state, earned income in the form of taxes on water and fishing and were maintained by a special set of engineers (Kenyon et al, 2006). Some were also donated to monasteries (Siriweera, 2002). For example, the Mahavihara monastery in Anuradhapura required villagers to supply labour or crops in return for irrigation and domestic water supplies (Gunawardana, 1971). Additionally, historical records document the existence of privately owned small tanks and canals beginning around the third century AD (Siriweera, 2002).

Although large tanks provided a reliable source of water for the populations surrounding the historic monasteries and royal complexes, most small villages still relied on small, community owned and managed tank systems. A strict set of guidelines regulated the maintenance and water management of these tanks. Village tanks were often interconnected in a cascade along a non-perennial stream or small river, requiring cooperation between several villages for sustainable management of water resources. A system of compulsory personal labour obligation helped guarantee the maintenance of these systems for many centuries. Each village had a system to divide up water during drought years (Kenyon et al, 2006), and an elected water-chief, who managed the village tank and irrigation system in return for a share of rice. These village systems were largely uncontrolled by the king.

Many of the small tanks established centuries ago are still used by rural villages for domestic and agricultural activities. Unfortunately, the traditional management system of community and government water resource systems was abolished in 1832 under British colonial rule (Panabokke et al, 2002). Now, as new water and sanitation facilities are constructed and existing systems are improved, the local government and villages must once again establish a long-term relationship to ensure operation and maintenance into the future. Whereas the traditional system was developed gradually over several hundred years, contemporary citizens and government authorities must adapt over a period of only a few years. To re-establish and foster the link with water managers, development projects such as WES must make it a priority to involve not only the community, but also local government (and inform all responsible government agencies about new facilities) in development activities.

Ultimately, communities themselves, or even aid organizations alone, cannot ensure a safe source of drinking water for all the rural communities in the country. Instead, the ability of communities to communicate their needs to the local government and manage local water resources will once again determine Sri Lanka's ability to sustain a healthy, productive agricultural economy. Practices stemming from the predominant religion also contribute positively towards this effect. For Buddhists, the Sri Lankan majority religion, sharing water is considered as gaining merit. Traditionally, fetching water from a neighbour's well or even contributing labour to help out a neighbour does not meet with objection.

Aside from the construction of new water and sanitation facilities, perhaps the most meaningful achievement of the WES project was the sharing of these ideas and values among neighbours. Although farmers in rural villages often meet to help each other with various agricultural or social activities, most had never discussed more intimate topics, such as their hygiene conditions. When questioned about the outcome of these initial meetings, many community members confessed that they found it novel to open up the dialogue on existing hygiene and water use habits, and to discuss how these habits differ between households. By discussing the social and economic value of high quality drinking water and sanitation at the family level, community members have taken the first steps towards prioritizing these commodities at the village level. If it is a high priority for the community to maintain water and sanitation facilities, it is more likely that village leaders and CBOs will become socially and politically active to make sure that these needs are met into the future.

The need for community empowerment, however, is a major challenge that extends well beyond Sri Lanka. Bell and Franceys (1995) relate development project success to the ability of all parties – including the government – to adjust to a new role in water resources management. Moreover, they make the observation that individuals generally look towards the government with expectations for the provision of basic human rights. Today, community-scale water facilities are often installed by outside parties such as COSI or an international NGO, instead of directly by communities or the government. Therefore, successful integration of the new facilities into local governmental oversight for operation and maintenance requires strong cooperation between the community, the organization implementing new facilities and the local government. From the outside party's side, this means both involving the government directly, and empowering the community to speak up when it needs support. Local governments rarely have the means to check up on rural community water supplies; therefore, the community itself must be motivated, informed and educated to organize maintenance and lobby its needs to local authorities.

The WES project produced mixed results with respect to community empowerment. In the communities where an appropriately motivated CBO was identified and strengthened, the future of community water and sanitation supplies looks good. In other communities, where many challenges arose, or community mobilization was delayed, the community's ability to work together for the sustainability of their new facilities is more questionable. COSI hopes to increase the success rate by offering to act on a community request for any kind of advice, even after project completion. This is possible because, unlike an international NGO, they are located geographically close to the communities served by the project. Additionally, COSI is involved in a network of water aid stakeholders on the island, including governmental representatives and acts as a community advocate in the local government.

Looking Ahead: The Future of Drinking Water and Sanitation Development in Sri Lanka

In 2000, when the Millennium Summit met and developed the MDGs, Sri Lanka fared better than most developing countries, with a safe drinking water supply already available to approximately 80 per cent of the population. However, this meant that meeting the 2015 goal of halving the population without access to this necessity would require an intensive effort to reach the poorest, most remote villagers. Will Sri Lanka meet this challenge?

The year 2007 marked the halfway point between the Millennium Summit and the goals outlined during it. The few reports that evaluate whether MDG countries are on track towards meeting their goals show that it is unlikely that the drinking water and sanitation goals will be met at the current rates of progression in most countries. Mara (2003) gives a more hopeful analysis; based on the accomplishments of the International Water Supply and Sanitation Decade in the 1980s, he predicts it should be possible to meet the drinking water MDG, though definitely not the sanitation goal. In Sri Lanka, the scarcity of reliable data and a wide variability of conditions make it difficult to estimate progress towards reaching the MDG goals (World Bank, 2005). However, in a 2007 publication, UNICEF reported that only 79 per cent of the Sri Lankan population had access to improved drinking water sources (UNICEF, 2007), suggesting that there is still a considerable amount of work needed to meet national goals by 2015.

The tsunami that struck the coast of Sri Lanka in 2004 caused widespread damage to the coastal areas. In addition to the many lives and homes lost to the floods, over 60,000 wells were either abandoned or made unpotable through saltwater intrusion (Hewage, 2005). An untold number of latrines were also destroyed, causing a grave sanitation problem and increasing disease rates. While the tsunami caused great setbacks in efforts to meet water and sanitation MDGs, it also focused world attention on Sri Lanka. If the monies promised by international donors are successfully implemented in this region, it could help to promote drinking water and sanitation in these communities; however, it remains to be seen whether this will be the case.

Another unknown in Sri Lanka is the stability of the current peace agreement between the government and the Liberation Tigers of Tamil Eelam (LTTE). From the mid 1980s, a violent ethnic conflict plagued the country and slowed development. A peace agreement, signed in February 2002, has led to relatively peaceful conditions. However, recent increases in bombing and conflicts are a worrisome reminder that relations between the government and the LTTE are less than optimal. The end of this conflict is a prerequisite for achieving universal access to drinking water and sanitation (Rijsberman, 2004). Thus, development of these sectors is inextricably tied to the prevailing social and cultural health of the country.

Internationally funded projects often leave little time for the community mobilization required to build long-term solutions to water and sanitation needs. Thus, cooperation between these large organizations and a local company such as COSI provides a distinct advantage. While the WES project was not without fault, we believe it produced acceptable results in terms of meeting community water needs and involving the community in the design and implementation process to ensure that their long-term needs are met. COSI could also establish a working relationship with local government more easily than an outsider might have, increasing the likelihood that the new facilities will be taken care of into the future. In conclusion, small, local organizations like COSI may be an essential ingredient in global efforts to improve water and sanitation access for all.

References

Amarasinghe, U. A., Mutuwatta, L. and Sakthivadivel, R. (1999) *Water Scarcity Variations Within a Country: A Case Study of Sri Lanka*, Research Report 32, International Water Management Institute, Colombo, Sri Lanka

Bajard, Y., Draper, M. and Viens, P. (1981) 'Rural water supply and related services in developing countries – comparative analysis of several approaches', *Journal of Hydrology*, vol 51, pp75–88

Bell, M. and Franceys, R. (1995) 'Improving human welfare through appropriate technology: government responsibility, citizen duty or customer choice', *Social Science and Medicine*, vol 40, no 9, pp1169–1179

Bessette, G. (ed) (2006) *People, Land and Water: Participatory Development Communication for Natural Resource Management*, Earthscan, London

Chambers, R. (1997) *Whose Reality Counts? Putting the First Last*, Intermediate Technology Publications, London

Cooke, B. and Kothari, U. (eds) (2001) *Participation: The New Tyranny?*, Zed Books, London

Dissanayake, C. B. (1991) 'The fluoride problem in the ground water of Sri Lanka – environmental management and health', *International Journal of Environmental Studies*, vol 38, pp137–156

Gleick, P. (2004) *The World's Water 2004–2005*, Island Press, Washington, DC

Gunawardana, R. A. L. H. (1971) 'Irrigation and hydraulic society in early medieval Ceylon', *Past and Present*, vol 53, no 1, pp3–27

Hewage, T. (2005) statement by Mr Thosapala Hewage, Secretary Ministry of Urban Development and Water Supply of Sri Lanka, high level segment of the 13th session of the Commission on Sustainable Development, 20 April 2005, United Nations, NY

Isham, J. and Kahkonen, S. (2002) 'Institutional determinants of the impact of community-based water services: evidence from Sri Lanka and India', *Middlebury College Working Paper Series 0220*, Middlebury College, Department of Economics, Middlebury, IN

Kenyon, P., Pollett, C. and Wills-Johnson, N. (2006) 'Sustainable water management practices: lessons from ancient Sri Lanka', *Water Policy*, vol 8, pp201–210

Mara, D. (2003) 'Water, sanitation and hygiene for the health of developing nations', *Public Health*, vol 117, pp452–456

Mendis, D. L. O. (1986) 'Evolution and development of irrigation eco-systems in ancient Sri Lanka', *Transactions of the Institute of Engineers Sri Lanka*, vol 1, pp13–29

Panabokke, C. R., Saktihivadivel, R. and Weerasinghe, A. D. (2002) 'Small tanks in Sri Lanka: evolution, present status and issues', International Water Management Institute, Colombo

Pickford, J., Barker, P., Elson, B., Ince, M., Larcher, P., Miles, D., Parr, J., Reed, B., Sansom, K., Saywell, D., Smith, M. and Smout, I. (1998) 'Water and sanitation for all: partnerships and innovations', selected papers of the 23rd WEDC conference in Durban, South Africa, 1997, Intermediate Technology Publications, London

Rijsberman, F. (2004) 'The water challenge', paper prepared for the Copenhagen Consensus project of the Environmental Assessment Institute, Copenhagen

Shand, R. (2002) *Irrigation and Agriculture in Sri Lanka*, Institute of Policy Studies, Colombo

Siriweera, W. I. (2002) *History of Sri Lanka from Earliest Times up to the Sixteenth Century*, Dayawansa Jayakody and Company, Colombo

Sooryamoorthy, R. and Palackal, A. (eds) (2003) *Managing Water and Water Users: Experiences from Kerala*, University Press of America, Lanham, MD

UNICEF (2001) *Progress Since The World Summit For Children: A Statistical Review*, United Nations Special Session on Children, September 2001, www.unicef.org/specialsession/about/sgreport-pdf/sgreport_adapted_stats_eng.pdf

UNICEF (2007) *UNICEF Humanitarian Action Report 2007*, United Nations, NY, www.unicef.org/har07/files/HAR_FULLREPORT2006.pdf

United Nations Development Programme (2006) *Making Progress on Environmental Sustainability: Lessons and Recommendations from a Review of over 150 MDG Country Experiences*, Environment and Energy Group, Bureau for Development Policy, NY, www.undp.org/fssd/docs/mdg7english.pdf

VanDerslice, J. and Briscoe, J. (1995) 'Environmental interventions in developing countries: interactions and their implications', *American Journal of Epidemiology*, vol 141, no 2, pp135–144

WHO/UNICEF Joint Monitoring Programme for Water Supply and Sanitation (2004) 'Meeting the MDG drinking water and sanitation target: a mid-term assessment of progress', United Nations, New York, NY

Wittfogel, K. (1957) *Oriental despotism: A Comparative Study of Total Power*, Yale University Press, New Haven, CT

World Bank (2005) 'Attaining the Millennium Development Goals in Sri Lanka: how likely and what will it take to reduce poverty, child mortality and malnutrition, and to increase school enrollment and completion?' South Asia Region: Human Development Unit, World Bank, www-wds.worldbank.org/external/default/WDSContentServer/WDSP/IB/2005/06/30/000160016_20050630090640/Rendered/PDF/321340LK.pdf

World Bank (2006) 'Sri Lanka at a glance', GenderStats Database, 13 August, http://devdata.worldbank.org/AAG/lka_aag.pdf

WSSCC (2000) 'Vision 21: a shared vision for hygiene, sanitation, and water supply', www.wsscc.org/fileadmin/files/pdf/publication/vision21.pdf

11

The Public's Role as a Stakeholder in the Yarqon River Authority, Israel

David Pargament, Richard Laster and Dan Livney

Introduction

Institutional design principles for water management have evolved from abstract concepts to application and acceptance. These principles are an integral part of major legislative initiatives, including the European Union Water Framework Directive. They include governance along hydrological boundaries (river basin or catchment basin), integrated water resources management (IWRM), stakeholder participation on all levels, transparency and accountability. No principle, however, can encapsulate the convoluted nature of an ongoing governing body, even one working within accepted boundaries and under clear policies and guidelines. The art of management requires that every institution understand its limitations, its strengths and weaknesses. All institutions are fluid; no government body works in a vacuum and they are all limited in time, personnel, resources and political support. For this reason, theorists, such as Kai Lee, who recognize the dynamic nature of institutions for the protection of natural resources, propagated adaptive management theory:

> Adaptive management is an approach to natural resource policy that embodies a simple imperative: policies are experiments; learn from them ... we do not understand nature well enough to know how to live harmoniously within environmental limits. Adaptive management takes that uncertainty seriously, treating human interventions in natural systems as experimental probes. (Lee, 1993, p9)

The Yarqon River Authority (YRA) was Israel's first attempt at governance of a river basin. When it was created in 1988 there were examples of river authorities operating in other parts of the world, but none in the region. There was also abundant literature about institutional structure and governance of natural resources both within and without Israel. Early literature from the 1970s steered policy makers towards several principles. One, suggested by Bruce Ackerman and James Sawyer (1972), predicts failure if there is separation of thinkers from doers. A single institution must develop a master plan for the basin and implement its provisions. A second, also suggested by Ackerman and Sawyer, is that decisions by a basin agency are political and 'decisions generated by the political process are generally accorded legitimacy in the contemporary polity' (1972, p423). Therefore a basin authority should be representative of those living in the basin. A second author of wide practical experience, Ian Sinclair (1985, 1994) refined these general principles into a guide for effective governance of a river authority. Sinclair posited that a basin agency, to be effective, must have knowledge of the basin, be open and transparent and have the power to determine the quality of the water resources in the basin. Building on these guidelines, Elinor Ostrom (1990) actually proposed design principles for common property resource management; that is, a system with clearly defined boundaries, decision-making in public with those using the resource involved in the decision-making process and the nesting of local institutions with other levels of decision-making.

These forerunners of the YRA inspired the management institution that has operated in the Yarqon Basin since 1988. The YRA actually adopted these principles and promoted the idea that the YRA be open to public debate. Nothing was held sacred that was not submitted to open debate, either within the managing directorate of the YRA, which itself is composed of representatives of the public, or with the general public at large. Issues that affected certain segments of the public were first presented to that segment for discussion prior to continuing with the programme. Examples of these will be brought forth in this chapter. It is clear, however, that the guiding principle behind the operation of the YRA was that openness, transparency and accountability are the basic principles of management that go hand-in-hand with the theoretical design of integrated water resource management. One cannot function without the other.

The public holds important information that can be useful and even critical to managing a river. And if the silent public, the trees, birds and other animals could talk, they would also add their voices to the vocal public. Yet due to their failure to make sounds that *Homo sapiens* can understand, the YRA stays attuned to ecosystems via research prepared by academic institutions familiar with ecosystem management. All major projects are first tested in an academic setting and knowledge soundboards before continuing, in order to get ecosystem feedback before embarking on a project.

In its early years, decision-making in the YRA did not involve local interest groups, both because few actually existed and because it was not considered important. The interests of the public were thought to be sufficiently represented by the local government authorities and the other representatives sitting on the YRA governing bodies. Social participation in the YRA, initiated by the YRA, began as a way of getting feedback from the public about its decisions. In 1994 the YRA turned to a group of the river's major polluters, the farmers who tilled the soil near the river's banks, and asked to hold a series of meetings to reduce nonpoint pollution of the stream. In addition, meetings were initiated with the residents of riparian urban neighbourhoods in Tel Aviv. As stream quality improved in the directions that served the local population, participation spread further. This in turn put pressure on the YRA as well as on local authorities and the national government to further fund river rehabilitation and increase amenity uses. The symbiotic three-way relationship between the YRA, government authorities and the public has created an upward spiral in the Yarqon River's environmental and social quality.

Today social participation in the YRA is comprised of an ongoing dialogue with several local interest groups and the scientific community, an open door policy, a multi faceted information dissemination system, involvement of the general public in developing a Yarqon Master Plan and a governance structure that includes representatives from government, non-governmental organizations (NGOs) and the general public. Implementation of these methods of participation requires constant learning and revision on the part of the YRA.

Setting the Scene – The Yarqon River

The vision of Israel, viewed from outside, is one of a semi-arid state located in the arid Middle East. This view misrepresents Israel as a desert state. Israel has a wet north and a dry south. Rainfall in parts of the north reaches 800mm a year (as much as London), but the southern port of Eilat has only 20mm annually. Yet unlike precipitation in Europe, rain comes only during the winter, and avoids the region during the summer months. This has led to a dependence on groundwater, including the sources that feed the Yarqon, as the major source of drinking water, so that the few rivers and streams that flow year-round are not used for potable water as they are in Europe and North America.

The Yarqon River was often mentioned in historical literature as one of the major routes for transportation as well as one of the most difficult barriers for merchants and armies coming from south to north through the Via Maris, the ancient caravan route between Egypt and Mesopotamia (Syria). This was especially true during the winter when the river often flooded. The base flow of $7m^3/s$ created a unique ecology and served as a source of energy for powering flour mills, adding to its importance as a source of water. This has all changed; the Yarqon headwaters have been captured and its flow has been reduced to a trickle; the underground water table has been lowered and the 2000ha of swampland in the Yarqon basin that existed a century ago are gone (Ortal, 1998).

The Yarqon River is geographically situated in the centre of Israel and flows through the country's most densely populated area. The river is the main channel of a watershed approximately 1800km² in size that begins at the Samarian Mountains and ends in the Mediterranean Sea. It rises in springs within Israel near the town of Rosh Ha'Ayin (in Hebrew, the headwaters), and flows westward for about 17 miles (28km) to the Mediterranean Sea where it exits in Tel Aviv–Yafo. About two thirds of the watershed lies east of the green line, the pre-1967 border, within the Palestinian Authority. This physical connection, seen mainly in runoff and sewage pollution, is not reflected in any official political cooperation between the YRA and the Palestinian Authority. The name Yarqon comes from the Hebrew word *yaroq* (green) because of its flora and fauna. Its Arabic name is Nahr el-'Auja (The Tortuous River) because of its winding nature; it curves around itself in several places (Figure 11.1).

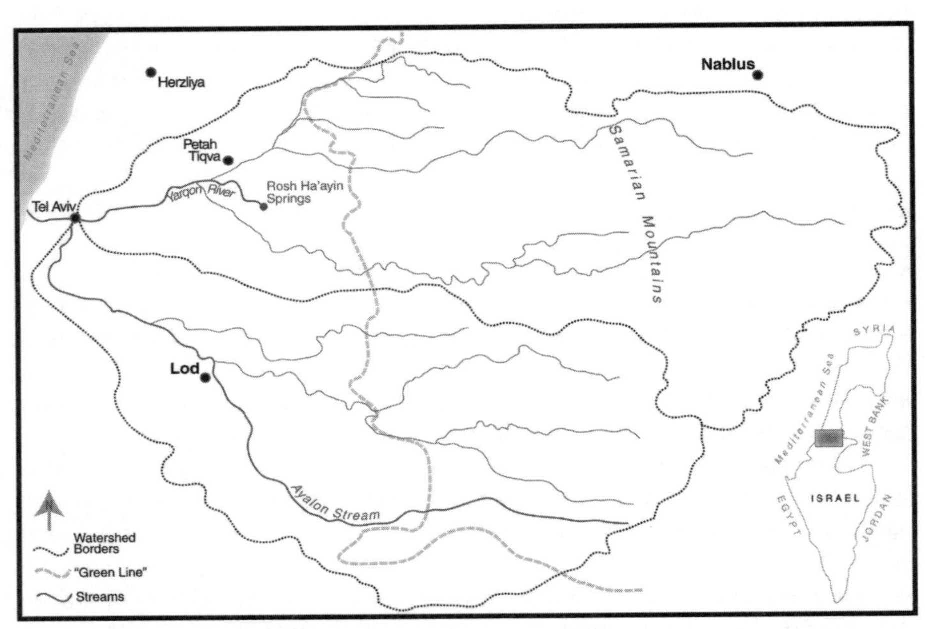

Figure 11.1 Map of the Yarqon river basin

Beginning 50 years ago, the headwaters of the Yarqon, at Rosh Ha'Ayin, were captured to supply water to the homes, factories and farmers of Israel for industrial, agricultural and domestic use. What was once a flowing stream containing over 220 million cubic metres of spring flow annually now contains less than 4 million cubic metres. On the other hand, the increase in population and the growing amount of raw sewage and low quality effluent discharged into the river resulted in destruction of habitats, the disappearance of many plant and animal species and turned the river into a general nuisance associated with mosquitoes and bad odours. This, augmented with the need to control mosquitoes by spraying, destroyed riparian vegetation and contributed to the poor state of the river and the negative feelings of the public towards the Yarqon. This known and expected vicious cycle was even more frustrating because there was no end in sight. Neither the public nor the municipalities had the means to address complaints. The increasing rates at which the springs were pumped contributed to the ongoing state of pollution and deterioration of the river, and this influenced urban planning processes in the surrounding cities to the extent that inferior land uses, such as industrial areas and major transportation infrastructure, were placed between the population and the river.

In order to stop the vicious cycle, the national government drafted a bill to protect the Yarqon. The Rivers and Streams Authorities Law was passed in 1965, and the Yarqon River Authority, Israel's first river authority, was created 23 years later.

The Legal Framework for the Protection of Surface Water in Israel

Israeli law was fortunate to have inherited Ottoman law for the protection of water sources. Ottoman law, which was in force in Palestine from 1516 until World War I, treated water resources as common property, not capable of private ownership. States under Ottoman rule never adopted a riparian rights doctrine or a prior appropriation doctrine found in the laws of most countries in the world. According to the *Majelle,* the Ottoman legal code, all major rivers and streams were considered free property that belonged to the entire community, and even groundwater was considered a community resource when it was needed for domestic use.[1]

The subsequent British Mandate over Palestine declared that all existing Ottoman legislation remain valid, unless amended or replaced by new legislation.[2] When the State of Israel was founded in 1948 there was little need to take water rights away from existing users, since due to the previous legal regimes for water, little of these existed. Rather, there was a need to regulate and prioritize water usage and to prevent flooding. This was done in a series of water laws from 1955 to 1959. The Water Measurement Law of 1955 required that all uses of water were to be metered. The Water Drilling Control Law (1955) required that drilling for water or using a well required government approval. The Drainage and Flood Control Law (1957) set up drainage authorities throughout the State to protect persons and property from runoff and storm water. The drainage authorities were independent bodies created to protect persons and agricultural interests from surface water, but not to protect surface water from persons and pollution.

In 1959 the Knesset passed one the world's most significant water laws, the Water Law (1959), which, in its preamble, creates the following conditions:

1 The water resources of the State of Israel belong to the people to be managed by the government.
2 Everyone has the right to the use of water but no one has the right to a water source, even those working or living on the banks of that source.
3 No one may degrade a water resource nor overdraw its source.
4 All water uses are connected to a permitted use, such as domestic, agricultural, industrial and public.

The Water Law created a Water Commissioner (today called the Director of the National Water Authority) with wide statutory powers to protect, enhance and manage the water resources of the State.[3] Using the powers invested in him, the Water Commissioner designed, built and supervised Israel's national water supply system. First a national water company was created, and then the Water Commissioner issued licences to regional and local supply companies. Israel has a well-managed supply system for water use regulated by a central authority. This, however, came at the expense of the surface and groundwater sources because the supply system developed without thought for sustainable development. The Water Law contemplated that the Water Commissioner would protect surface and groundwater as supply systems were developed; all the power was available by law, but the tools given the Water Commissioner, immense in scope, rusted in his hands. Due to this failure of governance, the rivers and streams became polluted and the aquifers suffered from over pumping. The Water Commissioner's reluctance to prevent water pollution or even resist over-pumping of groundwater led to the reduction of stream flows and an increase of pollution loads.

All of the 31 major streams and rivers in Israel deteriorated with little objection raised, but the Yarqon struck a special note with members of the Knesset, Israel's parliament. Beginning in 1964, the Knesset began debate on a law to protect the Yarqon River specifically, but the proposed law evolved into something wider – the Rivers and Streams Authorities Law of 1965. The law delineates a framework as to how a river authority should be set up and run. The Knesset gave the Ministers of Agriculture and Interior the power to implement the law. The two Ministers could decide which rivers or streams should have a river authority and what their powers should be. But by placing the law under the jurisdiction of two ministers, the national government delayed its implementation for 23 years. The Minister of Agriculture was in no hurry to grant power to a fledgling river authority that might interfere with higher policy set by the Water Commissioner. The Minister of Interior was in no hurry to take powers away from local authorities and grant them to a river authority. The creation of the Ministry of Environment in 1988 broke the deadlock. The Ministers of Agriculture and Interior finally agreed to create the Yarqon River Authority, and turned supervision of the law over to the Minister of Environment. Since then, only one other river authority has been created – the Kishon River Authority in 1992.

The law's stated purpose was to rehabilitate the Yarqon River and turn it into a source of recreational use rather than a conduit of sewage and solid waste. The law enabled the creation of a public body whose major stakeholders are representatives of the local authorities. The new law did not, however, change the contours of Israel's legislation for protection of water resources. This meant that the new authority would have to find its way through a bureaucratic maze if it intended to rehabilitate the river.

During the 23 year lapse between adopting the law and actual implementation, too much water had been pumped out of the headwaters of the Yarqon and too much pollution discharged into the river for it ever to reclaim its former glory. This, however, did not daunt the original members of the YRA who met in 1989 to begin reclaiming the Yarqon and its banks for public use.

The Yarqon River Authority: The Legal Background

As Ostrom writes in her design of institutions managing a natural resource, one major criterion is to nest the governing body with other levels of governance (Ostrom, 1990). The YRA is not the only government body concerned with the Yarqon River; it is, however, the focal point for all stakeholders concerned. This is achieved by statutory fiat; that is, the law creating the YRA established the pecking order for stakeholder involvement. Since water in Israel is under the auspices of five national government ministries, the order creating the YRA gives representation to all five government ministries on its executive council. All the cities riparian to the Yarqon are also represented, as are landholders and representatives of the Nature Reserves and National Parks Authority (NRA). A 2002 amendment to the Rivers and Streams Authorities Law added a requirement that river authority membership must also include three representatives of the general public, at least one of whom represents an environmental organization.

The YRA budget covers all routine activities including the creation of the master plan and many components of its implementation. The budget is funded by the 18 members of the YRA:

- seven municipalities;
- four government ministries;
- seven additional stakeholders.

In order for the Authority to work efficiently, the YRA has both a council and a directorate. On the council sit representatives of stakeholders from the national, local and regional government, plus NGO representatives, some 18 in all. It meets twice a year to set policy and budget allocations, debate proposed regulations and approve all major decisions of the directorate. On the directorate sit nine members:

- four representatives of local government;

- two representatives of the national government;
- one each from the Yarqon Drainage Authority, the NRA and the Organization of Towns for Sewage and Environmental Quality.

The directorate meets often, as much as once a month and more if needed. The public is represented in both the council and the directorate through representatives of the local authorities riparian to the Yarqon, who are the major stakeholders in both organs. Local environmental interests are served via representatives of the Nature Reserves Authority and environmental NGOs, but other local interests (for example, boaters, cyclists, fishermen and residents) have no direct representation other than their local government representative or through direct contact with the YRA.

The day-to-day running of the YRA is handled by a four-person staff that includes a general director, biologist, inspector and secretary. Other specialists such as engineers, ecologists, hydrologists, urban planners, project managers and lawyers are outsourced.

The order creating the YRA required it to operate under a master plan approved by the Council. In other words, the YRA had first to create a master plan in order to conduct all of its activities, this plan having to be approved by the full council. The creation of the master plan involved numerous representatives of the public and took one year to develop, as described in Case Study 1. The master plan mandated an action plan for implementation that is still under way. The two-phased action plan dictated a long, public, open process for protection of the Yarqon for future generations. Both the statutory plan process and its action plan for implementation require networking with statutory bodies, NGOs and the general public, as explained below.

The Concept of Social Participation in the Yarqon River Authority

In hindsight, the process that was initiated to create the master plan in 1995 became one of the most important assets for the future operation of the YRA, in addition to the plan itself. The process and plan, together with the publicly open and proactive mode of operation that the YRA adopted, serve as a basis for its activities. Within the framework of public participation, as perceived by the YRA, there are two pathways that the public can utilize for involvement in decision-making processes. The first is the statutory pathway and the second through an informal open door policy:

1. The statutory pathway does not just mean abiding by legal dictates. It includes the degree of openness by which the decisions are made by the statutory bodies for decision-making. In other words, are the statutory stakeholders of the authority a rubber stamp or is there a real process that results in decisions that have a wide base of support? In Israel, the technical part of the planning process that takes place in the regional planning boards includes representatives from the Society for the Protection of

Nature (SPNI). SPNI is the largest independent membership organization in Israel, aimed at protecting Israel's environmental assets and the SPNI's representative is always very active. In effect, this means that even before the part of the planning process that is fully open to the public, public representatives participate in planning and decision-making processes.

Many government entities, national and local, have statutory power over the Yarqon watershed and without the YRA would intervene and manipulate the watershed, each for its own interest. This is the natural state of governance, where government governs along political borders and nature operates along geophysical lines. Therefore, to be effective watershed management must include within the basin authority all the government authorities who have statutory powers affecting the watershed. The members of the YRA represent sections of public organizations from local and national government, including elected officials. At first glance this may appear to be only a governance structure, but it can serve as a benchmark for examining public participation (Ackerman and Sawyer, 1972). In YRA meetings, each representative must act according to needs and policies having to do with the river and not according to interests of the organization that each represents. In rare instances, government representatives acted in the interest of their proctors, but, overall, the YRA directorate looks beyond local interests to serve the basin as a whole. This proves the general conclusion that the actual success of public participation in general and local stakeholders in particular, in watershed management issues, requires the ongoing commitment of local organizations and representatives from the central government for the benefit of the basin and its constituency. Funding, confidence building, and an atmosphere of flexibility and openness facilitate collaboration (Ferreyra and Beard, 2007; Leach et al, 2002). In addition the central government aids basin management through funding, courses, workshops, technical aid and information transfers.

The effectiveness of collaboration efforts depends on the need and the willingness of the stakeholders to participate, and on the ability of the watershed manager to prepare and lead the process, in a way that will generate effective collaboration to navigate between the parties and the issues (Smutko et al, 2002). Conflicts among local representatives are to be expected, as well as conflicts and power struggles due to the basic inequality and lack of a common basis other than the watershed and the myriad issues associated with it. This will complicate the interaction between local stakeholders even with the best intentions (Ferreyra and Beard, 2007).

In addition to national and local government representatives, the YRA council includes two representatives from the public, nominated by the Minister for Environmental Protection as well as the SPNI representative. The YRA's adoption of this mode of operation coincides with its structure, which sets the agenda in a way that makes it clear that the authority represents the river's needs in its overall aspects as opposed to the needs of any particular municipality along the river or of central government authority. One way to test the success of this method is to see if the statutory decision-making process does not meet with the contemporary polity, in the words of Ackerman and Sawyer (1972). How many decisions were appealed to a court of law? How many overturned by the legislature? How many approved by the national government? The answer is that there has been only one court case against the YRA in its history (instituted by an environmental NGO) and this case ended in support for the master plan and its implementation. In a country as litigious as Israel, the lack of appeal to a court serves as an indication of public support.

Due to the recognition of the scientific community's importance to the work of the YRA and the health of the river, the YRA made a principle decision to tie itself with the academic community. The YRA works with scientists on several levels:

1. It aids researchers, scientists and students doing studies on natural resources.
2. The YRA hires scientists as part of collaboration with the Ministry of Environmental Protection. The YRA initiated a call for papers, funded by the Ministry of Environmental Protection, for research that would increase scientific understanding related to decision-making necessary for achieving the goals of stream rehabilitation. This ties into the fact that scientists, especially in the field of ecology, have a high environmental awareness and in many cases are involved in public environmental struggles.

2 The second component of participation needs elaboration. When the YRA first began operation, participation focused on statutory stakeholders: national and local government authority representatives and institutions like manufacturers and farmers associations, but not the general public. There were two reasons for this:

1. The poor water quality and resulting condition of the river over so many years caused people to give up and accept the Yarqon as a place to avoid rather than visit.
2. There was no precedent of social participation in Israeli water management institutions and very few public interest groups existed beyond a couple of nature protection organizations.

Integrated watershed planning and management traditionally relate to the technical issues and aspects of drainage, flood mitigation and ecology. But the real force that drives governmental organizations to implement watershed management planning processes is the protection of the different human uses. As the human footprint encroaches into the river basin, the need to protect these new uses arises. In addition, as the environmental quality of urban rivers like the Yarqon improves, the number and variety of uses increase. Seen together, this increased complexity requires the involvement of local stakeholders in decision-making processes in order to achieve proficient and integrated management. Participation in the YRA began as a sounding board in order to test the quality of its decisions. This was particularly apparent in the development of the master plan for the river, as illustrated in Case Study 1. Subsequently, as the quality of the river and its environs improved, amenity usage by the public grew. This in turn spurred the development of interest groups who demanded further improvements from the YRA and the relevant local and national government authorities. This symbiotic three-way relationship between the YRA, government authorities and the public has resulted in a steady improvement to the Yarqon River's environmental quality and social value.

Since the purpose of public participation in the YRA is to create consensus or maximize support for particular actions through collaboration and consideration of different views, the YRA adopted several principles, as illustrated in the case studies described below:

1 Decisions are made in informal consultation with a maximum number of stakeholders. This is relatively easy with existing organizations such as rowing clubs, bicycle clubs, individuals and any organization that plans activities in the river's corridor. Representatives from the YRA also meet with representatives from environmental organizations several times a year, as well as being at YRA council meetings. To date, all of the initiatives concerning public participation have been the result of a specific request either from a particular organization or were promoted by the YRA. In addition, the YRA has either initiated meetings or responded to requests to debate planned actions such as flood mitigation and the management of riparian vegetation.

   An example is the issue of large eucalyptus trees that grow on the banks of the River. Many of the majestic trees are between 50 and 100 years old and are identified by the public as part of the natural Yarqon scenery. Unfortunately the trees are the cause of several problems. Besides not being indigenous, they interfere with proper development of indigenous species because of allellopathy. They contain compounds that inhibit bacterial activity, thus interfering with the food chain. Furthermore, eucalyptus trees shed large branches without warning, endangering people using the river corridor for recreation. Even general maintenance of the trees, for instance, trimming overhanging branches, can be problematic. Therefore YRA general policy targets the removal of eucalyptus trees from the immediate river banks. On the other hand, many people feel that cutting the trees down, even for ecological reasons or as part of flood mitigation actions, is unacceptable.

   In order to address these objections and create a consensus, the YRA initiated meetings and field trips with representatives from the two prominent nature societies for the purpose of explaining the need for the removal of the trees and replacement with indigenous species. This YRA policy is not carried out at one time, but rather in small steps. The law requires a permit from the Inspector of Forests to cut down each eucalyptus tree. A permit is issued only after the Inspector weighs both scientific and professional aspects together with any public objections. Due to the general and justified sensitivity of cutting down trees, discussions between the YRA and NGOs and even private citizens are held in order to reach a consensus, sometimes even concerning a single tree.

2. Information is constantly available to the public through the YRA website, the press, verbal or written communication with local residents and signs explaining specific projects that are in progress. Educational projects involving local schools develop awareness of the Yarqon and its environment. Students conduct study assignments in different areas of the river in the fields of ecology, flood mitigation, urban planning and architecture and participate in projects such as tree planting, archeological excavations and clean-up campaigns.
3. Specific inquiries by the public are promptly answered by mail, email or telephone. Each year there are close to 100 requests for information on various issues concerning the river, at least as perceived by the public. Inquiries come from concerned citizens, students at all levels, professionals and laypeople with suggestions for improvements, as well as NGOs and other groups. Some of the queries relate to river issues such as vegetation rehabilitation, maintaining habitats for songbirds and kingfishers and the issue of eucalyptus trees. Most of the queries concern non-river specific issues such as the bicycle paths, pedestrian bridges and requests for specific information.

In recent years river quality has substantially improved, as have the services provided.The YRA put special emphasis on improving habitat conditions and increasing recreational activities. These changes include:

- rehabilitation of trees and vegetation along the river corridor;
- improved water quality;
- increased water quantity;
- developing an ongoing maintenance routine for trash collection, pathway upkeep and river clean-up and dredging.

The opportunity for making improvements increases if the achievements and advantages of integrative management, including collaboration with stakeholders, are identified, measured and expressed in terms of gains to the public. These gains include increases in public assets such as parks and recreational areas, as well as the expected increase in property values adjacent to rivers (Imperial and Hennessey, 2000). When the results of the work done in the Yarqon actually improved the conditions in and around the river in a manner visible to the public, negative pressures from the public dropped significantly, while positive pressures increased. Many of these public concerns focus on issues not exclusively within the authority of the YRA, chiefly the use and expansion of open spaces along the river corridor. This has extended the focus of public participation in the Yarqon River to the local municipalities, the National Park Service and other governmental authorities.

The following case studies demonstrate the public's role in the decision-making process and the extent of its influence on the improvements mentioned. Case Studies 1 and 2 analyse public participation in the YRA planning process. Case Studies 3 and 4 illustrate YRA interaction with boaters and farmers, two important stakeholder groups.

Case Study 1: Creation of the Master Plan

The master plan process was designed so that all of the members of the YRA and local stakeholders participate in the process. In the initial stage, the YRA's Board of Directors formulated the terms of reference (TOR) for preparing the master plan, thus creating a wide base of agreement among the members. The subsequent preparation of the plan itself required putting together a team that would study the river in depth in order to create a holistic overview of the river system and to understand how to integrate the factors so that the rehabilitation would answer the needs of both the ecosystem and the people. As planning progressed, consultations with local stakeholders provided feedback and direction for the planners.

The TOR was explicit in defining the relationship between the planning team and the steering committee. The planning team was instructed to consult with the members of the Board of Directors, who were part of the steering committee, at specific stages during the planning process. This arrangement served to familiarize the Board of Directors with the structural properties of the river and its environment, with the river's hydrological and ecological needs and to maximize their involvement in the planning process.

The goals that were set in the master plan reflect the tasks designated for the YRA, mainly to rehabilitate the river and make it suitable for recreational activity (Rahamimoff, 1996):

- To create and secure a green lung for the most populated region in Israel, providing a green oasis in the midst of Israel's largest urban area, as well as naturally cleaning the air of dust and toxins.
- To change the current public attitude toward the Yarqon from a backwater to an urban showcase.
- To rehabilitate the river's ecosystem and improve the water quality by solving the problems created by the discharge of sewerage and low quality effluent.
- To provide appropriate solutions for river regulation and flood hazard reduction.
- To suggest economic initiatives for river use compatible with principles of sustainable development.
- To improve the environmental and aesthetic values of the river and its adjacent corridor.

In effect, the intensive process, concluded in one year, became one of the most important interactions between the planning discipline, represented by city planners and decision-makers on all levels and between hydrologists and ecologists at the other end of the spectrum. This is true for all of the stakeholders at the level of the authority, the watershed and on the national level. The process of preparing the plan served as a workshop for educating decision-makers and all those involved including mayors, city engineers and other local and government officials, about the river‘s structure and needs. The regional planners, who operate according to planning and building law and who are key position holders in all planning processes, also adopted the principles described in the master plan and started to act according to them despite the fact that it is only a precursor to outline plans that are the statutory basis for detailed plans. This enabled the YRA to proceed with decision-making regarding the actions needed to initiate the river's rehabilitation.

After the YRA Board of Directors and Council approved the master plan, it was submitted to and approved by the Ministerial Committee for the Environment. This approval received the status of a Government Cabinet Directive in April 1996 and since then the master plan has served as a guideline document for the YRA's activities, which are presently focused on its implementation.

It is difficult to assess the direct contribution of non-YRA member stakeholders in developing the master plan because initially there were no meetings that were designed to elicit remarks. As this is the first master plan created for a river in Israel, when work began there were very few organized stakeholder groups that could be identified save for the farmers who irrigate with water pumped out from the Yarqon and the SPNI. Today it is possible to identify several groups of stakeholders as will be shown in the subsequent case studies and their participation is closer to the classical interpretation of public participation.

One particular early example of interaction with stakeholders was the YRA's work with cyclists. The master plan for the Yarqon created 'fingers', access routes from built-up areas to the open spaces and parks along the river, in order to encourage recreation and commuting by bicycle. Prior to the approval of the plan, the YRA produced a map of the available mountain bike paths around the Yarqon and distributed them, free of charge. The influx of large numbers of cyclists created a major stakeholder group promoting protection of the Yarqon and its environs, thus increasing public pressure on policy makers to rehabilitate the river.

Case Study 2: Creation of a Comprehensive Strategic Plan for Tourism

Preparing the river and its corridor for recreation is central to the YRA's activities and is an integral part of the master plan. However, it would seem that at this stage of the development of watershed management in the Yarqon, tourism is a necessary by-product but not a direct component of watershed management. In this context, the creation of a tourism plan may be ahead of its time but it is important to set such processes into motion, parallel to others.

The geographical position of the Yarqon, flowing through the most heavily populated area in Israel, serves as an incentive to provide quality open spaces and a substantial boost in the quality of life for over two million people who live within a half an hour radius of the river. The YRA initiated the creation of a tourism plan in collaboration with local communities with the aim of developing small businesses and jobs that relate to the activities along the river. Recently there has been a rapid increase in local eco-tourism, particularly in the rural areas of Israel. The idea is to transfer this idea to the urban setting. In order to do this, the YRA hired a specialist in eco-tourism and chose one particular regional council in order to develop a suitable model with the local residents. The head of the regional council supported the entire project that began with a series of workshops and lectures about the Yarqon River and the plans for its rehabilitation and development. These subjects were introduced on the local scale as well as on the watershed scale. Participants were asked to fill out questionnaires and to submit suggestions regarding the suitable concept for local economic endeavours as well as for possible recreational activities. The results were analysed and incorporated into the strategic plan that was the end product of this stage. Experts on starting small businesses were invited to talk to the attendants in the workshop and field trips were taken to areas in Israel where there are similar projects. The implementation of this has been slow mainly due to difficulties arising from the convoluted relationships between local authorities, who would like to see implementation proceed at a fast pace, as opposed to some national government agencies and NGOs who are slow to make decisions and wary of development even when considering eco-tourism projects. These conflicting approaches dampened the participants' initial enthusiasm, and after the development stage ended, the rate of the public's activity declined. Still, three years after the project ended, some of the initiatives are in the process of implementation.

Case Study 3: Interaction with Farmers

The farmers who grow crops and orchards along the river and especially those who extract water from it for irrigation are the stakeholders with the longest history of continuous use along the river. Even during periods of severe drought, the farmers' pressure forced the Water Commissioner to discharge spring water into the river for the farmers' benefit, thus preventing it from becoming a dry riverbed. The YRA recognized that these stakeholders had power, influence and a historical connection to the river. It was imperative to clarify to them that the YRA considers agriculture an important factor in preserving the heritage of the area and creating the desired atmosphere of open spaces within the urban setting, on the condition that it does not harm the river. Meetings with the farmers were held at the local branch of the Ministry of Agriculture and the Water Commissioner and his staff were involved.

Representatives from the YRA explained the master plan to the farmers and responded to the issues raised. Since these initial meetings with the farmers in 1995, there has been a generally positive relationship between the farmers and the YRA, with the following results:

- Changes were made to the weir system that enabled water diversion by building permanent weirs that serve both the farmers and the river. The new weirs were planned to perform ecological functions such as water aeration and habitat enhancement as well as to provide the necessary head for pumping. In addition, the weirs serve as passages from one side of the river to the other. This is important since there are not enough bridges that can serve recreational uses.
- In response to the YRA's request to minimize pollution from diesel fuel, pesticides and empty pesticide containers, farmers installed pollution prevention structures and cooperated with monitoring the use of pesticides, whether by air or land application. This has resulted in reduced pesticide residues reaching the river.
- After winter floods, the YRA removes debris from the river. At the request of the farmers, the YRA also removes sediment from the farmers' pumping stations so that they can begin irrigation when needed.
- The YRA provides water quality data to the farmers, a requirement for obtaining permits from the Ministry of Health to irrigate crops with river water.

It should be noted that not all farmers responded immediately to the requests made by the YRA. Some maintained an atmosphere of suspicion and in one instance, one of the weirs was purposely damaged in order to enable increased pumping. Even though this was one isolated incident, it was an important lesson to the YRA, which realized that its priorities are not always those of the farmers.

In another instance a conflict arose between the YRA and one particular farmer who for years had been issued an abstraction permit from the Water Authority at a point in the upper reaches of the Yarqon. The Water Commissioner decided to replace irrigation water pumped directly from the upper section of the river with water from the national water supply network. This decision resulted in a lawsuit brought by the farmer against the Water Commissioner. The YRA voluntarily joined the case as an additional defendant in order to work towards ending all pumping from the upper reaches. With the intervention of the YRA, the farmer dropped his demand for pumping from the river. It should be noted that the Yarqon River rehabilitation scheme calls for ending all pumping from the river in the future and replacing it with water from other sources. It is expected that this will create additional conflict, but the water situation in Israel cannot compare to the days when irrigation from the river was permitted. Under the final master plan concept, water flowing in the Yarqon will be pumped out for agricultural use only at the point where the river enters the Mediterranean. This will enable a double use of the river, once for amenity uses and again for irrigation, but irrigation will not be permitted directly from the river.

Partnerships between local stakeholders and government basin authorities need to be based on a win–win situation or at least on win–no lose. This does not mean that the partners, stakeholders and the organization managing the watershed agree on all subjects, nor that there is total agreement on projects central to watershed management (Imperial and Hennessey, 2000), but the start of discussions requires a uniform definition of the issues before the organizations and a division of authority and their degree of responsibility.

Case Study 4: Interaction with Boaters

There has always been boating on the Yarqon River, but activity increased in correlation with improved water quality. Measurements have shown faecal coliform bacterial counts dropped from 10^5 and 10^6 a decade ago to 10^2 and 10^3 today. The improvement brings bacterial counts to a range that is acceptable for rowing activities. The daily number of rowers has increased to hundreds and the number of active rowing clubs has grown from three to five. This led the Ministry of Health to demand that a bacterial quality standard be set for the river to protect the boating public. By law, however, river quality is the province of the Ministry of Environmental Protection and boating on the Yarqon, the province of the YRA. In addition, the Ministry of Transportation holds responsibility for transportation on watercourses. The YRA decided that due to the rising demand for rowing it would have to take the lead in drafting an ordinance to regulate the activity. Due to the number of stakeholders involved, however, drafting a simple ordinance became a major effort. The Ministry of Health has no responsibility for boating but it does set limits for public exposure to health hazards. The Ministry of Environmental Protection is responsible for setting water quality in rivers, and the Ministry of Transportation sets standards for boating. The rowing public, however, has been rowing in the Yarqon for over 50 years without a sign of sickness incurred as a result of rowing; their pressure to allow rowing year round forced the regulators to work together to solve this convoluted issue.

The process that took place began with a lobbying effort that was initiated by a rowing centre that, in coordination with the YRA, brought the issue to the Israeli Parliament's Environmental Committee. Members of the rowing clubs were invited to the deliberation in the parliament and they were allowed to state their case. Subsequently a committee was formed to set water quality criteria, balancing health considerations with boaters' demands.

At that point, the YRA began the process of drafting an ordinance. As the statutory process proceeded apace, the YRA recognized that it must encourage public debate in parallel. As mentioned previously, the management of the YRA recognized that adaptive management requires continual learning. This means that stakeholders with experience in boating must be heavily involved in the policy decision for rowing on the Yarqon River. In essence two types of stakeholders were engaged: representatives of the Ministry of Transportation skilled in boating regulation and the rowers themselves, including the commercial concessions that rent small boats.

In 2007 the YRA initiated contacts with the clubs and the commercial concessions in order to explain the process and to hear their reactions. The process began with a meeting held at the main boating club on the river. Present were representatives of clubs, commercial interests and individual boaters. A lively debate developed, in the spirit of adaptive management called bounded conflict, where conflict is welcome but confined to limited objectives. The wealth of knowledge in the room combined with the spirit of the rowers imbued the management of the YRA with the drive to pursue criteria advantageous to continued rowing. Knowledge gained from other rivers together with the active cooperation of the Yarqon rowing clubs led the management team to adopt a flexible rowing programme. No one would be forbidden from rowing; instead flags would be raised when water quality was below a certain bacterial level and rowers could row at their own risk. When the ordinance takes effect it will be studied, water quality will be monitored and studies will look at health effects among the rowers. Over time, quality levels and flag flying will be adapted according to study results.

Conclusions

Although social participation in the YRA seems quite tame when seen from afar, in many ways it is leading the way for participation in Israeli water management institutions. In its early years, the YRA used participation as a sounding board in order to test its decisions in developing the Yarqon River Master Plan. The main focus for participation was through representatives in the YRA's council and directorate.

As the first river authority in Israel, the YRA had no choice but to develop its own constituency. For some, this would mean working strictly within the statutory framework. Yet for the YRA, this would not include the silent public, those that benefit directly from the Yarqon and its environs, but do not get involved in public debate. These stakeholders had to be found and actually merged into a constituency, for example, boaters clubs and the farmers. There was no group of farmers created to deal with problems along the Yarqon until the YRA asked to meet with such a group, nor with the boaters and the bikers. In short, the YRA created participation forums to contribute to the goals of adaptive management of learning and adapting changing streams to changing streams of thought.

As these groups formed and grew, they began putting pressure on the YRA and the local and national authorities to further improve the river and its amenity uses. Today the YRA has an informal open-door policy that encourages the public to voice their opinions, and a management policy that is constantly changing in response to new information and ideas. While in many cases the classic mode of social participation is at most a technical issue or part of a checklist, in-depth and ongoing participation has become a deciding factor in the success of implementing policies in the YRA. What can be learned from the Yarqon experience is that principles of basin management are sound. Implementation of the principles, however, requires trial and error, with a learning curve built in:

> Adaptive management is field science; its laboratory is not a controlled setting, but a noisy, changing world of human actions and natural fluctuations. (Lee, 1993, p69)

No better sounding board for success exists than the public and, therefore, encouraging participation in the management process ensures that mistakes will be discovered; the response of management to those mistakes ensures improvement and further participation.

Channelling public debate from turmoil to effective management requires first and foremost an understanding of democratic institutions. It further requires a humility not often found in political institutions. More importantly there must be an understanding that no single organization, even a basin authority, can control the basin or manage a river and its environs. All rivers in Israel flow through local government entities. This means that getting the public involved must begin with the mayors of local and regional councils, who must recognize this fact and act accordingly. They must initiate processes by which the river will turn into an asset for public amenities and not public blight. Without a proper framework for operation, understanding adaptive management practices and public participation would not have occurred.

The involvement of so many interests in the welfare of the river has provided the YRA with the power to demand improved water quality and quantity from the Director of the National Water Authority and the relevant government ministries. The main result has been a change from a downward spiral of increasing pollution and neglect to ever-increasing improvements in river quality and public usage. For example, the National Water Authority is now allowing 3.6 million cubic metres a year of freshwater to flow down the river from the Rosh Ha'ayin springs. The Yarqon River and its environs are now among the most beautiful recreation sites in Israel, visited annually by millions of local residents and tourists. It was transformed from an embarrassment to a showcase for recreational use of rivers and streams in Israel within less than ten years of operation. The experience in managing the Yarqon River shows that integrated water resource management in a watershed, coupled with adaptive management practices and strong social participation at crucial stages, speeds the process of stream rehabilitation.

Notes

1 The Majelle:

- 'Waters flowing in the bowels of the earth are not the property of any man.' §1235
- 'Oceans and large lakes are ownerless.' §1237
- 'The many rivers that are no man's property and are not separated into rivulets, i.e. they don't enter into channels that are the property of a recognized group of people, are ownerless ...' §1238

2 Article 46 of the 1922 King's Order in Council proclaimed that:

> The jurisdiction of the Civil Courts shall be exercised in accordance with the Ottoman Laws in force in Palestine on 1st November 1914, and such Ottoman Laws as have been or may be declared in force by Public Notice, and such Orders in Council, Ordinances and Regulations as are in force in Palestine at the date of the commencement of this Order, or may hereafter be applied or enacted.

These laws were subject to the enactments of the new legislator, the British High Commissioner, to the Orders-in-Council issuing from the King of Great Britain and to Acts of the British Parliament.

3 Water Law, Article 124s. As of 2006, the Water Commissioner is now called the Director of the Government Authority for Water and Sewage (Director of the Water Authority).

References

Ackerman, B. A. and Sawyer, Jr, J. A. (1972) 'Scientific factfinding and rational decisionmaking along the Delaware River', *University of Pennsylvania Law Review*, vol 120, no 3, pp419–503

Drainage and Flood Control Law 1957, 12 LSI 5

Ferreyra, C. and Beard, P. (2007) 'Participatory evaluation of collaborative and integrated water management: insights from the field', *Journal of Environmental Planning and Management*, vol 50, no 2, pp271–296

Imperial, M. T. and Hennessey, T. (2000) 'Environmental governance in watersheds: the role of collaboration', *8th Biennial Conference of the International Association for the Study of Common Property (IASCP)*, Indiana University, Bloomington, IA

Leach, W. D., Pelkey, N. W. and Sabatier, P. A. (2002) 'Stakeholder partnerships as collaborative policymaking: evaluation criteria applied to watershed management in California and Washington', *Journal of Policy Analysis and Management*, vol 21, no 4, pp645–670

Lee, K. N. (1993) *Compass and Gyroscope*, Island Press, Washington, DC

Ortal, R. (1998) 'Marine and wetland conservation in Israel', *Nature and National Park Protection Authority Internal Report,* Jerusalem, Israel

Ostrom, E. (1990) *Governing the Commons: The Evolution of Institutions for Collective Action*, Cambridge University Press, Cambridge

Rahamimoff, A. (1996) *Master Plan for the Yarqon River*, Yarqon River Authority, Ramat Gan

Sinclair, I. C. (1985) 'Institutional and legal aspects of water management', International Seminar on Institutional and Legal Aspects of Water Management, Madrid, Spain, unpublished

Sinclair, I. C. (1994) *The Water Industry in England and Wales*, International Association for Water Law – Italian Section Proceedings 15–16 March

Smutko, L. S., Klimek, S., Perrin, C. A. and Danielson, L. (2002) 'Involving watershed stakeholders: an issue-tribute approach to determine willingness and need', *Journal of the American Water Resources Association*, vol 38, no 4, pp995–1006

Water Drilling Control Law 1955, 9 LSI 88

Water Law 1959, 13 LSI 173

Water Measurement Law 1955, 9 LSI 85

Part V

Participation and the Politics of Governance

12

Water Rights and Rule-Making Justice as Fruits of Social Struggle in the Ecuadorian Andes

Rutgerd Boelens

Introduction

In these times of growing scarcity and competition regarding access to water resources in the Andean region, water control – and in particular water rights – have become pivotal issues in the day-to-day struggle of local communities and indigenous and peasant organizations to defend their livelihoods and secure their future. Water in Andean communities is often a very powerful resource. Apart from being a foundation for productive, social and religious practice and local identity, the particular collective nature of water almost by definition forces people to build strong organizations: in most cases, the resource can be managed only by means of day-to-day collective action. Collaboration, instead of competition, is the only way to survive and secure water rights in this extremely adverse environment.

Seemingly, current water policy proposals coincide with this local challenge. Most water policy and irrigation or watershed intervention proposals in the region include strategies such as user participation, decentralization and transferring management to local forms of government. At a first glance, these new orientations constitute important steps toward empowering water user collectives, by granting them greater autonomy, decision-making power and water tenure security, and by respecting local water resource rules and rights according to their own needs and capabilities. However, radical state downsizing in the Andean countries' water resources administration during the last decade is likely to be a more important reason for the 'participatory' policies that were installed in many places. These include the abandoning of fundamental public tasks and cutting back on public spending for water management (Cremers et al, 2005; Guevara Gil, 2006; Hendriks, 2006; Perreault, 2006). The facilitation of private sector water control appears to be another major element of the political agenda. As in most regions of the water world, social participation is, in many instances, politically conceptualized as reinforcing state water control at local management levels. This may be through decentralized local government agencies, presumably autonomous but in fact state-dependent user organizations, and participation of private water enterprises, commonly through the so-called public–private partnerships (Assies, 2003, 2006; Bauer, 1998; Gleick et al, 2002; Hendriks, 2006; Meinzen-Dick and Pradhan, 2005; Mollard and Vargas, 2005; Swyngedouw, 2003; van der Ploeg, 2006; Zwarteveen, 2006). The latter – the word is telling – typically work for but not through and with local common property institutions. New alliances of the public with the private often profoundly exclude the commons from water policy design and project decision-making.

Any serious consideration of social participation faces the need to understand the cultural and political rationality behind the local forms and norms of water control and the struggles they entail, besides addressing power differentials. In the Andean region, land and water are among the most basic, essential elements of peasant households' livelihood strategies. And indeed, the extremely skewed distribution of these land and water resources has been and continues to be a source of recurrent frictions and struggles. Water rights stand central in the local and national conflicts. Here, struggles over water rights do not just refer or restrict themselves to questions of material resource distribution. The multi layered local notions of water rights in Andean indigenous and peasant communities involve rights of access to water and system facilities, but also recognize that water rights are to be considered as authorized claims to control decision-making about water management (Boelens and Doornbos, 2001).

As a result, as this chapter will argue, social participation in water development affairs involves a struggle among multiple actors to establish water resource distribution, the contents of rules and rights, the legitimacy of authorities and the discourses that defend these three key aspects of water control. These notions are closely related to power, identity and cultural politics in the Andes. In fact, the struggle over water rights is simultaneously a battle over the material control of water control systems and over the right to culturally define, politically organize and discursively shape and control their existence: a battle to exist as water control communities. This chapter, therefore, also challenges the notions of participation as either management decentralization, inclusion of local beneficiaries in national water frameworks or water user consultation and involvement in state or development agency projects. I argue that it is not coincidental that the words 'participation' and 'inclusion' are fundamental pillars of neoliberal water ontologies and policies, aiming for the inclusion and disciplining of those who are, in neoliberal terms, not sufficiently included in the market system (Boelens, 2009).[1]

All water knowledge and water truth are situated immanent to the networks in which they operate and develop, and my position in this paper reflects this. Researchers cannot observe norms and rights in action without interpreting and, thus, using their defining power and network relations. My focus is on how the dynamic working mechanisms and resilience capacity of Andean collective water rights repertoires are threatened, thereby endangering the self-sustained reproduction of the very water management collectives themselves. It concentrates on the position of those many user groups with less influence in the Andean water world who find social and political-economic shelter and livelihood security precisely within collective water rights systems.

The chapter starts with an analysis of local irrigation water control in peasant and indigenous community systems; the diversity and embeddedness of rules, rights and rationalities and the ways water rights are dynamically related to local water organization and infrastructure. The next section takes the case of Licto, Chimborazo Province, Ecuador, as an illustration. User communities, forced by the national state agency to accept its rules and regulations, challenged this framework by mobilizing its member families and constructing their own water rights and organizations. When the productive and effective practices of farmer-managed water rights and rules were not able to convince the bureaucracy, again, social mobilization was (and remains) necessary to explain the farmers' concept of social participation to the state agency. Throughout the years, decentralization (and privatization) of water control in the Andean countries, inspired by neo liberal policies, has led to bizarre consequences for local communities, who see the need to mobilize massively in order to challenge the products of participatory and decentralized water governance institutions. Rather than depending on and waiting for development and state agencies, it is precisely the water user organizations who have to monitor the water development process in order to guarantee social justice. As the final section concludes, these water user organizations and federations seek to invert the notion of social participation, to make the state and development agencies participate in their water projects and programmes, and not the other way round.

The Art of Local Water Control and the Creation of Property

Water law and policy-making face particularly difficult constraints in the Andean region, owing to the existence of a huge variety of water use systems in a context that is extremely diverse. All countries have great differences in ecological and climatic regions as well as in institutional and technological environments, and also show impressive, historically evolved diversities with respect to organizational and political structures, cultural backgrounds and production rationalities. Socio-economic stratification and the large variety of management forms create even greater complexity, since rules and regulatory mechanisms must necessarily adjust to the actors and issues at stake.

Water rights are an expression of agreement about the legitimacy of the right claimed by holders. To be functional, they need to be certified by an authority that is able to enforce the rights considered legitimate by the users, and users need to refer to this rights framework as orienting their behaviour and thus build it into their actual social relationships (Beccar et al, 2002). It does not always mean that this authority will be recognized by all. Because most Andean zones have more than one authority – representing different (socio)legal systems or frameworks, whether at the local or national level – there are often diverging positions as to recognizing the legitimacy of different claims to water use (Boelens and Hoogendam, 2002). Indeed, a diversity of water rights exists in conditions where rules and principles of different origin and legitimacy co-exist and are encountered, in one and the same water use system, community, watershed or territory. In general, state law as well as local law, ancestral rights systems, the rules and regulations of water development projects and other frameworks interact (Benda-Beckmann et al, 1998; Roth et al, 2005). Such different rights regimes can complement, overlap or even contradict. As a direct consequence, legitimate authority in Andean water management is not restricted to the state agencies only, nor do rights and rules refer to only those emanated by state law (Roth et al, 2005; Zwarteveen et al, 2005). And it comes as no surprise therefore that, behind the struggle for distributing water and water rights in the Andean countries, an intensive battle is going on in order to establish both the contents of water rights and the authorities with decision-making powers about these rights. This section takes a closer look at some aspects of Andean community water control, since comprehending their local water uses and rights is a necessary first step to build pluralist policies.

Local water control and Andean community organizations

The art of community irrigation is not confined to simply planning optimal amounts of water and scheduling its distribution, defining methods of application and building conduction canals. At the level of irrigation systems or watersheds, beyond the analysis of certain individual practices or techniques in isolation, it is above all water control – in its socio-technical sense and situated in institutional landscapes, cultural frameworks and power structures – that merits central attention. Water users who take part in planning, implementing, operating and maintaining farmer-managed irrigation systems share, beyond the technical and production-related demand for water, a social demand. Water is the liquid that makes their plants grow, but it is also the fuel for the zone's organizational engine, the blood in the veins of the rural livelihood system and the heart of survival and co-existence of many peasant communities. As Jaubert de Passa (1846) observed long ago, irrigation is a practice as miraculous in its effects as vulgar in its means. Being a fundamental pillar of many local communities' production and livelihood systems, control over their own water reality, under their own norms and decision-making power, is a major challenge for local communities.

Direct control and accountability are also common features inside the system: users can hold local leaders responsible for their actions, and control mechanisms and penalties in cases of abuse are applied in a relatively straightforward manner, compared to accountability relations with bureaucrats, politicians, development agents and others. In community water meetings and general assemblies this often results in fierce, open critiques as well as apparently dramatic debates and quarrels – particularly when dealing with issues of water rights and obligations. Also, members as co-owners of the system commonly co-decide aspects of its management. This includes self-mobilization and direct action on the basis of social control, collective monitoring and collectively elected and rotating leadership as characteristics of the involvement of all recognized members in water control affairs. Leadership turnover (yearly or every other year) and the sharing of positions among most water community members, however, cannot be taken as proof that social and political differentiation are in check. First, generally the same group of leaders tends to surface when crucial decisions are to be made. These election processes tend to manifest local structures of prestige, re-distributive power and local legitimacy (Mayer, 2002; Salman and Zoomers, 2003). Second, social participation in system control does not necessarily mean that all have the same class, gender or ethnic background. Class and ethnic differences, for example, may be an obstacle to strong user organizations, but it is crucial to also recognize the organizing potential of heterogeneity. Heterogeneity can reinforce the forms of cooperation based on interdependence and complementation of capacities and resources. However, relatively large economic power differences commonly work as impediments to sustainable organization, particularly in those systems where the distribution of benefits and burdens is considered to be inequitable, or when market integration erodes traditional relations of (symmetrical and asymmetric) exchange (Gelles, 1998; Mayer, 2002; Pacari, 1998).

An adequate, sufficiently consensus-based definition of collective and individual water rights[2] is essential to make a user-managed irrigation system work. Apart from the issue of who will have water rights and who will not, other basic questions that define water rights are:

- What is the local conceptualization and the precise facets or contents of water rights in the particular case?
- What mechanisms are collectively recognized as legitimate to obtain and maintain water rights?
- How will water benefits be divided?
- How will contributions and disadvantages be divided?
- Who will be entitled to participate in decisions about management, acceptance of new members and any changes in future system ownership?
- How, and with what results, can different users activate and materialize their water rights in practice? (Beccar et al, 2002)

Unlike government-managed irrigation systems in the Andes, where state authorities formally establish and enforce rules and specialized managers (often at various levels) are supposed to carry out most management tasks, in farmer-managed systems the roles of water authority, water managers and water users are much more integrated. Despite the above heterogeneity, in most of these small-scale systems it is common to see that water users and water authorities share a similar social and cultural background, being members of the same kinship or community organization; also both sets of roles and responsibilities circulate among the same group – issues that are entirely different from bureaucratically managed systems in the region. Generally, (rotating) water authorities are simultaneously water users. Moreover, although formal lines of command and conduct are often presented to the outside world, internally the norms, tasks and penalties tend to be clear but also much more flexible, as long as there is conformity to the community (re)production objectives. Informality and local social relationships play a decisive role. Operational rules and tasks may be divided into the following main categories, which are coloured and shaped according to time, space and context. These tasks relate to:

- regulation and authorization;
- operational water management;
- internal organization;
- (re)constructing infrastructure;
- mobilizing and administering resources;
- alliance-building;
- networking and ritual tasks, according to the system's embeddedness in its supernatural environment. (Boelens and Hoogendam, 2002)

The latter tasks concern activities related to maintaining and reproducing reciprocal relationships with deities in order to secure adequate water delivery and productive sufficiency.

There are as many ways to organize for water allocation and distribution as there are water uses, types of water (Hendriks, 2006), water user groups and cultures and organizational levels within Andean society, from the family level to the inter-family group, community, ethnic group and watershed network. As a result, there are various entities within a community to allocate, distribute and watch the water and each community member commonly belongs to several of them. Here informal organizational patterns and agreements play a highly important role, confirming the need to go beyond just formal user associations (board, committee or others) and their formalized rules and practices.

The highlands experience unpredictable climates and frequent drought periods, unstable geophysical and agroecological conditions causing constant canal breakdowns and erosion hazards, as well as adverse power conditions and powerful encroachment interests. Here, irrigation is grounded in mutual dependence and intrinsic obligations for intensive cooperation among users more than almost any other economic activity. Irrigation forces people to operate collectively every day. It is impossible to manage a system in this context with just a group of individuals; intra- and inter-community collaboration and their respective collective agreements are indispensable prerequisites to ensure access to water. Because of this obligatory reciprocity required to operate and sustain the system and because of the common ownership of the system in which the rights of each user are created, recreated and embedded (see below), users identify with the system and relate to each other. This is at the heart of collective action in water control and, jointly with the historical struggle for water, collective defence of community authority and development of the community's own rules and customs, it reinforces context-specific hydraulic identities. As I will show below, far from being just an ideological construct based on presumed Andean and indigenous solidarity, the material creation of this hydraulic property that links individual action and property to the collective water property owners group (infrastructure, rights and organization) is largely based on a logic of defence and reproduction of water community in the harsh Andean geographical and political context (Boelens, 2008).

The creation and re-creation of the rights–organization–infrastructure triangle

As the foregoing sections make clear, water rights embody social relations among the actors involved: water rights are intimately linked to the existing social and cultural organization and relations of authority and power. In day-to-day political encounters, diverse interest groups in water management – both users and non-users[3] – challenge the construction, application and reproduction of rights and bring the different rules and powers of their socio-legal systems into interaction and confrontation. Rather than participatory policies or user-oriented strategies, it is this fact that gives social participation in water control its concrete shape. This interaction among interest groups relates both to fierce contestations and to forms of active collaboration. Consequently, both these aspects importantly influence water rights definition and distribution. Here I will concentrate on one important element of this Andean water rights development process.

Before and during system construction, and during system use, water user collectives develop their local rights framework: a set of norms that guide system creation; administration and maintenance; allocation and utilization of water; and relations among users. Thereby, the mechanism of user investment for acquiring rights is often extremely important for Andean communities managing their own irrigation systems or doing so jointly.[4] The development of irrigation infrastructure simultaneously establishes property relationships among the system creators, as Coward (1983) argued. By investing in the facilities, users create their hydraulic property, a common ownership of the system, which is the factor bonding irrigators together and driving their collective action. This forms the foundation that guarantees realization of the different operation and management activities required by a user-managed system. The mechanism also guarantees peasant and indigenous communities, as collective bodies, that they will have effective control over the development and application of their own norms for managing their system (Boelens and Hoogendam, 2002).

So, the appropriation of families' individual rights directly coheres with the appropriation of the group's collective rights, and these water access and control rights are directly connected to collective infrastructure and underlie collective system management. In many indigenous systems, families obtain irrigation rights not only through their own contemporary investment in building the collective facilities, but also as an inheritance of the investment made by their ancestors. Moreover, for most indigenous communities this act of ancient water system creation is entwined with their myths of origin and belonging, and so relate to important sources of water culture and current identity formation; in part, water rights are a loan borrowed from the deities. This is confirmed through rituals rooted in irrigation practice.

After generating rights, users must consolidate them. They do so by fulfilling their obligations within the irrigation system, which generally also takes the form of user investment. Participation in collective work, payment of dues, attendance at meetings, and the like are important obligations, both to conserve one's rights and to keep the irrigation system itself working.[5] Thus, the conservation of water rights plays a key role in effective irrigation system management: maintenance of the system means maintenance of water rights.[6] In other words, users' investment in construction (or rehabilitation) of the system is grounded in the logic that creates individual and collective water rights, while their investment in maintenance reaffirms and recreates them. The precise quantitative and qualitative relationship between investment contributions with rights creation and re-creation, and the perception of whether this is equitable, usually differs per system. This is a pivotal issue for local decision-making and, not uncommonly, a matter of intensive struggle and negotiation, internally as well as with third parties (Beccar et al, 2002).

Despite the obvious influence of infrastructure modification on existing water rights, many development projects do not explicitly address it – let alone understand the relationship between prior and new user investments and collective hydraulic property creation. This denial of local systems' foundations often give rise to the following situation: local contributions are called for, but without stating that these inputs are individual investments in a co-owned system (which by definition requires prior clarification about the relationship between contributions and benefits for each user, as well as a common understanding about the ownership of the system and the collective water rights). When this disorganized investment is over, no solid foundations are left for the users to organize water distribution, much less maintain their channels as collective property (Beccar et al, 2002; Gerbrandy and Hoogendam, 1998).

The confusing of existing property relations lies at the heart of many intervention failures of development (Gerbrandy and Hoogendam, 1998; Urteaga and Boelens, 2006). That is why, when one visits a post-project irrigation system, it is not unusual to find the users' group still bitterly arguing about conditions for access to water and the corresponding maintenance obligations. Particularly problematic are the many cases in which external agencies intervene and invest in existing irrigation facilities, founded on local rights agreements and based on prior user investment. These new, agency-led investments often destroy existing collective and individual property rights and, thereby, the necessary collective action to sustain the system (Beccar et al, 2002; Coward, 1983; Gerbrandy and Hoogendam, 1998).

Common notions in most irrigation development programmes – paternalistic or participatory – present social participation and local contributions as a mechanism to create a feeling of ownership in future beneficiaries. In contrast, peasant and indigenous users in a community-managed system do not claim *a sense* of ownership but *factual relations of ownership or co-ownership*, establishing precise access and control rights (Beccar et al, 2002).[7] This also shows the political nature of water rights. Water rights are the object of struggle and the product of power relations. Both their contents and their distribution in society and the way they dynamically adapt to new situations, reflect the prevailing power structures and the way they are contested.

Struggles over Water Rights and Legitimacy in the Ecuadorian Highlands

Although fundamental class and ethnic contradictions form the background in the Andean region, classically they have largely been ignored in official water policy and law-making processes and in agency-led irrigation interventions. Historical inequities created by colonial encroachment and post-colonial hacienda-based rights monopolization have never been tackled effectively, rather, they have worsened. For instance, in Ecuador, land reform that started in the 1970s has had poor outcomes, reconfirming historical concentration of access and control rights over land and water in the hands of bigger landholders and forcing the indigenous peasantry to occupy the marginal lands. Consequently, water rights distribution is enormously skewed towards benefiting a small minority of powerful enterprises and landlords. According to Galárraga-Sánchez, 88 per cent of the irrigators who are small farmers (*minifundistas*) have access rights to only in between 6 and 20 per cent of the available irrigation water, while the big landlords (who comprise 1–4 per cent of the irrigators) have rights to 50–60 per cent of the water (Galárraga-Sánchez, 2000). Recent Ecuadorian governments, so far, have paid mere lip-service to the social water drama, as in the case of the distribution of other resources.

This also is the backdrop of irrigation development in the area of Licto, located in the central Ecuadorian highlands in the Chimborazo province. The area encompasses more than 20 indigenous rural communities. Licto is also the name of the main town.[8] In these highland communities demographic pressure has led to the rapid degradation of natural resources. Consequently, subsistence agriculture, carried out primarily by women, does not meet basic needs. Intermittent migration and wage labour, especially by men, is a necessary complement to local production. Some 13,000 people live in the

Licto area, of which some 90 per cent self-identify as indigenous and 10 per cent as mestizo (mixed). The latter generally reside in the town of Licto. As in many Andean regions, Licto has a long history of white-mestizo landowners and other local powerholders subordinating the indigenous communities. Although the forms and relationships of exploitation have shifted in the last few decades, poverty and ethnic and class-based discrimination endure.

Such extreme poverty provides fertile ground for various top down development programmes and politicians who advocate social participation in the name of, among others, (neo)liberal equality. As I will detail later in this chapter, the latter typically promote participation as the inclusion of the poor and indigenous in society. But, first, they neglect the fact that, obviously, these marginalized groups are already (and always were) included in society, but in a subjugated position. Second, neoliberal participation characteristically refers to efforts to make these groups participate in a water world that is defined by the main elements of the neoliberal model itself: private rights, the reallocation of water rights through market transfers, the redefinition of user and rights identities through commodity-based arrangements, all elements that are linked by means of a naturalizing economic–scientific water policy model and discourse (see the final two notes of this chapter for the materialization of this policy in the Andes).

In Licto, communities have suffered during recent decades, not just from liberal and neoliberal policies, but equally from radical or socialist–bureaucratic policy orientations, which all had in common that they denied the very needs, solutions and resource management rationalities of the local population itself. There have been countless promises from outsiders and development institutions to help these indigenous communities, and just as many disappointments. Therefore, when local communities, through the rural indigenous Corporation of Rural Organizations of Licto (CODOCAL), were invited in 1989 to take part in an ambitious irrigation project, many local indigenous residents were wary. Indeed, it was precisely the white-mestizo people living in Licto town who had promoted this project through their contacts with the Ecuadorian Institute of Water Resources (INERHI),[9] at that time the governmental irrigation agency. Nevertheless some indigenous leaders from the communities, and many poor women from the town of Licto, also saw this project as a potential means to alleviate their poverty and to challenge existing power structures.

INERHI had been working on the studies and execution of the system since 1974. It was a classic example of a vertical design and implementation process that excluded the rural population from decision-making. In their offices in Quito, technical staff who were unfamiliar with rural Licto provided the technical and organizational designs, which were presented in 1990 as the final designs. According to the plans, the main canal would first benefit the town's mestizos, some 500ha at the upstream part of the system. This would, yet again, weaken the situation of the indigenous communities located at the downstream end of the system, thus reinforcing prevailing power structures.

Contrary to local norms concerning the creation of water rights, the design would distribute water to irrigators according to the size of their holdings, regardless of their active labour investment in building or maintaining the system. Moreover, no night storage reservoirs were included in the design, thus mandating nocturnal irrigation. This implied severe problems for the great majority of future irrigators, women. The difficulties of getting permission to go out at night, of leaving little children behind and the threat of sexual violence would make it difficult, if not impossible, for women to enact their water rights. Finally, the canals and hydraulic sectors were designed purely on the basis of technical and geographical criteria, ignoring the boundaries of the social units, the peasant communities. This top-down blueprint was bolstered by an organizational and legal design grounded in national law and uniformly enforced nationwide. Hydraulic sectors and new government-dependent water leaders would form structures parallel to the existing communities and their leadership structures. As has happened elsewhere in the region, these imposed structures would have interfered with communal institutions that assure collective survival. Moreover, the proposed regulations set a single blanket fee that everyone would pay for water service, in this way ignoring user labour and organizational contributions, as well as the broad, national diversity in systems and their productivity and profitability context.

The techniques of governance of INERHI are powerfully illustrated in this combination of organizational, legal and technical designs. The state agency, as well as the local elites, would now have an even stronger influence on the daily running of water affairs. Most communities that joined the project did not question the socio-technical designs since these were portrayed as being 'normal', that is according to the official standards, as well as modern and efficient, based on expert knowledge. Nevertheless, the design of the system would thus further strengthen state control by the so-called inclusion of the indigenous communities in exogenous, modern management. These uniform rules from outside would deny local control over decision-making; they would also facilitate extraction and intensify external domination over local livelihoods.

However, the socio political panorama changed dramatically in Licto, when in the course of 1992 and 1993 CODOCAL elected new, strongly committed leaders. They wanted to bring the communities together within a unified indigenous organization. They also established a water user organization, the Irrigation Directorate, within CODOCAL's inter-communal structure and rooted it in community organization – not based on the hydraulic sector formula prescribed by law. An Ecuadorian non-governmental organization (NGO) joined the water development efforts of the local communities and supported their claims vis-à-vis the government agency. In late 1993, INERHI sent the official regulations for administering and organizing governmental irrigation systems.

But rather than accepting the organizational blueprint and simply paying for the water at national rates, which was the official approach that in practice favoured some powerful white-mestizo families, the users refused and made a better, context-adapted regulation. In this alternative regulation, the communities decided to establish their own rules and rights based on a fundamental principle: the right to water and management decision-making is earned by those who work in the communal labour work parties, who participate in the water user organization and who pay their dues according to collectively established contribution rates. The crux of the indigenous peasants' protest and proposal was that rights cannot be purchased, they must be earned. As outlined in the previous section, this water rights creation mechanism is in keeping with notions of equity and reciprocity found elsewhere in the Andes. At the same time, the communities, with support from the NGO, carefully analysed the technical, organizational and normative designs and discussed their implications and probable consequences; they then redesigned the system.

However, the government agency's technical and administrative officers would not agree to such fundamental changes in their system. In addition to having to discard the designs and norms in which they were professionally heavily invested, they would have to agree to the design criteria of outsiders, indeed, that of Indians, which was entirely counter to the prevailing racial logic in Ecuador. Nevertheless, after two years of fierce struggle, social mobilization and negotiations, in which the countervailing power of the indigenous communities grew strong, the agency had to accept that the project was destined to fail and, even worse, was in danger of becoming a political fiasco; social peace, public image and election votes were at risk.[10] INERHI was forced to redesign the system with the participation of the communities and the NGO. Even though they accepted changing the infrastructure design as far as possible, they refused to discuss organizational and normative proposals proposed by the communities that they considered illegal. But the communities prevailed; in spite of the fact that the state agency was unwilling to ratify the rules formulated by the indigenous organization, the latter quickly enforced those rules as the guiding principles of their own system.

According to these principles, paying fees to the state was not enough to obtain rights to water and decision-making; rather these rights would be created during system construction and then recreated and consolidated through the users' participation in operation and maintenance. This ancient Andean norm was often abused by power elites throughout the Inca, Spanish-colonial and Republican times. Under so-called pacts of reciprocity they appropriated the collective labour force of local communities, who had to work in irrigation system construction and maintenance while most of the fruits of their labour were delivered to the rulers. But now this norm of collective labour investments in order to create rights to access water as well as decision-making power, was reclaimed by the communities. It became the solid foundation for the system's management, and, indeed, much more. The norm became an all-purpose tool that indigenous communities could use to challenge both the state as well as the town's white-mestizo families. On the basis of this criterion of creating rights, there has been a consciousness-raising process in which communities built a new water cultural identity, a hydraulic identity, based on their own specific rules. This way, in a first phase, downstream indigenous communities were united in an indigenous irrigators' association with its own principles. In the next phase, they were on solid ground with clear, strong norms and the capacity for negotiation, and this organization invited the town to join them.

The white-mestizo townsfolk of Licto, thus, had to accept the equitable criteria already established by the communities themselves. As one might expect, this organizational strategy and transformation at first faced great resistance from white-mestizo power groups in the town of Licto. But the strategy of the inter-community irrigation organization, structured within the indigenous peasant organization CODOCAL succeeded; it has grown to become the strongest political force in Licto.

Indigenous communities united with lower-class strata in the town of Licto itself and together they successfully faced down the abusive power relationships in the region. Moreover, through sound critiques and counter-proposals regarding system design and construction and massive uprisings to defend their rights and technical and organizational proposals, they also convinced the donor SDC (Swiss Agency for Development and Cooperation, who funded a large part of the project) and imposed their arguments upon the state agency. The communities took over the management of the system, and since that time, through a series of ups and downs, the Licto irrigators' organization has earned increasing respect from outside development and state agencies, as well as recognition from more and more of the local mestizos, who have been forced to make the best of a bad situation and who have now applied for membership in the irrigation organization.

But this was not the end of Licto's struggle for a well-functioning and more equitable irrigation system. The struggle to claim and materialize social participation, as co-designers of rules, rights, infrastructure and organization, is part of an ongoing battle as new threats gain momentum. Although the current government, compared to the previous ones, is more respectful toward peasant and indigenous communities' own management practices, threats to their collective property linger on. For instance, new initiatives to bring about the privatization or reconcentration of water rights and the individualization of collective water control are certainly not imaginary. And despite the fact that the indigenous and peasant inter-community organization of Licto has created an important and solid base to counter such threats to their very existence as a collective, private enterprises and governmental agencies – commonly under the banner of social participation and decentralization policies – will continue to challenge their rule-making authority in the future.

Indeed, in the background of the Licto water development process described above, a new tradition of rule-making in Andean irrigation, under the auspices of decentralized decision-making and user-controlled water management, has been implemented, and the transition has been swift since the early 1990s.[11] But rather than peasant and indigenous water user organizations taking the lead in this transition, national authorities and elites were advocating what they called modernization through state withdrawal from bureaucratically controlled systems. It is striking to see how this transition was not so much a national affair or a process restricted to the Andean region but a global move, in which basic rules were set by international policy-making institutions.

Clearly, models for rules and rights in irrigation water management are not system-, country- or region-specific, but are international in scope.[12] As part of the global wave of structural adjustment policies, starting two decades ago, an intense process of state and policy reform has been put in motion in Ecuador. Decentralization, modernization and privatization endorsed new modalities for public entities and a different political culture. In the process of decentralization in Ecuador several functions, tasks and responsibilities were handed over to local government bodies in order to enhance overall government functioning and efficiency. Tellingly, the process was enforced by the law of state modernization (1993), the law of state decentralization and social participation (1997), and the national plan for decentralization (2001). However, local government capacity building was neglected and many local governmental institutions have inherited the technocratic water control legacy (Cremers et al, 2005). Although INERHI, the former powerful, central state agency for water management, was dismantled in 1994, its biases and disciplines have been largely reproduced in reduced, decentralized government institutions. Mono-disciplinary engineers (civil and hydraulic) and planners (economists) are mainly the ones in charge of large-scale irrigation and water development. Even provincial governments often interpret their new tasks as just building large-scale irrigation infrastructure, neglecting manifold opportunities to support small-scale highland community systems (Cremers et al, 2005; Hendriks et al, 2003; Pacari, 1998). Although civil society platforms and water policy alliances, such as the National Water Forum, provide a growing counterweight through interactive training of young water professionals and by organizing policy debates in all regions, engineers' expertise in water design and technology development continues to be reified in the main offices. Therefore, despite the new pluralistic constitution, issues such as democratic water control, organization building, legal pluralism and recognition of locally particular rules, rights and management frameworks are still difficult to get into the everyday practice of the official water development sectors. In Ecuador, the modernization discourses, policies and practices have dismantled not just state bureaucracy, but also its capacity as a framework to creatively help local government and water user organizations manage their own water affairs (Hendriks et al, 2003).

Peasant and indigenous water user organizations and federations do not remain silent and have protested fiercely. This new water policy is viewed as a new attack on both their water resources and irrigation infrastructure. Since 1998, though facing ups and downs in their organizational empowerment, Licteños have united in new, supra-system alliances, and new federations are rising up to demand and actively construct social participation. Ecuadorian peasant and indigenous water users decided to curb top-down policies and abusive water rights allocation practices (Dávila and Olazával, 2006), through such inter-system networks as Interjuntas (Chimborazo) and the Federation of Irrigation User Groups of Cotopaxi (FEDURIC), among others. These organizations realize that, in the Andean region, the vast majority of public investment in water management so far was for the benefit of those areas and stakeholders who were already well-to-do and better organized. Governmental institutions (whether national or local) and public action – even worse when delegated to private market players – are by no means neutral. Clearly, social participation is not a gift and is often not based on open, transparent dialogue among equals in a setting of harmony. Prevailing societal relationships and power structures provide the backdrop for consensus building. Social struggle and mobilization in contexts such as Licto's seem fundamental to achieve or guarantee sufficient political democracy, distributive justice and respect for indigenous–peasant rights in water management.

Social Participation: From Alignment to Mobilization for Water Rights Justice

Within the unequal power structures in water development processes, different user groups – as well as non-user stakeholders – set their strategies to demand, defend and materialize their own interests. They establish alliances, gather their powers, capacities and resources and confront each other. Here, groups with less political and economic bargaining power commonly face huge obstacles to get their rightful share and materialize their interests.[13] These struggles are also complex, since in the Andes water rights and benefits involve conflicts about the access to and withdrawal of this fundamental resource as well as about control over its management and recognition of the respective authorities' legitimacy. Cases such as the one of Licto make clear that, for peasant and indigenous communities all these levels have great importance, since it is precisely the authority of indigenous and peasant organizations that is increasingly being denied, their water usage rights that are being cut off and their control over decision-making processes that is being undermined. Often this is defended by modernist policy discourses and contested by various counter-discourses of the user groups affected.

The importance of local and indigenous rights regimes is often overlooked by official policies and intervention strategies in Ecuador. Moreover, as in most other Andean countries, decision-making power of state irrigation institutions is often based on undemocratic principles and an unequal representation of local communities. Thereby, both centralist and neoliberal water policy models not only assume and desire universal laws but also actively establish them (Boelens and Zwarteveen, 2005). Co-existence of a great diversity of rules, rights and obligations is actively discouraged since such diversity would obstruct inter-regional and international transfers and trades, which require a uniform legal framework. In most Andean countries, market policies do not replace bureaucratic policies, as is commonly suggested in decentralization discourses, but act as allies since both perceive the need to discipline and counteract pluralism of water rights repertoires. Recognition of the diversity of water authorities and rules would undermine the power and rule-making capacity of both national bureaucrats and powerful market players. State bureaucracies, therefore, impose their uniform rules or are reformed to provide and enact regulations that allow markets to emerge. Communities and collective rights systems that do not fit a state or neoliberal model are defined as inefficient and backward. Therefore, they are either doomed to wither or get included in the state or market model, to join the game on unequal terms, precisely in line with the externally imposed politics of participation and inclusion (Assies, 2006; Boelens and Zwarteveen, 2005).

Nevertheless, the state has the potential to contribute to a more equitable, sustainable development of irrigation through, for instance, legal and operational backing for peasants' own management of irrigation systems. But the lack of interactive strategies and user-oriented investment policies, together with the absence of gradual water rights redistribution efforts, only increases the current problems. As long as peasant and indigenous rights systems do not have a formal legal support, there is a possibility for them to be overruled. The recognition of collective water rights, assigned to common property groups, rather than allocating water rights merely to individuals as most modernization policies in the Andean region propose, would increase security, productivity and stability of water use systems.[14] National legislative efforts should aim to create a strategic space for the operation and strengthening of these local normative systems, without codifying their precise contents; for strong and internally consistent irrigation systems it is crucial that the local system's collective rights are in the hands of user organizations. In case local community institutions fall short, non-local, higher-level institutions for dispute resolution have to be in place and backed by national law and administration. Fundamentally, local user groups establish their internal rules and rights, within the framework of broader legal principles of justice, democracy and sustainability.

Such a process is essentially political and cannot be given as a handout or legally engineered: it takes place in an arena in which conflicting water interest actors negotiate, compete and struggle. Here, bottom-up processes of awareness creation and capacity building may support but cannot supplant grassroots claim-making, collective action and mobilization. While at the central government level, political will and an enabling political environment is required, commonly in Andean countries, such political acceptance and accountability processes do not appear out of the blue. They result from collective pressure from below.

Although water development often accelerates social differentiation, the case illustrates that it can also be a vehicle for empowering the less privileged groups. Water is power so changing water is changing power. In Licto, for example, some representatives of marginalized groups were the first to recognize the validity and challenge of this statement. We have seen that support processes cannot plan – in a linear sense – irrigation development or the new constellation of powers, but this does not deny the importance that outside support may have in influencing the (re)definition of control over water and positions of power. Identification of interest groups; analysis of problems; needs and potential assets of less-favoured groups; participatory research; facilitation of networks and horizontal linkages (alliance-building capacity); facilitation of negotiation forums that are accessible to the least powerful groups preparing them (demand- and proposal-making capacity); and institutional backing in the platforms are all important elements in a strategy of empowerment.

As Evo Morales argued in the publication, *Water and Indigenous Peoples,* it is time that, rather than having farmers participating in the projects of development institutes, the latter understand that they need to participate in the projects and strategies of marginalized water user groups, local communities and grassroots organizations 'We indigenous peoples do not want to be research objects, but fellow combatants in the struggle' (Morales, 2006, p22). Thus, a process of assisting in local empowerment does not require the generalized processes of 'McParticipation' or 'McDecentralization', as if empowerment could be planned and materialized through universally applicable strategies that lack a deep understanding of context and historical background and a critical-constructive positioning in the water policy and politics world. On the contrary, such a process is to be founded on critique of the power games, analysing governance and rights constructions in particular water control cases as a movement through which water user groups and their members are subjugated and controlled (Galeano, 2009). This is done by means of power mechanisms and instruments that are fundamentally associated with a truth constructed and diffused by dominant players. Assisting local critique of such domination means facilitating subjugated water user families and communities to question these truth claims, and to analyse the effects of water control, challenging prevailing power structures and their claims to rationality, efficiency, adequacy, democracy and equity.

Notes

1 In other works I have elaborated on the capillary, bottom-up, equalizing powers of participatory domestication and disciplining in water control. My book, *The Rules of the Game and the Game of the Rules. Normalization and Resistance in the Andes* (Boelens, 2008) details how this process takes place through cultural politics, hydro-policy modelling, technological dream scheme development, expert-dominated constructs of local rights, and subtle power games that seem to recognize local water users but in fact deepen their subjugation. This is part and parcel of both neoliberal projects and top-down state politics (see also the multiple studies of the WALIR Water Law and Indigenous Rights programme that are referred to in the references list below).

2 Collective rights are the demands by the user organization vis-à-vis third parties, and also determine the collective forms and conditions for access to water and the prerogatives and burdens assumed as a group. Individual rights establish the relations for access to water and the distribution of other privileges and burdens inside the users' collective, that is, among the right-holding persons or families.
3 Divergent claims from users refer both to the demands by different (consumptive and non-consumptive) water user sectors (irrigation, drinking water, hydropower, industry, mining, logging and others), as well as to sector-internal struggles over water and rule-making. Examples of non-users involved are politicians and donor agencies, as well as the downstream watershed inhabitants and other groups affected by water use.
4 Apart from being an important mechanism for obtaining rights (allocation), it is simultaneously an important principle for scheduling water access once rights have been allocated.
5 Although labour input is a vital element to create and conserve rights, other factors are also important, for example:

   - money;
   - goods (agricultural produce, materials, instruments and others);
   - intellectual inputs and organizational efforts, such as at meetings;
   - operational contributions to water distribution and communal cultural investment, which is present in the communities' collective memory (for example, joining ritual irrigation activities and remembrance of the blood, sweat and tears and casualties invested in the system).

6 The user investment rationality that is needed for sustainable, autonomous systems does not call for romantic approaches. In Arroyo and Boelens (1997) we have analysed the consequences for female irrigators, already overburdened with work, in communities with much male migration.
7 Then, the issue is not to organize contributions themselves, but to define specific contributions by each party, their relationships with rights, privileges and with other right-claimants (Beccar et al, 2002). Nor is it a question of getting them involved in the project but for the users themselves to generate, modify and maintain their own collective ownership arrangement, at the same time generating and conserving their individual rights.
8 This section is largely based on Boelens and Gelles (2005) and Chapter 7 in Boelens and Hoogendam (2002).
9 Created in 1966, INERHI was linked to the Ministry of Agriculture and Livestock and made large investments in mainly large-scale hydraulic projects. During the decentralization and privatization years it was replaced by the National Water Resources Board (CNRH), recently transformed into the National Water Secretary (SENAGUA).
10 See also the film *The Right to be Different* (Wageningen University-Agrapen, 2003).
11 As a reaction against state intervention, over-regulation, protectionism, and state-installed bureaucratic water management systems, this policy fosters irrigation management transfer (IMT) to the local government or user organization, elimination of subsidies, deregulated management and a self-regulating water rights and/or water service market, privatization of system construction, individual initiative and involvement of private (and currently also public–private) enterprises.
12 Despite the fact that neoliberal proposals to decentralize and commercialize water resources management in Andean countries are disguised as a national system for public ownership, in order to quell popular protests, these water reform proposals

and actions in the Andean region blatantly link local community systems and resources to international domains of rule-making and resource control and exploitation (Assies, 2006; Swyngedouw, 2003). Not only do these policies aim to decentralize (supposedly on the basis of subsidiarity principles) but also to get the government to smooth interaction and exchange among local, national and international levels, by making the rules and rights the same for all. It is also astonishing to see how, over the last two decades, these international policies, claiming to democratize water management and decentralize decision-making, have spread through the Andean region, aggressively running roughshod over most water users, totally beyond their power to curb or control them. Furthermore, it is important to note that, at the moment, the water policy directions in the different Andean countries are increasingly divergent.

13 See also the Andean country contributions of Evo Morales, Armando Guevara Gil, Rocío Bustamante, Pablo Solón, Francisco Peña and Paulina Palacios to *Water and Indigenous Peoples* (Boelens et al, 2006). See also Berry, 1998; Gelles, 1998; Getches, 2005; Hendriks, 1998; Meinzen-Dick and Pradhan, 2005; Pacari, 1998; van der Ploeg, 2006; Zwarteveen et al, 2005.

14 As many Andean studies have shown, privatization and individualization of collective water rights, particularly when below the system level, seriously threaten the survival of common property systems, and peasant and indigenous communities as a whole (Gelles, 2000; Hendriks, 1998, 2006; Mayer, 2002; Pacari, 1998; Urteaga and Boelens, 2006).

References

Arroyo, A. and Boelens, R. (1997) *Mujer Campesina e Intervención en el Riego Andino. Sistemas de Riego y Relaciones de Género, Caso Licto, Ecuador*, CAMAREN, Quito, Ecuador

Assies, W. (ed) (2003) *Gobiernos Locales y Reforma del Estado en América Latina*, El Colegio de Michoacán, Zamora, Mexico

Assies, W. (2006) 'Reforma estatal y multiculturalismo latinoamericano al inicio del siglo XXI', in R. Boelens, D. Getches and A. Guevara Gil (eds) *Agua y Derecho. Políticas Hídricas, Derechos Consuetudinarios e Identidades Locales*, WALIR, Abya Yala and IEP, Lima, Peru

Bauer, C. (1998) 'Slippery property rights: multiple water uses and the neoliberal model in Chile, 1981–1995', *Natural Resources Journal*, vol 38, pp110–155

Beccar, L., Boelens, R. and Hoogendam, P. (2002) 'Water rights and collective action in community irrigation', in R. Boelens and P. Hoogendam (eds) *Water Rights and Empowerment*, Van Gorcum, Assen, The Netherlands, pp1–21

Benda-Beckmann, F., von Benda-Beckmann, K. and Spiertz, J. (1998) 'Equity and legal pluralism: taking customary law into account in natural resource policies', in R. Boelens and G. Davila (eds) *Searching for Equity. Conceptions of Justice and Equity in Peasant Irrigation*, Van Gorcum, Assen, The Netherlands, pp57–69

Berry, K. (1998) 'Race for water? American Indians, Eurocentrism and western water', in D. Camacho (ed) *Environmental Injustices, Political Struggles: Race, Class, and the Environment*, Duke University Press, Durham, NC

Boelens, R. (2008) *The Rules of the Game and the Game of the Rules. Normalization and Resistance in the Andes*, Wageningen University, Wageningen, The Netherlands

Boelens, R. (2009) 'The politics of disciplining water rights', *Development and Change*, vol 40, no 2, pp307–331

Boelens, R. and Doornbos, B. (2001) 'The battlefield of water rights. Rule making amidst conflicting normative frameworks in the Ecuadorian highlands', *Human Organization,* vol 60, no 4, pp343–355

Boelens, R. and Gelles, P. H. (2005) 'Cultural politics, communal resistance and identity in Andean irrigation development,' *Bulletin of Latin American Research*, vol 24, no 3, pp311–327

Boelens, R. and Hoogendam, P. (eds) (2002) *Water Rights and Empowerment*, Van Gorcum, Assen, The Netherlands

Boelens, R. and Zwarteveen, M. (2005) 'Prices and politics in Andean water reforms', *Development and Change*, vol 36, no 4, pp735–758

Boelens, R., Chiba, M. and Nakashima, D. (eds) (2006) *Water and Indigenous Peoples*, UNESCO, Paris

Coward, E.W. (1983) *Property in Action. Alternatives for Irrigation Investment*, report prepared for the workshop on *Water Management and Policy*, University of Khon Kaen, Thailand

Cremers, L., Ooijevaar, M. and Boelens, R. (2005) 'Institutional reform in the Andean irrigation sector: enabling policies for strengthening local rights and water management', *Natural Resources Forum*, vol 29, pp37–50

Dávila, G. and Olazával, H. (2006) *De la Mediación a la Movilización Social: Análisis de Algunos Conflictos por el Agua en Chimborazo,* AbyaYala, Quito, Ecuador

Galárraga-Sánchez, R. (2000) *Informe Nacional sobre la Gestión de los Recursos Hídricos en el Ecuador*, Foro de RRHH, Quito, Ecuador

Galeano, E. (2009) Cover text of the Aguas Rebeldes book (R. Boelens and R. Parra, eds), Imprefepp, Quito, Ecuador and IEP, Lima, Peru

Gelles, P. H. (1998) 'Competing cultural logics: state and "indigenous" models in conflict', in R. Boelens and G. Dávila (eds) *Searching for Equity. Conceptions of Justice and Equity in Peasant Irrigation*, Van Gorcum, Assen, The Netherlands, pp256–267

Gelles, P. H. (2000) *Water and Power in Highland Peru: The Cultural Politics of Irrigation and Development*, Rutgers University Press, New Brunswick, NJ

Gerbrandy, G. and Hoogendam, P. (1998) *Aguas y Acequias*, Plural Editores, Cochabamba, Bolivia

Getches, D. (2005) 'Defending indigenous water rights with the laws of a dominant culture: the case of the United States', in D. Roth, R. Boelens and M. Zwarteveen (eds) *Liquid Relations: Contested Water Rights and Legal Pluralism*, Rutgers University Press, New Brunswick, NJ, pp44–65

Gleick, P. H., Wolff, G., Chalecki, E. L. and Reyes, R. (2002) *The New Economy of Water: The Risks and Benefits of Globalization and Privatization of Fresh Water*, Pacific Institute for Studies in Development, Environment and Security, Oakland, CA

Guevara Gil, J. A. (2006) 'Official water law versus indigenous and peasant rights in Peru', in *Water and Indigenous Peoples*, UNESCO, Paris, pp126–143

Hendriks, J. (1998) 'Water as private property. Notes on the case of Chile', in R. Boelens and G. Dávila (eds) *Searching for Equity, Conceptions of Justice and Equity in Peasant Irrigation*, Van Gorcum, Assen, The Netherlands, pp297–310

Hendriks, J. (2006) 'Legislación de aguas y gestión de sistemas hídricos en países de la región andina', in *Derechos Colectivos y Políticas Hídricas en la Región Andina*, WALIR, Abya Yala and IEP, IEP, Lima

Hendriks, J., Mejía, R., Olázaval, H., Cremers, L., Ooijevaar, M. and Palacios, M. (2003) *Análisis de la Situación del Riego en la República del Ecuador*, Mission Report, WALIR, Quito, Ecuador

Jaubert de Passa, F. (1846) *Recherches sur les Arrosages chez les Peuples Anciens*, 6 parts in 4 vols, 1981 reprint, (Collection 'Les Introuvables'), Éditions d'Aujourd'hui, Paris

Mayer, E. (2002) *The Articulated Peasant. Household Economies in the Andes*, Westview Press, Boulder, CO

Meinzen-Dick, R. and Pradhan, R. (2005) 'Analyzing water rights, multiple uses and intersectoral water transfers', in D. Roth, R. Boelens and M. Zwarteveen (eds) *Liquid Relations: Contested Water Rights and Legal Pluralism*, Rutgers University Press, New Brunswick, NJ, pp237–253

Mollard, E. and Vargas, S. (2005) 'Introducción', in S. Vargas and E. Mollard (eds) *Problemas Socio-Ambientales y Experiencias Organizativas en las Cuencas de México*, IMTA and IRD, México, pp9–23

Morales, E. (2006) 'Message on behalf of the indigenous peoples', in R. Boelens, M. Chiba and D. Nakashima (eds) *Water and Indigenous Peoples*, UNESCO, Paris, pp22–23

Pacari, N. (1998) 'Ecuadorian water legislation and policy analyzed from the indigenous-peasant point of view', in *Searching for Equity. Conceptions of Justice and Equity in Peasant Irrigation*, Van Gorcum, Assen, The Netherlands, pp279–287

Perreault, T. (2006) 'Escalas socioespaciales, reestructuración del Estado y la gobernanzaneoliberal del agua en Bolivia', in R. Boelens, D. Getches and A. Guevara Gil (eds) *Agua y Derecho. Políticas hídricas, Derechos Consuetudinarios e Identidades Locales*, Abya Yala and IEP, Quito, Ecuador and Lima, Peru, pp285–320

Roth, D., Boelens, R. and Zwarteveen, M. (eds) (2005) *Liquid Relations. Contested Water Rights and Legal Complexity*, Rutgers University Press, New Brunswick, NJ

Salman, T. and Zoomers, A. (eds) (2003) *Imaging the Andes: Shifting Margins of a Marginal World*, CEDLA Latin America Studies 91, Aksant/CEDLA, Amsterdam

Swyngedouw, E. (2003) 'Privatising H_2O: turning local water into global money', *Austrian Journal of Development Studies*, vol 19, no 4, pp10–33

Urteaga, P. and Boelens, R. (eds) (2006) *Derechos Colectivos y Políticas Hídricas en la Región Andina*, WALIR, Abya Yala and IEP, Lima, Peru

van der Ploeg, J. D. (2006) *El Futuro Robado. Tierra, Agua y Lucha Campesina*. WALIR, Abya Yala and IEP, Lima, Peru

Zwarteveen, M. (2006) 'Wedlock or deadlock? Feminists' attempts to engage irrigation engineers', PhD dissertation, Wageningen University, Wageningen, The Netherlands

Zwarteveen, M., Roth, D. and Boelens, R. (2005) 'Water rights and legal pluralism: beyond analysis and recognition', in D. Roth, R. Boelens and M. Zwarteveen (eds) *Liquid Relations. Contested Water Rights and Legal Pluralism*, Rutgers University Press, New Brunswick, NJ

13

Water Management Practices on Trial: The *Tribunal Latinoamericano del Agua* and the Creation of Public Space for Social Participation in Water Politics

Carmen Maganda

Water Justice and Social Participation

Current research on water distribution demonstrates that water politics are significantly affected by notions of ethics and power (Petrella, 2001). Lack of public information concerning water distribution combined with the political influence of economic organizations have led to inequitable distributions of water resources throughout the world along with changes in physical/climatic conditions and social inequity.

According to the 2006 Human Development Report the world is not running out of water. The report states that globally there is more than enough water to meet human needs for water for consumption, agriculture and industry (United Nations Development Programme, 2006). Thus, the crisis of water is predominantly a management issue. However, the consequences are, and still can be, dramatic. In addition to the negative environmental impacts tangibly observed in aquifers and water reserves, the United Nations Development Programme (UNDP) reports 1.2 billion people without access to safe water, 2.6 billion without access to sanitation and about 2 million deaths (mostly children) due to water quality-related illnesses, such as diarrhoea and cholera (United Nations Development Programme, 2006, Chapters 2 and 3). Of course, most of the people lacking access to water and sanitation services are poor and characterized by socio-economic vulnerability.

The current water panorama is so negative that in 2000 many governments established the fight against the so-called global water crisis and they strengthened their common support for the Millennium Development

Goals (MDGs), which include a reduction of the proportion of people with no access to safe water by 2015. The 2006 UNDP Human Report states that the world has the technology, finance and the human capacity to end this blight and take the opportunity to unleash another leap forward in human development. However, the problem cannot easily be solved with the progress currently being achieved through international meetings, declarations or accords. All nations face different water management challenges and they set their own limits of what they consider to be fair distribution of water. Moreover, states do not always follow paths that address the limits of ecological sustainability (defined as the long-term maintenance of environmental resources) and most of the time this contributes to the justification of unequal distribution of these resources.

Due to some violations of environmental values and standards related to human well-being, particularly in poor sectors, there is growing acceptance of international norms concerning a human right to water.[1] There is an international recommendation by the UNDP for recognizing water as a human right as a way to enable poor people to expand their entitlements through legal channels. However, the human right to water is not a self-standing human right and in the international political arena water is being discussed more as a need than a right. Fortunately, states commit in many other ways and there are a few examples of national legislation, most of them African (Congo, Ethiopia, Gambia, South Africa, Uganda, Zambia), and some Latin American (Ecuador, Panama, Uruguay, Venezuela and Bolivia in 2009), that include the human right to access clean and safe water. Despite these efforts, globally there is still unequal access to water in disadvantaged communities, and this problem leads to environmental injustice and inequity. In fact, a main problem regarding environmental injustice (particularly regarding water) is the lack of measures on which to base claims judicially. Procedural justice is a fundamental part of environmental justice and the judicial vacuum in international water politics creates a formidable obstacle for the equitable distribution of this resource.

According to the National Center for Human Rights Education (2006) in the United States, environmental justice is defined as the right to safe, healthy, productive and sustainable environment for all.[2] Water is not an exception. Following the Fourth World Social Forum in Mumbai, India, in 2004, a water justice organization was created to promote 'effective, democratic and equitable ways' to solve the global water crisis (www.waterjustice.org). There is also a relatively new water justice theory (McLean, 2007) based on environmental justice approaches,[3] which focuses on the analysis of the ways that different water cultures are valued in their views on access to water services (this involves an examination of the norms that dictate acceptable behaviour in water distribution processes).

In fact, the reference to the notion of justice in water issues introduces an interesting element for framing social participation in water politics. According to David Schlosberg (2006), an approach to the term justice, and more specifically environmental justice, requires the need to see the interrelations between distribution, recognition, capabilities and participation. In this chapter, I want to focus specifically on participation because historically (particularly in Latin America), water management problems have led to calls for improved access to water resources that are directly related to the lack of plurality in social participation in water decision-making processes. For example, many hydroelectric dam projects are built without social consensus, which then leads to protest and social resistance affecting both the environment and cohesion in local communities (Cana Brava in Brazil, Yacireta in Argentina and Paraguay, Urra in Colombia and Parota in Mexico are a few examples).

The negative environmental effects that we already have (pollution, inequitable access, health risks and illness) highlight the inefficiency of regulations in terms of the application of environmental laws, as stakeholders easily violate these measures due to a lack of governmental authority or binding declarations in this field. In the arena of environmental justice, there is an existing impunity in the world and water is not disconnected from this. Many social actors and environmental non-governmental organizations (NGOs) do not believe in simply waiting until environmental promises from decision-makers can be delivered in the future, so they have decided to create their own social institutions as public spaces. Public spaces can be defined as the arena in which democratic discussions occur, which allow for observing, participating in, and even judging environmental and water policy-making. Creating public spaces within new institutions is meant to apply pressure on decision-makers to improve the protection of water rights.

International Spaces for Social Participation in Water Management

Like other transnational issues, such as global warming or human rights, water management is characterized by a gap between international discourse, sometimes presented in global summits, and political action. The evolution of international political commitments focused on water starts with the United Nations Conference on the Human Environment held in Stockholm in 1972. It continues through to the official World Water Forums that have taken place since 1997 along with their analogous People's World Water Forum (Forum Mondiale Alternativo dell'Acqua held in Florence in 2003), to the MDGs adopted in 2000 and the 2006 United Nations Human Development Report. Most of these conferences, such as the World Water Forum, aim to discuss the principles of ethics and equity as well as concrete actions at the national and sub-national levels where decisions are made concerning the distribution of water and where elected officials are often pressured to protect economic interests or favour urban contexts over rural areas. Effective political mobilization is rarely fostered by increased international attention because the structures that could allow the opportunity for social movement remain restrictive. Scholars of border water issues, such as Maria Garcia-Acevedo and Helen Ingram, have illustrated this point with research that has shown that domestic political structures are more significant frames of reference in cross-border debates than are international norms (Garcia-Acevedo and Ingram, 2004).

In fact, this chapter argues that a distinction must be made between social mobilization in water politics and social participation. Social mobilization is characterized by reactive protest against unjust water policies. Social participation is defined as democratic involvement in water policy-making. Conflict over water resources is not seen as negative unto itself but it needs to be institutionalized in democratic debates that include representatives of government, stakeholders and NGOs.

Moreover, we can also identify two categories within social participation: institutional social participation (or participative planning) and communitarian social participation. Institutional social participation works as a mechanism that looks to create co-responsibility between citizen input and state planning in environmental management. This evolution transforms participation from political protest to legal and political consultation. By contrast, communitarian social participation refers to collective action against unpopular political initiatives (Orr, 2007).

In Mexico, for example, this type of co-responsibility has been limited. According to the Mexican Ministry of Environment and Natural Resources (SEMARNAT) social participation is an instrument that can widen the reach of environmental policies. This is already a limiting concept because social participation is considered more of a mechanism for the evaluation of this ministry's programmes instead of a space for democratic discussion where policies are defined. Also, the promotion of social participation in environmental management has been limited by the institutional profile of sharing responsibility in environmental planning with society. This issue is important in Mexico because by law 'every person has the right to petition federal government [SEMARNAT] with requests related with the management of the environment and natural resources' (Article 8, Political Constitution of Mexico). So, why don't people usually do this? Because even when the law exists, there are no real ways to make this kind of request. On one hand, there is no publicly accessible information on water. On the other hand, institutional consultation forums are not an option as a representative space for social participation in Mexico for a number of reasons:

1 they do not circulate relevant information;
2 the discussions are not broad-based as they focus in academic, professional and technical sectors;
3 their session formats are rigid.

Such institutional forums have not promoted democratic participation in the environmental decision-making process in Mexico.

The political distance between environmental institutions and civil society has generated, and continues to generate, water conflicts in Mexico and Latin

America. The keys to social participation include accurate and accessible public information on water resources, the creation of public space in which water policies can be openly debated and transparent laws in the field of water management. Traditionally, each country has its own mechanisms to address water management and water rights. However, during the past three decades water has become an international political issue. In 1992, the International Conference on Water and Environment in Dublin stated, as a second guiding principle on water and sustainable development, that 'water development and management should be based on a participatory approach, involving users, planners and policy-makers at all levels'.[4] International organizations have implemented specific initiatives (programmes, forums and commissions) to address global water management issues.

The largest and most important international body in global water debates is the World Water Council (WWC), which was established in 1996. It is composed of international multi-stakeholders and plays an interesting role in promoting awareness, building political commitment to and triggering action on critical water issues in the global community. Since 1997 the WWC has organized the World Water Forum (WWF) every three years in close collaboration with the authorities of the hosting country. The WWF is the best example of institutional social participation as it attempts to:

- raise the importance of water on political agendas;
- support the deepening of discussions towards the solution of international water conflicts;
- formulate concrete proposals and bring their importance to the world's attention;
- generate political cooperation.[5]

The Fourth WWF, held in Mexico City in 2006, had the specific objectives to 'actively promote the participation of all stakeholders during the preparatory process and the Forum itself'.[6] Even more, the main theme of the forum was local actions for a global change. The forum was supposed to engage the public in debates concerning water scarcity and the need to find new methods to distribute water resources more equitably. The existence of the forum itself was hailed as a sign of increased social participation in international water politics. However, starting with the price of US$600 to participate in the WWF, this event was dominated by the circle of elites in water politics (governments, international NGOs, water management corporations and academics). Moreover, even though debates focused on the role of ethics in water management, few concrete mechanisms were proposed to improve social participation and ethical behaviour in water distribution. It is true the WWF is the world's largest water policy event but we may question if it is the most appropriate platform to discuss the world's water problems in a broader perspective, including political pressure on the diplomatic scene to keep pushing for the human right to water.

In 2003 the second major forum for the cultivation of social involvement in international water politics took place in Florence, Italy, in response to the participation problem, which had been apparent since the first WWF. The first People's World Water Forum was organized as a social and academic alternative to the WWF with the objective of guaranteeing access to drinking water for all world citizens and the recognition of water as a common good in the time frame of a generation. This Alternative World Water Forum focused specifically on:

- elaborating a series of legislative, political-institutional, financial-economic and socio-cultural actions aimed at creating a right to water for all;
- promoting campaigns in support of the right to water for all;
- contributing to the reinforcement of synergies among social movements, civic groups and associations active in water issues, environmental protection, democracy and human and social rights;
- promoting increased responsibility in social and individual behaviour and the participation of citizens in water management.

All these objectives are directly inspired by principles elaborated in the *Manifesto for a World Water Contract* and more specifically in the *Manifesto Italiano per il Contratto Mondiale dell'Acqua.*

The People's World Water Forum and its related initiatives directly focus on social participation in water debates. These initiatives aim to promote international reforms from local action by facilitating communication between various movements throughout the world. However, while these strategies have progressive objectives and ties to the World Social Forum, their influence and impact are not widespread. These movements are known in Europe but their impact has been minimal in developing states, especially in Latin America. The Alternative World Water Forum has not even received publicity in Latin American academic circles where one finds leading experts on water issues. For example, recent water conferences in Colombia that were sponsored by the Parliament of the Andean Community did not address any of the main points of this alternative participation framework. Moreover, academic seminars do not mention these initiatives and of even greater concern is the fact that libraries in many Latin American universities do not have materials on the alternative World Water Contract. If Latin American water specialists do not know about these initiatives, what impact can they have on the public? Thus, the governance problem in focus is: if the World Water Forum is dominated by elites and if the influence of the People's World Water Forum is restricted to Europe, then what strategy can be developed to improve social participation in water politics in Latin America? This is the focus of the following section.

Is There a Need for Water Courts?

As stated above, international efforts to promote social participation in water management have not been very effective in Latin American contexts. This region has its own problematic water background due to complex stakeholder scenarios and a general weakness of democratic performance. With few regional exceptions where there is abundant water, water management in Latin America has caused significant competition for the resource due to:

- social inequities;
- technical problems with hydraulic design and networks;
- short-term policy solutions;
- vertical decision-making regarding hydraulic infrastructure;
- economic problems because poor people lack the means to pay for water services as well as other reasons.

Because of the lack of participation spaces, public protest has been the common answer to these problems. This is emblematic of traditional difficulties with democracy in Latin America, and scholars such as Evelyne Huber (1990) and Wayne Cornelius (2005) have identified the contentious relationship between authorities and citizens as a defining characteristic of many political systems.

In my opinion, these forms of mobilization are not the answer to social participation in water management:

1. As stated above, they are reactive measures rather than proactive ones.
2. They are often organized by radicals who may hinder mainstream acceptance of social participation movements more than they help because of the chosen *modus operandi* (strikes and barricades) for protest.[7]

For these reasons, regional initiatives, such as the *Tribunal Latinoamericano del Agua* or Latin American Water Tribunal (LAWT) offer interesting alternatives for the improvement of social participation in water debates. The specific question that this chapter addresses is: what potential do schemes such as the LAWT have in making a significant impact on water decision-making processes? I argue that this initiative is innovative and despite its limits, which will be addressed below, it offers an avenue that could lead to the significant improvement of social participation in water management processes.

The LAWT focuses on legal justice. Unlike the political protest approach described above, the basis of this initiative is the legal interpretation and application of social rights. The unequal access to water resources that characterizes Latin American water politics is not only a social problem but it reflects a crisis of environmental legislation that focuses more on the symptoms of water scarcity than on its structural causes, which include inefficient jurisdiction and the aforementioned lack of participation. Water injustices, as stated in the introduction, are present in all these terms and it seems that a kind of ecological impunity seems to reign 'if an environmental law was of relevance as part of transparent jurisdiction, an ethical water tribunal would not be needed' (Helfrich, 2006).[8] Thus, the LAWT's first contribution to improving democratic participation in water debates results from its focus on creating a quasi-legal context in which participation can be fostered.

Internationally, there are some historical examples of environmental civil society tribunals where water injustices have been exposed and resolved. For example, in a public hearing in 1983 covered by the mass media in Rotterdam, The Netherlands, an environmental tribunal analysed environmental damages caused to the basin of the river Rhine. Some effective recommendations were made concerning water contamination safeguards in the tribunal's resolutions. Similarly, in 1992, severe water contamination cases from Asia, Africa, America and Oceania were considered in Amsterdam. While governments and international corporations were held responsible for the allegations, the impact of this hearing was less influential than the first as no positive results were attributed to the resolutions. One of the problems regarding the effectiveness of this hearing was related to the fact that the Dutch court did not have jurisdiction in other parts of the world. Despite these limits however, the court was able to expose environmental problems in a legal context.

Like the European examples cited above, Latin America also has had some NGO-run water courts. In 1983, Brazil's National Water Tribunal held its first public hearing in Florianopolis to review cases on mining, radioactive and agro-chemical contamination as well as cases related to large-scale hydroelectric generation projects. The Central American Water Tribunal (CAWT) was created in 1998 with the purpose of contributing to the resolution of conflicts related to water ecosystems in Central America. It responded to what its founders (activists, jurists and academics) called the democratic deficit in water management and the environmental impunity situation. It was inspired by international environmental instruments that recognized the vulnerability of water and the need to ensure its responsible use, such the Fresh Water Treaty (1993), the Dublin Statement (1992), the Convention on Biological Diversity (1992) and the San Jose Declaration (1996). The CAWT was:

> an ethical institution supported by the social endeavours of citizens committed to water preservation for human consumption for current and future generations. Its legitimacy was based on the moral nature of its resolutions and its juridical foundations, including covenants, declarations and international agreements over environmental protection. (Picolloti and Crane, 2005, p462)

The CAWT was established to:

- create new justice settings;
- resolve water-based conflicts;
- promote the use of environmentally cleaner technologies;
- promote adequate water resource management.

The Tribunal scheduled workshops, seminars and hearings in which water crises in Central America were discussed, thus implementing the goals outlined above.

According to Juan Miguel Picolotti and Kristin Crane (2005), the CAWT represented an example of social and environmental justice created by a coalition of NGOs from Central America in collaboration with an interdisciplinary team of lawyers and scientists from around the world. The innovation of these tribunals went beyond their mere focus on environmental jurisdiction and legal interpretation of social rights. Their main function was to create a public space for democratic participation in water debates for groups that do not usually have access to traditional avenues for justice and those directly affected by the degradation of this resource (Picolotti and Crane, 2005). Local actors were able to utilize these courts as arenas for public debates on water policies that had regional significance.

After two years and five public hearings held in different capitals of Central America, which caught the attention of the international media, in 2000 the founders of the CAWT created the LAWT in order to increase the impact of this body throughout Latin America. Its objectives are the same as the CAWT: to contribute to the resolution of conflicts related to water ecosystems in Latin America by providing an alternative platform for social participation, mediation and environmental justice. Thus, the LAWT, like its predecessor, aims to create a framework for social participation that is pro-active rather than reactive and conciliatory rather than accusative.

Of course, the LAWT's impact is significantly limited because it is not a judicial tribunal in the constitutional sense. Its decisions are not binding and it does not impose sanctions. Its creation, however, is significant for two reasons:

1 It represents an autonomous international judicial body where legal discussions concerning water management and ethical decisions are handed down. These efforts introduce the principle of ethics into international water debates both in form (through the rule of law) and substance (through the content of the tribunal's rulings).
2 The Tribunal hears cases brought by grassroots organizations; it links international water debates to sub-national mobilization.

This chapter analyses the activities and significance of the LAWT, with the aim of addressing the following questions:

- What is the social benefit and/or political impact of this court?
- How plural are the processes and who is participating in this water court?
- Can this body be a new regional answer for promoting real social participation in water politics?

The Latin American Water Tribunal

As the World Water Forum was taking place in March 2006, the LAWT simultaneously held its first public hearings in Mexico City involving ethnic communities, rural populations and other civic groups contesting inequitable water management practices. In fact, the hearings faced some serious logistical problems, including the Mexican refusals to issue visas to some jurors and a change of venue that apparently resulted from political pressure to stop the tribunal's activities. Local and international mass media were already following the development of the LAWT before the first hearing took place.

This section specifically focuses on the LAWT's proceedings. Even though the tribunal's rulings are important, this chapter argues that the LAWT's most important contribution comes from its proceedings that encourage forms of institutional social participation (as defined above) that extend beyond simple protest. Because the LAWT's verdicts are not binding (no sanctions are implemented), the body's major innovation lies in the fact that it convenes all actors in specific cases and attempts to mediate their conflict impartially. Hence, the tribunal's activities create public space within the context of water management. How does this happen in practice? The following sections explain the LAWT's formal procedures.

The complaint

According to the LAWT:

> every person, group of persons or organizations aware of threats to the sustainable use of water or that undergo the consequences caused by the water resource mismanagement or abuse, may file a complaint before the Tribunal. The allegations must be properly supported with scientific and technical evidence proving the possible impacts on the environment.'[9]

Every person or industry whose activities may contaminate, misuse or threaten water resources can be accused in the tribunal. Those public institutions whose actions or omissions favour the above practices can also be accused. While it is true that the call for complaints focuses more on negative environmental impacts than it does on water resources, in practice the complaints also address socio-economic impacts in vulnerable communities as a result of vertical decisions regarding the construction of large hydraulic infrastructure (such as dam projects).

Like other international courts, such as the European Court of Human Rights, the LAWT hears cases after local and national legal remedies have already been exhausted. This means that plaintiffs must first demonstrate that they had followed all national legal and administrative procedures aimed at protecting environmental and civil rights as well as the aquifers in question. Nevertheless, the LAWT recognizes that in some cases this rule is not applicable given the predominant impunity that characterizes environmental problems in the region, the overload of cases in national justice systems, the political and financial interests of the agents named in complaints, the socio-economic vulnerability of plaintiffs and the holes that exist in matters of environmental regulation in most Latin American countries. Thus, the tribunal's first contribution to social participation is its attempt to offer recourse to parties excluded from decision-making concerning water rights.

Often, poorer stakeholders do not have the opportunity to participate in decision-making procedures because water management decisions are often made by technocrats who were not elected and who do not make information readily available to the public (Aboites Aguilar, 1998; Maganda Ramírez, 2004). These processes shield managers from public pressure but they also restrict transparency. The LAWT improves social participation because it sheds light on distributive decision-making in public hearings (see below) and it includes new actors in policy discussions by sending experts to communities to inform citizens of their rights and educate them on avenues for involvement in democratic processes.

The cases

The technical/scientific commission of the LAWT is responsible for selecting the cases, followed by a thorough evaluation of them. This commission consists of professionals and technicians who carry out a study in order to select the most representative and best-sustained cases before bringing them to a public hearing.

Once the case is received, the LAWT's technical commission reviews the lawsuit and the supporting evidence. If the accusation is accepted, the LAWT formally notifies the opposing party accused of committing the environmental damage of its right to respond to the allegations in a public hearing. The chosen cases for this public hearing are those that pose a serious threat to or put in danger significant populations, or important water bodies and their basins. This is particularly meaningful if such water bodies are critical to human life.

The cases that do not reach the public hearing are also evaluated and addressed through a conflict resolution agenda, based on mediation, in which the LAWT comes to an agreement with the opposing parties. This is achieved through technical advice from the LAWT. Experts, in fact, play both formal and informal roles in the LAWT's activities. Formally, they compose the technical committee that reviews cases, as well as the jury that decides them. Informally, environmental lawyers participate in the deliberations and make recommendations to the parties involved. This point is very relevant because the court does not just listen to one party while ignoring the claims of accused water agencies. Instead, the court aims to reconcile the differences between both parties. This demonstrates the tribunal's commitment to democratic dialogue over social protest. Social participation is nurtured through mediation and institutionalized communication.

The jury

Once a case does go to a hearing, it is heard by a jury of water experts. These juries are composed of people with different professional backgrounds. In geographic terms, most jurors come from Latin American countries, although some from other continents also participate. The main requisite to become a member of the jury is the acknowledged ethical background of the candidate. The jurors are selected by the LAWT's general coordination body, and at least one third should be female. They are committed to attending the hearings, analysing the accusations, issuing the verdicts and formulating recommendations.

The jury of five to nine members is always installed one day before the start of the public judicial hearings. In the act of installation the jurors appoint a president and a vice president from among their ranks. In lieu of a judge, the president has the responsibility of directing the sessions and organizing the testimony and general course of deliberations. The vice president replaces the president in his temporary or permanent absence. The jury also appoints a secretary of minutes, who is charged with recording an accurate registration of the deliberations, events and testimony. The jury has the right to revise the documentary evidence and to cross-examine the witnesses.

The public hearing

The public hearing is the most significant activity of the LAWT. It is a public event in which cases where human activities may threaten water availability or where water mismanagement is involved are presented and evaluated. The allegations are put forward by civic organizations from Latin American countries (through their legal counsel) and are assessed by the aforementioned jury.

During the hearings, the parties involved in the conflict have the opportunity to be heard and to state their viewpoints. If plaintiffs cannot attend the hearing then their cases are withdrawn (Bogantes and Borrero, 2000). However, if the accused do not attend the hearings for whatever reason without giving just cause, then it is understood by default that they renounce the right to an oral defence and the hearing continues without defence witnesses and legal counsel. This point is important because governments often ignore invitations to defend their policies because it would contribute to the legitimization of the LAWT as a judicial body, which may be viewed as adversely affecting their sovereignty.

The event's protocols are similar to those of an ordinary trial. The secretary of minutes and a bailiff assist the jurors with proceedings. The bailiff announces the entrance of the jurors and she will be responsible for general public order throughout the proceedings. The secretary of minutes has two assistants who help record all of the events of the trial.

Throughout the hearing jurors examine the evidence and interrogate witnesses. As soon as they finish hearing witness testimony and cross-examination, the jurors announce that the hearing has finished and like any other trial, the jury then withdraws to deliberate. Their verdicts are announced at the end of the hearings. Once they have begun the case, the jurors order the secretary of minutes to read a summary of the corresponding actions that occurred in controversial cases.

Testimony also follows the pattern of a normal trial. Once the hearing begins, the court gives the plaintiffs the opportunity to present their case. Once they have finished, the accused are given an opportunity to present their defence. Each side has the right to speak for 30 minutes. Once both sides have presented their cases, the jury hears testimony from witnesses and notaries during a 20 minute period in which cross-examination occurs. The court then accords ten minutes to each party to present summaries and conclusions, then they once again present the formal accusation to the jurors who retire to deliberate and pass judgment.

The LAWT and the jurors demonstrate a strong commitment to guaranteeing equality and fairness to both parties in the hearings. They work to ensure equal treatment in terms of testimony, responses to evidence presented, publicity, the right to opportune information, the right to a proper defence and other legal guarantees included in international treaties, constitutions, laws and state regulations. In short, the LAWT is not merely a showcase for the accusation of public officials and the organization of protest. The tribunal works to ensure fairness like any other international court and aims to guarantee environmental and social justice. During the hearings each side can freely express its points of view with respect to the facts of the case that is in dispute. Those claiming to be victims of unfair water policies are usually represented by lawyers from environmental NGOs. State representatives are also invited to participate or be represented legally, but until now, only one city government in Mexico has chosen to do so, in a single case.

The verdicts

The LAWT does not have formal judicial power because its verdicts are not binding and its judgments do not derive from any legitimate legal document. Therefore, it cannot apply administrative, penal or financial sanctions. In their place, their judgments, especially the condemnations, constitute true moral sanctions, as well as moratoriums and social refusal of harmful conduct against the water resources of Latin-American citizens. To ensure valid and legitimate verdicts, the tribunal's decisions are produced after hearing testimony.

As soon as the two parties finish presenting their cases according the protocol described above, the president declares an end to the hearings and sends the jury for deliberation. The juries of the LAWT have absolute liberty and unlimited time to develop their verdicts. These verdicts do not attempt to assign guilt or designate responsibility to one party or the other. Instead, with the objective of contributing to the maintenance of environmental justice, the juries' verdicts contain recommendations for both sides that explain in detail the responsibilities that each party assume in order to resolve the water conflict presented at the hearing. This point is very important because the LAWT's goal is not to blame water authorities, political actors or economic stakeholders for injustices. Instead, its goal is to bring parties together in order to resolve grievances. Thus, the LAWT fosters social participation in water decision-making as defined above rather than organize social mobilization that is one-sided and conflictive. These verdicts, which aim to improve social participation, are then announced to the media and environmental NGOs in order to guarantee impartial monitoring. If anything, this monitoring aspect of the tribunal's work is its Achilles' heel. This is the focus of the following section, which covers the LAWT's involvement in water-related conflict in Guerrero, Mexico, where national and state authorities are trying to displace communities in order to build the Parota hydroelectric dam.

Parota Dam Case

As stated in the previous section, the LAWT seems to demonstrate potential as a facilitator of social participation in water politics because it acts as a quasi-judicial body that focuses on mediation and conflict resolution. This section demonstrates the LAWT's approach in real terms through an examination of the Parota Dam Project case, which was heard by the tribunal at its first public hearings in 2006.

Located in the southern Mexican state of Guerrero, the city of Acapulco has become an internationally known tourist resort that has grown tremendously since the 1950s into a large metropolitan area with more than 700,000 inhabitants. Its market share currently represents 40.5 per cent of Mexico's tourist industry.[10]

The development of the city's tourist industry has attracted many migrants from surrounding areas in Guerrero, in addition to tourists. This demographic and economic growth has placed pressure on strategic resources in the city. Periodically, local residents lose water pressure and electricity. In response to these problems, and in order to guarantee electricity to the city's large hotels, local, state and national officials have proposed a plan to build a large hydroelectric power plant called La Parota on the nearby Papagayo River. Of course, urban inhabitants applaud this measure and authorities have invested significantly in public relations campaigns, including public figures such as Jorge Campos, the ex-goalkeeper of the Mexican national soccer team who comes from Acapulco. Conversely, rural populations would be affected adversely by this project. Specifically, a number of communities would be displaced and, already, state officials have attempted to force these people off disputed land. Thus, the LAWT has become involved in the case.

This case has been filed by the *Consejo de Ejidos y Comunidades Opositores a la Presa la Parota* (CECOP, or Council of Cooperatives and Communities Opposing the Parota Dam) who are the plaintiffs attempting to prevent the construction of dam. These plaintiffs represent more than 20 communities concentrated in a rural area (from the municipalities of Acapulco, Juan R. Escudero, and San Marcos) in the state of Guerrero, which covers approximately 37000ha. The most prominent localities are Aguas Calientes and Cacahuatepec where most of the CECOP's meetings were organized.

As stated in the hearings, the accused include the State Government of Guerrero, the Federal Electricity Commission (CFE), the SEMARNAT and the Agrarian Attorney General's Office. These authorities were chosen because the CFE is carrying out the project with the support of the other agencies. One of the most interesting aspects of this case regards NIMBY (not in my backyard) theories of politics. One of the reasons that the project was pushed through, despite the human consequences in the above-mentioned communities, is that it received widespread support in Guerrero. Most citizens in the State have opposed the CECOP because they disagree with their tactics, which include organizing strikes and marches that block traffic, and forcefully taking control of water containers, temporarily cutting off water supplies to many communities. Thus, those who would be affected by the Parota Dam project have turned to radical and sometimes violent forms of protest whereas those whose backyards are not in the affected area have supported the project out of self-interest. This has increased public pressure on the CECOP to accept the dam.

The LAWT's intervention is especially significant given NIMBY considerations. Public support for the Parota Dam project has not been democratic. Interviews with citizens of Guerrero have demonstrated a clear disinterest in issues related to rights, governance or responsibility.[11] Instead, they are interested in guaranteed services for their own communities. Of course, water resources should not be distributed on such considerations, should sustainability be a policy goal. By shifting water policy from the subnational to the supranational arena, the LAWT has the potential to introduce democratic themes into local policy debates and lessen the impact of NIMBY or even clientelist politics. For example, the decision to construct the Yesca Dam in Nayarit was quickly made in 2006, in part due to the social conflict that the Parota Dam created in Guerrero. The purpose of the LAWT is to harmonize water policy-making principles so that such practices would have to be reconsidered. At the very least, the LAWT places local projects into regional political contexts.

In fact, the Parota project has a wider background that gives it regional significance. It is grounded in the Latin American Puebla-Panama Plan. This plan, signed in 2001 by the presidents of Mexico, various Central American states and Colombia, represents a multi-billion dollar investment in large-scale infrastructure projects in the region, and includes significant investments in transportation, telecommunications and energy sectors. Part of the plan in Mexico includes the proposed construction of the Parota hydroelectric dam. When it was conceived, this was the largest dam proposal in Mexico, which, when constructed, would have flooded 17,300ha with a 192m high dam.[12] According to the opposition parties, this would affect 25,000 farmers in five districts in Guerrero: Acapulco, San Marcos, Juan R. Escudero, Tecoanapa and Chilpancingo. *Espacio de Derechos Economicos, Sociales y Culturales* (*Espacio DESC*),[13] a confederation of 13 civic organizations, including academics active in human rights (including environmental rights) recently stated:

> The project entails significant cultural, economic, social and environmental repercussions that extend beyond the local level alone. The Parota project has provoked opposition among local and indigenous communities in the affected areas and a mounting level of conflict between opponents and supporters of the project. There is now a plethora of social actors involved in the Parota conflict, including state and federal levels of government in Mexico, as well as local, national and international indigenous and civil society organizations concerned not only with the potential detrimental impacts of the proposed project, but also with the abuse of the human rights of individuals and communities opposing the Parota project. (Habitat International Coalition, 2007)

The conflict surrounding the Parota project began in 2003 when the CFE started construction (opening of roads, introduction of heavy machines, perforations (CFE workers were making holes in the ground as a base for the dam's construction), logging and the building of construction worker camps) in the communal land of Cacahuatepec, supposedly with no previous communications to the community about this procedure and with no authorization from SEMARNAT to make changes in the use of land. The community opposed the CFE's entrance into their lands and there were reportedly some damages to communal land as well as human rights violations.

The CFE has attempted to buy land for the construction of a dam project that it argued was necessary to guarantee energy resources to Acapulco. The main problem is that the majority of the members of the cooperative that owns the land do not want to sell and be relocated. The CFE has been accused of utilizing corruption, particularly within the Agrarian Attorney's Office, in order to buy land under the market value or to take it outright. There is a long history of accusations of illegal assemblies supposedly approved by the Agrarian Attorney General but later cancelled when dam opponents could show proof of illegal activities based on corruption (purchase of votes, offering of food and improper military presence).

The local communities in the region where the dam has been proposed have created a CECOP. This group has answered political pressure from the CFE and the state government with protest and mobilization. These actors resorted to numerous actions that promoted illegality, a climate of harassment and violence amongst the populations that oppose the project. The escalation of the political conflict has led to physical attacks in the area, and state government has repressed political protests through heavy-handed police action. Numerous protesters have been injured and, to date, there have been four deaths related to these conflicts concerning the proposed dam project. Due to the repression of the CECOP movement, the local communities have also turned to violence attacking the police and taking the water containers used in Acapulco by force.

In 2006, the LAWT accepted this case, brought to them by the CECOP, in order to attempt to mediate between the two sides. Attempts at reconciling the opposing positions failed because the CFE, SEMARNAT, the State of Guerrero and the Agrarian Attorney General's Office did not send representatives. The LAWT verdict ruled that the project should be at least temporarily halted because:

1 authorities did not consider the damage to public health or to the quality of life of the local population when they examined the project's environmental impact;
2 authorities did not evaluate the project's impact on the supply and quality of water to the city of Acapulco or rural localities that depend on the Rio Papagayo and their study lacks the basin management hydrographic criteria established in national water law;
3 the technical justification for the project is ambiguous and sometimes contradictory and it never mentions the energy needs of affected rural communities;
4 the expropriation of cooperative and communal lands at very low prices contradicts the 27th Constitutional Article;
5 the presence and intervention of the CFE in the region has contributed to the violation of human, civil and political rights, and it has caused the discontinuation of inter-community relations that resulted in clashes between the followers and opponents of the project; and
6 the CFE, until March 2006, had not yet presented a resettlement plan or reparations package to those affected by the dam project.

In general, the LAWT verdict noted that democracy had been violated. Specifically, it stated that the Agrarian Attorney General's Office should guarantee an adequate environment of legality and liberty in agrarian matters for democratic decision-making and that the CFE should avoid carrying out illegal assemblies and disturbing social peace. Also, the state of law should be respected in the region in terms of the rights to:

- information;
- consultation and participation;
- free determination;
- housing, water, food, land and territory;
- development.

The LAWT's specific recommendations in the verdict were:

- To do an integral study of the social, environmental, economic and cultural impacts associated with hydro electric megaprojects.[14]
- To consider social impacts in the environmental impact evaluation.
- To create a new mechanism at the national level (through the Secretary of Social Development, the Commission for the Development of the Indigenous Towns and the National Population Council) to promote public consultation and social participation in decision-making processes for projects of public interest.

- To create a new inter-institutional citizen commission where affected people and social organizations can be represented to promote dialogue and peaceful solutions to conflicts around the dam construction.
- To re-submit the Parota Dam project for evaluation by the National Water Commission, *Consejo de cuenca del Pacifico Sur* and *Comité de cuenca del Rio la Sabana y la Laguna Tres Palos*. It is recommended that these institutions create specific spaces for participation of affected people.
- To exhort federal, regional and local institutions to protect the human rights and the procedural rights of affected people.

Also during March 2006, CECOP and the *Espacio DESC* organized a visit to the affected region for a number of activists and journalists participating in both the International Forum in Defense of Water and the Fourth World Water Forum in order to deepen understanding of the Parota project and its impact on local communities. While the LAWT has sided with these communities in terms of the improprieties involved in the construction of the Parota Dam, it has also criticized them for resorting to violent means of social mobilization. These facts are discussed in the conclusion.

Conclusions

The LAWT represents a constructive criticism of the lack of social participation in environmental politics in all of Latin America and it has the potential to significantly improve social participation in water politics, thus, in my opinion, improving the democratic quality of public water debates.

One could argue that its activities link interest groups in the field of water management so as to facilitate transnational social movements (Brooks and Fox, 2002). However, this is not the goal of the LAWT. Its real innovation regards the quality of debates. Most scholars of social movements (Tarrow, 1989) and urban politics (Orr, 2007; Stoker, 2003) argue that even successful popular movements have limited long-term impacts because they do not have the resources to sustain their activities. The key to long-term success in creating equity is affecting the nature of systems rather than winning battles within them. The LAWT helps institutionalize social participation through its quasi-legal approach. The framework improves the quality and dissemination of public information on water management in Latin America. For example, following CECOP's decision to bring a case to the LAWT, the lawyers representing this group informed the movement's leaders of both the environmental laws that had been violated and the environmental impact of the proposed project. This specific information was unknown to the group, most of whom are poor rural residents. The lawyers diffused their findings publicly, which created more informed opinions among both group members and citizens of Guerrero.

Through these educational activities, the LAWT creates a public space for democratic debate of water issues as participation in these discussions by all parties, including governments, NGOs and stakeholders, is encouraged. While much of the literature on environmental politics at the global level encourages open debates and social participation, it often ignores institutional necessities without which democratic discussions cannot take place. The public space literature argues that citizen participation cannot occur without concrete institutional settings. The LAWT provides more than a forum for NGOs and activists to stake their claims. It facilitates democratic discussions because all parties are invited to participate in public hearings. Of course, the effectiveness of this setting depends on the will of those invited. As Mexican officials ignore the Tribunal, their own positions are not heard, thus, limiting the effectiveness of the LAWT. However, one Mexican city government (Zihuatanejo) recently participated in a hearing. This provides hope that the Tribunal will be able to convince other officials to participate in the future. In fact, the representative from Zihuatanejo was not attacked by anyone from the LAWT or from the hearing's audience. He was treated with respect and his participation in the hearing was greatly appreciated. This may serve as a positive example to other officials.

The LAWT is not just an NGO attempting to publicly denounce unfair water practices. Even though its hearings are public, its initial efforts in the cases accepted are aimed at quietly mediating local disputes. Thus, it is institutionalizing conflict and providing a mechanism for healthy democratic debate. Moreover, all parties are invited to participate in the public hearings and voice their own positions in water disputes. Such actions in other fields, such as human rights, have shown that international mediation of local conflicts often provides the basis for long-term conflict resolution whereas public activism forces interested parties into more radical positions (Korey, 2001). Thus, mobilization may bring about short-term victories but systemic problems are rarely addressed in a sustained way. The case of the Parota Dam project demonstrates this point.

The events of March 2006 clearly indicate that the model proposed by the LAWT holds more potential for improving social participation than those organized within the framework of the World Water Forum. CFE temporarily stopped work on the Parota Dam until the CECOP could explain their demands to them. There are still many rumours that the project has stopped permanently and also that an alternative dam project called 'la Yesca' has started to be built in central Mexico.

Unfortunately, violence between supporters and opponents of the project has been a constant and it has impacted the international context. The LAWT has acted with significant political weight on the international diplomatic scene because, after its verdicts, the Parota project has also been condemned by Amnesty International and the United Nations. The most recent news from Guerrero, however, also demonstrates important achievements of the LAWT. On 10 February 2009 an article was published in a local newspaper stating that the CFE announced the 'eventual' cancellation of the Parota project due the lack of validity of temporary territorial occupation, cancelled assembly acts and suspended negotiations.[15]

The LAWT's focus on mediation and institutionalized dialogue provides an avenue for improved social participation in contentious water debates. Nonetheless, the impact of its decision seems to have temporal limits. While it was effective in stopping violent and contentious behaviour in 2006, the non-binding nature of its decision ultimately led to the refusal of this mediation and a return to animosity one year later. Also, while reports have surfaced that the Parota Dam project has been permanently abandoned, no official declarations have been made, thus rendering the true long-term effects of the LAWT on this case difficult to measure at present.

Nonetheless, the impact of the LAWT's involvement in the Parota case should not be considered a failure. Instead, it should be seen as a temporary success on which to build in order to ensure social participation in Latin American water debates in the future. Even though violence has returned to Guerrero, the opposing parties that are active in the Parota debate met for the first time in two years in an organized forum in May 2007. This demonstrates that institutionalized dialogue persists despite the increase in political tension, and this gives hope for an eventual peaceful solution. Moreover, the LAWT sent recommendation letters based on their verdict to the Mexican government and the state and local institutions involved in the case and these letters were consulted during the period when the project was suspended and planning was revised. Ultimately, a 'human right to water', like all human rights, can be considered a 'right' only when judicial vehicles exist for recourse when an individual feels that his rights are violated. The LAWT is beginning to open this possibility and, consequently, it is spearheading an avenue for the creation of a 'human right to water' in the future.

Should the tribunal's verdicts ever be considered binding by nation states, then Latin America could one day potentially claim the world's first working legal regional water distribution regime. This strategy seems to offer more potential for real social participation in democratic debates concerning water than political protests organized at the international level by NGOs and social movements. These protests denounce abuses and expose unfair water distribution strategies. Nonetheless, the keys to further introducing equity in the world water system involves introducing ethics into international water debates and mediation between conflicting actors who would hopefully adopt these principles and operationalize them in their water policies. The LAWT pursues all of these goals. Thus despite its present limits, it does broaden regional discussions that are characterized by social participation, and the Tribunal mediates conflict, including the encouragement of direct dialogues between citizens, stakeholders and public officials. This is the very definition of social participation.

Notes

1 The United Nations approved General Comment 15 in 2002, explicitly acknowledging a human right to water but this document does not hold any binding

authority. However, the right to water was not mentioned in the 4th and 5th World Water Forum's Ministerial Declarations, whereas water was recognized to be of critical importance for sustainable development and, as a human need but not as a human right, despite the request from several delegations.

2 Where environment is considered to include ecological (biological), physical (natural and built), social, political, aesthetic and economic environments (www.nchre.org)

3 According to Plumwood (2004), the concept of environmental justice has been employed especially to focus on the distribution of environmental risks, harms and benefits among human populations. One of its major theorists, Robert Bullard, writes that environmental justice raises questions of:

> differential exposure and unequal protection ... the ethical and political questions of 'who gets what, when, why and how much'. (Bullard, 1999, p35)

Because socially privileged groups can most easily purchase alternative private resources (clean water, for example), they have the least interest in maintaining in generally good condition collective goods and services of the sort typically provided by undamaged nature.

4 The Dublin Statement on Water and Sustainable Development, www.wmo.ch/pages/prog/hwrp/documents/english/icwedece.html

5 World Water Forum home page at the World Water Council website: www.worldwatercouncil.org/index.php?id=6&L=0, accessed March–May 2008

6 Objectives of the Mexico 2006 World Water Forum, www.worldwaterforum4.org.mx/home/genwwf.asp?lan=

7 In *Democracy and Disorder*, Sydney Tarrow (1989) sets out a theory of social movement cycles in which he argues that movements begin when radicals call for social mobilization. However, once the mainstream gets involved in the political issue, these radicals actually hurt the movement more than they help because their ideological inflexibility and sometimes violent behaviour causes the movement to lose mainstream support.

8 The author was the director of the regional Office of the Heinrich Böll Foundation Mexico, Central America and Caribbean.

9 Tribunal Latinoamericano del Agua, 'Justicia para el Ambiente', p5. Descriptive LAWT document distributed during the first hearings in Mexico City 2006.

10 Source: Royal Caribbean International, based on data from SECTUR (Statistic Information System DATATUR).

11 The author conducted interviews with citizens of Guerrero about the proposed Parota Dam in December 2005 and March 2006.

12 La Yesca Dam project (www.cfe.gob.mx), in construction since January 2008, is now the largest dam under construction in Mexico, with a 220m high dam curtain and covering almost 4000ha. It is expected to be one of the three largest dams in the world.

13 *Espacio DESC* is an NGO established in 1998. It is formed of 13 civil society organizations dedicated to human rights (economic, social, cultural and environmental), and development in Mexico. *Espacio DESC* has taken on the case of La Parota as one of its priorities since 2003.

14 In the case of the Parota project it would be essential to integrally evaluate:

   1 the risks associated with the seismically active location for the project;
   2 ecological failures;
   3 the collateral damage of the destruction of the forest on the aquifer cloaks and hydraulic basins in the area;

4 the irreversible damage to the flora and fauna, as well as to public health;
5 the violation of human rights of the 25,000 inhabitants affected in an indirect way by the project.

15 www.lajornadaguerrero.com.mx/2009/02/10/index.php?section=sociedad&article=005n1soc, accessed 21 April 2009.

References

Aboites Aguilar, L. (1998) *El Agua de la Nación: Una Historia Política de México, 1888–1946*, Centro de Investigaciones y Estudios Superiores en Antropología Social, México

Bogantes, J. and Borrero Navia, J. M. (2000) 'Documento de procedimientos', Tribunal Latinoamericano del Agua

Brooks, D. and Fox, J. (eds) (2002) *Cross-border dialogues: US–Mexico Social Movement Networking*', Center for U.S.–Mexican Studies, University of California San Diego, La Jolla, CA

Bullard, R. (1999) 'Environmental justice challenges at home and abroad', in N. Low (ed) *Global Ethics and Environment*, Routledge, London, pp33–46

Cornelius, W. (2005) 'Mexicans would not be bought or coerced', in G. M. Joseph and T. J. Henderson (eds) *The Mexico Reader: History, Culture, Politics*, Duke University Press, Durham, NC, pp684–686

Garcia-Acevedo, M. R. and Ingram, H. (2004) 'Conflict in the borderlands', NACLA report on the Americas, vol 38, no 1, July–August, www.nacla.org

Habitat International Coalition (2007) 'La Parota dam follow-up. Guerrero´s struggle against the construction of the Parota hydroelectric dam (Mexico)', www.hic-net.org/articles

Helfrich, S. (2006) 'Reasons to support the Latin American Water Tribunal', www.boell-latinoamerica.org/download_en/2006-04-05artikel2TLA-silke-ingles_corr.pdf

Huber E. (1990) 'Democracy in Latin America: recent developments in comparative historical perspective', *Latin American Research Review*, vol 25, no 2, pp157–176

Korey, W. (2001) *NGOs and the Universal Declaration of Human Rights*, Palgrave Macmillan, New York, NY

McLean, J. (2007) 'Water injustices and potential remedies in indigenous rural contexts: a water justice analysis', *The Environmentalist*, vol 27, no 1, pp25–38

Maganda Ramírez, C. (2004) 'Disponibilidad de agua, un riesgo construido.vulnerabilidad hídrica y crecimiento urbano-industrial en Silao, Guanajuato, México', PhD dissertation, Centro de Investigaciones y Estudios Superiores en Antropología Social, Mexico

National Center for Human Rights Education (2006) 'Environmental justice and human rights fact sheet', http://nchre.org

Orr, M. (2007) *Transforming the City: Community Organizing the Challenge of Political Change,* University Press of Kansas, Lawrence, KS

Petrella, R. (2001) *Il Manifesto dell'Acqua. Il Diritto alla Vita per Tutti*, Edizioni Gruppo Abele, Torino, Italy

Picolotti, J. M. and Crane, K. L. (2005) 'Access to justice through the Central American Water Tribunal', in C. Bruch, L. Jansky, M. Nakayama and K. A. Salewicz (eds) *Public Participation in the Governance of International Freshwater Resources*, United Nations University Press, Tokyo, pp460–473

Plumwood, V. (2004) 'Environmental justice, in institutional issues involving ethics and justice', in R. C. Elliot (ed) *Encyclopedia of Life Support Systems (EOLSS)*, UNESCO and Eolss Publishers, Oxford, www.eolss.net, accessed 20 September 2007

Schlosberg, D. (2006) *Defining Environmental Justice: Movements, Theories, and Nature*, Oxford University Press, Oxford

Stoker, G. (2003) *Transforming Local Governance*, Palgrave Macmillan, New York, NY

Tarrow, S. (1989) *Democracy and Disorder: Protest and Politics in Italy, 1965–75*, Clarendon Press, Oxford

United Nations Development Programme (2006) 'Human Development Report 2006: beyond scarcity, power, poverty and the global water crisis', http://hdr.undp.org/hdr2006

14

The Local Application of Global Sustainable and Participatory Development Norms in Turkish Dams

Stéphane La Branche

Introduction[1]

The research presented here is based on a 15 year research programme on sustainable and participatory development in several countries. Two main questions are raised:

1 How do actors translate and implement global participatory and environmental norms (or not) into local projects, and behaviours and values, at the level of the individual?
2 What is the link between participatory democracy and environmental protection?

Democratic procedures are not necessarily ecological, and thus one could see conflicts between globally defined environmental norms and local democratic ones emerge. Indeed, cases abound of civil society's refusal to adopt ecologically friendly energy-producing methods (such as wind power), ecological taxes or changes in behaviours associated with transport by car. Yet, the argument that participation as the basis of sustainable and participatory development (SPD) is the best way to achieve environmental protection, is put forward by the Report of the World Commission on Dams (WCD) and by many new water management and water governance approaches. One of the questions I addressed in my research was: since environmental and participatory democratic norms are not the same, even though they are linked to each other in actors' discourses and in public policies, then are these two sets of norms adopted to the same degree by opponents to dams? What are the implications of this for the depth of environmental and democratic values in actors? I will develop this further on.

Drawing from research on dams undertaken for the French Energy Council with comparisons between France, Quebec and Turkey, this chapter will first present a summary of the theoretical framework, a constructivist version of regime theory, which addresses the issue of ideas, norms and beliefs as well as legitimacy and their role in international regime construction. This approach deals with the question of how global norms and rules, and in this case, SPD norms, are diffused – or not – at the local level into, for example, public policy regarding citizens' participation and environmental protection in the decision-making process and then, how these norms are – or not – translated into values. Such values would imply that the population may want to go further than institutionalized procedures and develop a desire for more participation in countries where these institutions do not necessarily exist. In the case of Turkey, opponents' arguments and strategies are linked to global environmental and participatory governance norms, protocols and institutions in their local efforts at blocking the construction of dams. But what is the weight of environmental values relative to democratic ones? The aim of the research on dams was to understand the process by which international SPD norms, both participatory and environmental, are translated or not into local practices and individual beliefs.

For the purpose of this chapter, the SPD notion is used rather than that of integrated water management (the differences and common points are beyond this paper's scope) as the latter was not used by Turkish actors. Indeed, whether in their texts or in discussions, dam opponents refer explicitly to sustainable development and do not use the model of water management, integrated or not. In addition, dams are not as much part of a water governance framework in the national public policy scheme, as they are inscribed in an energy policy strategy. Also noted is that the notion of sustainable development is used throughout the Report of the WCD while that of integrated water management is conflated within this approach (World Commission on Dams, 2000). Created by the World Bank following a seminar on controversies related to dams, the WCD published its famous report in 2000 after a large-scale comparative study of the environmental, social and economic impacts of dams. Interestingly, the WCD no longer exists; only its report and internet site remain, as it has not undertaken any studies since publishing its results. Yet its report continues to play a role in propagating these norms in dam projects, as is revealed by the large number of international and national non-governmental organizations (NGOs) opposing dams and making reference to it, as uncovered in our study. In the WCD report, SPD is put forward as the only solution, the argument made being for its integration into all technical and political aspects and into all phases of a project. However, the report did not address some fundamental problems, among which were the tensions between participation and environmental protection, such as a population-level refusal or ignorance of ecological imperatives and of international environmental protection objectives. The problem is that we are far from understanding all the expected and unexpected effects of SPD but it has become the dominant regime in matters of national and international development everywhere, including that of dams. SPD is the core framework within which solutions to conflicts over water and dams are offered, and so cannot be left out of the discussion; yet nothing in SPD guarantees that it can resolve conflicts to the satisfaction of all, nor in favour of the environment. But before we go any further with this discussion as it applies to Turkey, it is necessary to present the theoretical framework.

Regime Theory and SPD in Dams

Of all international relations (IR) theories, regime theory best explains the process by which norms may become a relatively well coordinated set of rules supported by institutions and procedures: that is, a regime (Rittberger, 1993).[2] Indeed, it attempts to explain why and how there is cooperation on the international level and the effects of this cooperation on national and local public policy. However, even regime theory fails to explain how and why legal or political norms become legitimate to individual actors, in other words, how they transform themselves into individual level beliefs. My general working hypothesis, which this study only partly addresses, is that in the final analysis the persistence, efficacy and efficiency through time of a regime depends on the actors' perception of its legitimacy, that is, the belief that the norms and principles put forward by the regime are true, ethical or good – and one might add – capable of being implemented (La Branche, 2003a). The overall question raised in our case study is: are global participatory and environmental SPD norms translated into national public policy and, if so, what is the role of civil society in this process? Then, do democratic and environmental norms and principles have the same importance and role in civil society in their opposition to dams? To answer these questions on the empirical level, several theoretical and methodological issues arise:

1 The problem of multilevel analyses. Indeed, the hypothesis implies that we should be able to observe a relation between individual-level beliefs and the global level. This difficulty cannot be avoided, because of the very nature of the problems raised by dams and SPD: environmental issues are inherently multilevel as well as multidisciplinary; they involve different types of actors as well as the physical and natural world, animals, people and technology.
2 The issue of values and ideas. IR theory is wary of psychological factors (Rosoux, 2001) and has not on the whole developed the tools to integrate these into its analysis. This explains in part why even constructivist regime theory, even though it focuses on the role of ideas and values in international relations, still suffers from a serious lack of empirical research to demonstrate its points (La Branche, 2003b). Yet, this school has shown

> that epistemic communities, that is, scientists who organize around a common, agreed-upon body of knowledge and ideas and who try to influence political decisions and public opinion to solve a problem, have played a key role in the construction of environmental and participatory regimes. At the same time, the conceptual and methodological difficulties linked to the issue of ideas and opinions and their role in constructing regimes explain why so few internationalists have attempted to deal with the issue of legitimacy. In other words, my argument is that the legitimacy of global norms rests upon individual or local actors' perception of the truth or of the ethical weight of these norms rather than on the capacity by a powerful actor to impose them from the top.

If one is to find solutions to water-related problems then, it is necessary to engage in this type of theoretically grounded empirical research. Indeed, concentrating on the issue of legitimacy in dams allows us to go beyond a coercive explanation of power as a force that allows an actor to oblige others to obey him. Environmental issues are indeed a good illustration that coercion and domination are not the only forces, and actually may not even be the most significant in a regime's construction, persistence and demise. This hegemonic view is therefore faulty because, among many reasons, it cannot account for the fact that less powerful global and local actors play an essential role in the emergence and diffusion of SPD generally and in dams, specifically (Young, 1989). As we will see, the Turkish university professors, engineers, archaeologists and lawyers who lead the opposition to dams, constitute such an epistemic community and they have played key roles.

A major empirical problem for regime theory and for this research on SPD in dams is this: how can one show that rules and norms derived from a regime have power and influence of their own, that is, that they are legitimate? For Hurrel (1993), a regime is based on its legitimacy, which comes from a shared sense of belonging to a community, whose rules serve as a link between actors and institutions. There is, however, a problem of level of abstraction with this definition since institutions and rules cannot be the origin of legitimacy; they can only be its object. Legitimacy can only rest, fundamentally, on individual beliefs, whether scientifically true or not. As Hurd (1999) argued, it is a relational quality between actors, defined by their perceptions of an institution or a rule. A norm becomes, at the institutional level, a rule, and it can then influence behaviour and contribute to the actors' definition of interest (Hurd, 1999). Legitimacy is thus linked to values but not to norms. For the sake of our present argument, a norm is a rule or an institutional culture imposed by an actor on others who then follow it because they must, usually for fear of retaliation, cost, scandal, bad publicity or fines. A value is a belief by actors that a norm, behaviour or an idea is good or true, and, thus, that it should be obeyed because of its intrinsic worth, that is, it is directly related to legitimacy.

In the case of Turkish oppositions to dams, are actors putting forward environmental arguments because they believe they are right, good or true or because they seem a good way to achieve their goal? The latter could include 'not in my backyard' (NIMBY), more democracy, or better financial or material compensations. I put forward the suggestion that the SPD regime's diffusion and internalization at local and individual levels in part depends on the actors' internalization of these norms (the processes by which a norm is transformed into belief at the individual level), that is, its legitimacy. Thus, legitimacy becomes fundamental to theory since without actors' beliefs, one can only talk of an imposition of rules, a process that is highly likely to engender opposition and resistance. Is this taking place in Turkey in the field of dams? If opponents' goals are more money or better material compensations, such as new housing, then one could conclude in part that the democratic and environmental principles inscribed in the SPD regime are not the main reasons for opposing dams and, thus, that this regime is not perceived as legitimate by actors. Instead, it would be used instrumentally.

Two more points need to be made before turning to the Turkish case:

1. The discussion on legitimacy has considerable import for empirical research on international regimes, since it implies one has to understand individual values and strategies, for example, through participant observation and interviews as well as organizational and discourse analysis. Are Turkish opponents to dams referring to global or European norms in their arguments? Do they make use of international networks to gather support and pressure their government into abandoning the project? Finally, what is the importance of their environmental arguments and values in their strategies: smokescreen, strategic or founded on deep values? The field research first drew a portrait of the actors' characteristics: position, objectives, strategies, arguments and beliefs related to the environment and participation. I concentrated on two opposition Turkish associations: the Committee for the Evaluation of Historical Impacts of Dams (CSBCT), a national NGO involved in opposing large dams,[3] and the Foundation for the Protection of the Camlihemçin Valley in the Firtina region, which opposed a very small dam. In each case study, I focus on the links between the global and the local and on the weight of the environmental arguments and values relative to democratic ones, with actors involved in the controversy and opposition strategies.
2. Participation does not only take place through formal political mechanisms, such as participatory procedures or politically militant action; social actions are also possible, through information campaigns and teaching opposition strategies. For the sake of this discussion, I will distinguish political participation from social participation in that political participation takes place through formal procedures and institutions while the latter rather refers to informal and non-institutionalized means. While regime theory explains the global diffusion of formal participation norms related to institutionalized political participation (rather than social participation) rather well, their translation into social practices is more

difficult to evaluate in that framework. However, for the purpose of this paper, this question can be put aside, since the significant actors opposing dams in Turkey mostly use formal and institutionalized methods rather than informal social participatory methods and strategies, for reasons explained later. Turkish local actors (motivated by NIMBY) and other local/national people do take action but they seem to have little direct impact on the final decision, although they do have an indirect effect through their ability to gather international and European support. These international allies consider that the local demands are based on democracy and environmental values, thus they support local actions, hence increasing pressure on the Turkish government and dams constructors. It should be noted that these national and international coalitions are organized and headed by an elite composed of activists, academics and lawyers, indeed making up an epistemic community in the classical sense described by regime theory.

The Environment and Turkish NGOs

While Turkey was for a long time an authoritarian country, the last two decades have seen an explosion of local NGOs involved in matters of local democracy and environmental protection. This has been in part supported by Turkey's efforts at becoming a member of the European Union (EU). Nevertheless, improvements in democratic structures, culture and procedures are recent and the Turks interviewed still do not have a high level of confidence in their police and military forces when it comes to rights (Dorronsoro, 2005). On the other hand, they do seem to trust the judiciary system (judges and legal court proceedings) and they refrain less and less from criticizing their own government in public and in daily newspapers. This democratization process has played a significant role in the rise of oppositions to dams, aimed either at cancelling or modifying a project or at getting better compensation.

Today, there are over 300 dam projects proposed in Turkey, which is not surprising since they play a key role in economic and energy development strategies, especially with high oil prices and corresponding increases in dependence on other types of energy resources. But beyond energy needs and on a more subtle level, dams are also integrated into a more general economic, political and technical modernization process and they are linked to the process of constructing a modern Turkish national identity. Dams are thus a source of national pride and are part of a visionary national project for a wealthy Turkey on a par with European countries. This political discourse and the drive for economic development also help to explain why democratic procedures in large-scale infrastructure projects came so late in Turkey; democracy was not a high priority. Yet, in the last few years, participatory procedures have come to be increasingly recognized as being a legitimate way of reaching a decision by both government authorities and civil society. At the same time, local NGOs are also gaining in legitimacy and in legal and political recognition, as demonstrated by their capacity to exert pressure in a number of projects.

In the case of Camlihemçin, a local NGO brought the central government to the administrative court and won, with the assistance of independent local media and international NGOs. While it is not easy to counter a nationalistic debate, opponents argue that Turkey is producing more energy than it needs and that the social costs, in particular, population displacements with over 300,000 people a year migrating to Istanbul, as well as the environmental costs, are not worth it. Technical arguments have also been put forward as many of the projects were drawn up so long ago that they are not technically appropriate and, in other cases, construction started so long ago that the dams will be obsolete by the time they start functioning. Clearly, NGOs opposing dams play a role in both the democratization process and environmental protection, but what are their strategies and arguments? We will see that these local NGOs make use of the global level in their opposition to dams, making reference to global sustainable development norms, but what is the weight of the environment relative to economic compensation, democratization or NIMBY in their arguments and strategies? An unexpected aspect that also emerged has been climate change. The EU's lead on climate change in particular has had direct and at times contradictory effects on government energy policy, its discourses, as well as NGOs discourses and arguments in their social and political participation strategies.

The Effect of the EU's Climate Policy on Turkey's Environmental and Energy Policy

While it is not bound by international nor European agreements in matters of climate change, Turkey is well aware of the importance of the general environmental question for its EU candidacy and has thus adopted a strategy aimed at moving closer to EU environmental norms since 1998. But this raises a difficult issue, putting the Turkish government in a contradictory position. On the one hand, environmental protection, including policy implementation, is a strong EU policy, but dams have serious negative impacts on local ecosystems. At the same time, Turkey must show the EU that it is making efforts to reduce CO_2 emissions, and dams offer a solution since they produce very few greenhouse gases. Thus on the one hand, the EU gives Turkey a bad grade on environmental issues, criticizing its dam development programme, which, on the other hand, is a valid response to the fight against climate change.

The EU is not alone in facing this contradiction: local opponents and the national government are also caught in the same paradox: while dams have a negative local impact, they are also a form of renewable energy that can contribute to the fight against global climate change. Some Turkish actors who oppose dams recognize this issue. When pressed about it in interviews, almost all individuals ran into a cognitive dissonance, that is, they faced contradictions between different sets of values or between values and behaviours, which they resolved by focusing on the local impacts. Only actors with a high degree of education or those involved in environmental controversies on a continuous basis (with a full-time salary in environmental protection or with lengthy involvement) acknowledged the difficulty while pressing for a reduction in consumption pattern and changes in living styles for the wealthy classes and countries.

These numerous national controversies and local opposition to dams are supported by international NGO networks, such as Greenpeace or the World Wide Fund for Nature (WWF), which played a role in getting the Turkish government to accept the study of the World Commission on Dams of its Aslantas project in the Ceyhan region. At the same time, the government created public information and consultation mechanisms, such as the publication of environment and social impact evaluation reports on the internet as well as a forum for answers to questions, which are then used for information campaigns in daily media. As for whether this national public opinion is taken into account, the debate remains open. But the Aslantas region, as well as the large Greater Anatolian project, raised another issue, connected neither to democracy nor to the environment: the flooding of Roman and Hittite archaeological sites, some nearly 3000 years old.

The Archaeological Dimension in Opposition to Dams

The archaeological aspect in controversies over dams emerged in Turkey well before environmental concerns and it brought to the forefront a specific set of actors: archaeologists and the national chamber of archaeologists. These actors have an advantage over most NGO activists, as several told me in interviews. Being university professors, they are not in danger of losing their income and are thus less vulnerable to pressures from state agents and dam constructors. They can, thus, get involved in fighting dam projects without great economic risk and they have a strong scientific legitimacy, which they strengthen through references to international conventions on historical sites and the United Nations Educational, Scientific and Cultural Organization (UNESCO). With respect to regime theory, this intellectual elite, along with lawyers and the national chamber of engineers leading the opposition movement, corresponds closely to the definition of an epistemic community. For example, it played the major role in getting the Illisu Dam project cancelled, by convincing European banks to withdraw their loans.

Planned in the 1990s, the Illisu Dam would have been the second largest after Ataturk. The government was well aware that problems would be raised and that there would be opposition so there was an attempt to respond to these issues from the plan's inception. Thus, it implemented measures to save some archaeological sites such as Hasankeyf with the help of foreign scientists, by moving parts of it and sending some pieces to museums (NGOs are not the only ones making use of foreign allies). The opponents felt this was not sufficient and called for assistance, launching a European-wide anti-Illisu campaign, based on three main arguments:

1 the dam would have caused the displacement of about 65,000 people, most of them Kurds;[4]
2 it would cause some irreparable ecological and archaeological damage;
3 the quality of the water flowing to Syria and Iraq would have been seriously damaged and so could create inter-state tensions.

In 2005, the largest Swiss bank, USB, along with Balfour Beatty (Great Britain) and Skanska (Sweden), all cancelled their loan offer for the Illisu Dam project, three years after having signed the agreement. The rejection was based on the proposed dam's probable social and environmental impacts, as argued by a Western European ecological lobby, including Greenpeace and WWF, that had come to the aid of Turkish associations. Local Turkish actors called upon those at different levels in the EU to continue to pressure other European national and private actors in order to have the project cancelled. And in this, they succeeded.

The Arguments and Strategies of the Committee for the Evaluation of Historical Impacts of Dams

One of the major actors in this movement is the Committee for the Evaluation of Historical Impacts of Dams (CSBCT). A national NGO founded in 2000, its main role is to follow and evaluate all steps of a dam project, from conception to post-construction management. It was especially involved in the Greater Anatolian Project (GAP) and Ceyhan Aslantas Project (CAP) dam schemes. The organization aims to find a workable integration between electricity production, social issues, and environmental and historical protection through a multidisciplinary approach. Its members, essentially archaeologists, engineers and hydrologists, argued against the negative ecological impact of dams on flora and fauna on the regional ecosystem, as well as on the social impacts on farmers' and fishermen's livelihoods. This epistemic community thus seems to put the needs of local communities ahead of its own needs and professional training.

Its major position paper drafted in 2002 stressed the negative social, natural and historical impacts of dams as well as financial overruns. In order to address these issues, the association, which worked as a true epistemic community, offers scientific solutions coherent with an SPD framework. It proposes:

- A basin-based, multi-actor management approach to dams (similar to terminology used in an integrated water management approach) that includes environmental and archaeological protection.
- On the political level, a geographically local basin authority, to be integrated into the regional public administration but coordinated at the

national level. Its role would be to select priority zones, both natural and archaeological, to be protected.
- To inscribe its SPD approach in a legal framework so as to make it binding on all actors, including the government.

Thus, all phases of a dam project would be based on a multiprofessional, multidisciplinary, multi-institutional, equalitarian and democratic framework. This would imply engaging in pre-project social, historical and environmental impact evaluations and, when necessary, the government would be under the obligation to guarantee enough time and funding to investigate archaeological sites, even moving them to another location, which is a time- and money-consuming operation criticized by dam promoters.

As one can see, the solutions proposed by the CSBCT go much further than simple criticism, requiring real competences and skills not necessarily available to all civil society actors. Also, it went even further by making reference to international protocols in its arguments, such as the Malta Convention, in which all large public works from inception to construction must respect natural and cultural heritage and be implemented in association with all pertinent and competent stakeholders. The association also constantly reminds the central government of its own international obligations in matters of signed agreements and protocols over water, SPD and dam issues. According to these agreements, Turkey must finance archaeological protection measures, including moving historical sites to safe areas and it has to make public all pertinent information regarding a dam, including the rules and laws regarding participation and archaeological and environmental protection. European directives and legal frameworks are also used in the association's arguments, going so far as to suggest that if the EU does not implement its own rules by pressuring the Turkish government into respecting its own legislation, then the association would consider taking it to the European Court of Justice or the European Court of Human Rights.

Legal arguments are central to the CSBCT pressure strategies. This necessitates specific and high-level skills and knowledge regarding constitutional, international and national law. Thus, it is not within everyone's reach, which raises the question regarding the breadth of the democratic process of these controversies. However, the NGO does not have much freedom since taking to the streets and demonstrating does not appear to them as a 'very wise nor successful strategy', as they put it, since they still fear police crackdown. This is in large part why opponents to dams make use of legal and administrative tribunals in Turkey. Nevertheless, these difficulties did not prevent the Association for the Protection of Camlihemçin, a small local association, from democratically participating in and even leading a successful fight against a dam.

Camlihemçin: A Small NGO Takes on the Turkish Government

Camlihemçin is a small river flowing in Firtina, a deep and narrow gorge that attracts tourists, and is connected to the Black Sea. It is a humid ecosystem unique in the country and home to an endemic species of salmon. The government decided, without consultation, to build a very small dam. The project soon faced opposition by local fishermen and tourist-related businesses that depend on the valley's aesthetics. This project was very different from the GAP and CAP projects opposed by the CSBCT, as it was very small, inexpensive, and appeared to have only small and localized social and ecological impacts. The dam was to produce 592.7 million kWhrs thanks to several tunnels, which would, according to government numbers, funnel 96 per cent of the Firtina and 97 per cent of the Ayder River – the strict minimum for energy production. With a small local population, mostly composed of farmers and fishermen, the government was not expecting much opposition. Nevertheless the state followed the prescribed legal phases, including a social, ecological and economic impact study. It was this study that became the focal point of the controversy and the basis for the opposition, eventually taking on an international dimension.

The official evaluation report

It started with the publication of the evaluation report, adopted by the Ministry of the Environment but contested by a local NGO, the Foundation for the Protection of the Camlihemçin Valley, whose members are located mostly in Ankara. Several events took place at once. First, informal pressures by the Vice Prime Minister on local elected officials became public, which led to the public's counter-reaction, led by the local media that then started investigating the project closely. They discovered that, in all probability, the dam would have permanent and very detrimental effects on the local aquatic ecosystem, especially the unique salmon species that could be endangered or made extinct.

At the same time, the Foundation received the complete official report from a contact within the government. Many of its members have a high level of social skills, education and networks, which would play a significant role in later opposition strategies. Foundation members realized that the impact study was scientifically unsound as they discovered that the report and the underlying analysis were the work of a young, unqualified biology student. In addition, the student undertook the work as a summer job and was paid by the company planning to build the dam. Yet, according to law, such impact evaluation must be undertaken by neutral and competent specialists. The Foundation thus decided to undertake its own counter-study, which requires several types of specialization: sociology, biology, hydrology, economics and also legal expertise since the Foundation decided on the basis of its own report to file charges against the government in an administrative tribunal.[5] The counter-report exposed several serious methodological problems in the original report, as well as its serious legal and scientific failures, and reached very different conclusions. On this basis, the judge accepted the counter-report.

As in the case of the CSBCT, such a strategy requires skills that are not necessarily widespread. How did this small local NGO gather such competence? Expertise was mostly the result of the inclusion of members from the national chambers of engineers and hydrologists; several of its members were also lawyers. This association has become rather well known for the high level of education and training of its members, which are not entirely representative of the Turkish population. Furthermore, the use of efficient strategies reflected the skills of its members.

The Camlihemçin case in the administrative tribunal

The Foundation's main approach was to use legal arguments and procedures, integrating environmental and participatory democracy along with national legal norms. The ethics of democracy and environmental protection were almost left aside in court, not being applicable in this context. In court, the Foundation's lawyers put forward that neither the principle of public information and participation nor national and international environmental protection norms had been respected by the government. The court considered that the Foundation's claims were sufficient to warrant opening a court case, in part based on the counter-report. At the same time, the NGO did not limit its actions to the judiciary: it launched into a public information campaign that was supported by local newspapers. Our analysis of several months' worth of daily articles appearing in the local press shows that the discourse used by the NGO was far more ethically-oriented than judicially-oriented.

Up to this point, only the Ministry of Energy had been involved, the other ministries, such as that of the Environment, remained neutral. But at the first tribunal session, the government was taken by surprise: the Foundation was represented by eight lawyers, assisted by scientific experts: hydrologists, engineers, biologists and other specialists. The government asked the court for an adjournment, which it received. When it came back to court, the Ministry of Energy was assisted by others: Environment, Tourism, Economy and Public Works, all having received an order from the Prime Minister's office to get involved. All experts and lawyers representing the NGO side were volunteers, most coming from the Camlihemçin area. To the government's economic and legal arguments, the Foundation responded with its own legal, democratic and environmental arguments, attacking the legal foundation of the official socio-environmental impact report.

The Foundation first presented a scientific evaluation and analysis of the report, undertaken by three university professors born in the region, assisted by the lawyers who associated the scientific analyses to legal and constitutional arguments. While the Foundation questioned the very existence of the project and the necessity of the dam, it insisted that it was not against all dams per se but only against the use, value and pertinence of this one specific Camlihemçin dam, a typical NIMBY argument that has also been used by other opponents of dams in the country. In its court argument, the Foundation also referred to national, European and international democratic and environmental norms, arguing that the government was not respecting not only these supra-national norms but its own as well, and more specifically its own constitution.

As regards the general population and the media, our interviews and our analysis of newspaper articles strongly suggest that some of the other local NGOs opposing dams, those who did not play a major role as they did not get involved in the court case, had organized themselves as ecological associations only for the controversy, most disappearing after the case was closed. Most local businesses and local decision-makers were far more interested in the economic impact on fishing and tourism than on ecological protection. Seville Ulcer, archaeology professor at the University of Istanbul, feels that for the most part, Turkish opposition groups to dams are not interested in the environment as such. In her involvement in many controversies and as a former head of the National Chamber of Archaeologists, she often noted how local opponents who use ecological arguments cease all form of opposition if enough money is offered as compensation. She feels that it is 'rather natural since these populations are often poor or disadvantaged'. She also notes that the environment is not included much in daily conversation. In general, the daily media do not raise the issue, except in the case of local oppositions to dams and usually from the angle of democracy. These local media are often left wing and very pro-local democracy and they often play a significant role in raising popular consciousness about environmental issues even if this is not their main goal.

The Foundation made use of these media and of popular pressure through public meetings. While this did not influence the court, it did put pressure on the government, forcing it to develop its strategy and defence, which is significant considering that the project was not considered to be important. The Foundation's arguments to the public differ somewhat from those used in court: less scientific and less environmental, placing more emphasis on democracy, the valley's beauty, its tourism, ancestral ways of living and the negative economic impact through the loss of livelihoods. It reminded the population that it was never consulted on the project, that it had been imposed, and that it was not even necessary in terms of energy production for the region or for the nation. In court, however, the environment took a fundamental place in the argument.

One of the first ecological arguments was tied to global actors. The Foundation reminded the court that both WWF (which also gave its assistance in gaining international public attention through a broad campaign) and the UNEP included the valley on the list of the 100 ecosystems to be preserved on the planet. In a way, even the Turkish government agreed to this since a large part of the valley is located in a national park while another section is a protected site. The Foundation thus emphasized that building a dam, with the roads and power lines associated with it, would go against this protection. The court agreed to take the environmental arguments into account, but on the basis of their scientific foundations, not for ethical reasons, recognizing de facto the value and legitimacy of the counter-evaluation report submitted by the Foundation. The court also acknowledged serious flaws with the official report. For example, it recognized that the trees cut for the dam's construction would not be replaced by other trees, but rather by small brush that would very probably contribute to a decrease in the area's humidity, negatively effecting the local ecosystem. As for the aquatic ecosystem, it concluded that it would simply disappear, including the endemic species of salmon. The court also agreed with the Foundation that the initial report largely underestimated the impact of the construction process, with its roads, power lines and human presence that would have necessitated the cutting of an extra 45,356 trees not accounted for in the initial report, nearly the equivalent of those felled for the dam itself.

My analysis of both the counter-report and the court's reasons for its decision suggest that the environment did not simply appear as a secondary argument or as one that is used as a means to justify other goals, such as disguising NIMBY arguments. The environment was indeed the basis for the decision to cancel the project, in a legal framework of course, as this is the court's function. The court went further with the recognition of this dimension of SPD than in many other instances. This appeared in the court's reference to judicial texts and constitutional articles on environmental protection, national parks and natural reserves. On 11 February 2001, the court handed down their verdict. The judges declared that, in view of such an unreliable environmental impact assessment, the probable environmental cost was in all likelihood too high and unacceptable so it cancelled the project.

The Global and the Local in Turkey

The preceding analysis shows that the two associations analysed, CSBCT and the Foundation for the Protection of the Camlihemçin Valley, both acted as genuine epistemic communities. They made regular references to the global arena both in terms of protocols mentioned but also in terms of allies sought. They also put emphasis on the environment but in an unequal fashion. Indeed, while the CSBCT does mention the environment as an issue, its emphasis really is on archaeological protection, the environment being secondary. For the Foundation, the environment appears as the most important issue, as there are no archaeological sites in the area. So, what does the analysis teach us regarding the depth of environmental values among the actors involved and the links between the global and the local/individual levels?

The organizations studied made references to global protocols (environmental, historical or democratic) for two main strategic goals:

1. Putting pressure on their government by reminding it of its global responsibilities and forcing it to respect these norms, often translated into national rules and laws, some of which are even inscribed in constitutional frameworks.
2. They increased the legitimacy of their position and actions relative to the media and population. Global norms are indeed well appropriated by some of the opponents. There is a feeling among interviewees that not having recourse to these global norms and allies would be detrimental to their objectives and strategies, 'it would not look serious'; 'people would not trust us as much'; 'government representatives would not listen to us with as much attention'. Thus, it would seem that reference to the global level is both seen as an opportunity but also as a strategic obligation. This also indicates qualitatively that the global regime is in part seen as being legitimate by local actors, thus contributing to my idea that parts of the global SPD norms do indeed have roots at the local level.

As mentioned earlier, these references to the global level have implications for the type of actor involved. Even when one knows these norms, knowing how to handle the language and acquiring the necessary resources and skills to get the information is not within everyone's reach, even in the age of the internet. Actors thus need to be relatively well educated, with a high level of social skills, but this is especially true in Turkey where the level of formal education among the general population is lower than in more industrialized countries. In Turkey, organized opponents form an epistemic community in the sense given by regime theorists; these lawyers, engineers and archaeologists are scientists who structure themselves as a pressure group in order to influence political decisions. But the national political context also plays a role. In Turkey, street demonstrations in opposition to dams are not very feasible, as there is a potential for a strong police crackdown, so it is necessary for opponents to use more formal, institutionalized means.

Finally, the economic context needs to be mentioned as a factor playing a role in the process by which a global norm is translated into local strategies, arguments and values. The relatively low level of economic development and material security in Turkey means that potential local opponents without sufficient or secure incomes will hesitate far more and will be more ready to accept economic compensations offered by dam constructors, even preferring these compensations over more abstract rights to participate to the decision-making process or environmental protection. To end our discussion, a last point needs to be made concerning a central actor in the process of national diffusion of SPD norms and their implementation: the state.

We have seen that the environment and democracy are strongly present as arguments in opponents' discourses as are references to international environmental protocols and agreements, but these are also used by the government (and even dam constructors). Indeed, in response to its national and international critics the government puts forward the view that dams do not contribute to climate change, which is a top EU priority. As the Turkish state reminds its detractors, the EU plays a key role in climate change negotiations and the Kyoto Protocol, arguing for an eight per cent reduction in CO_2 emissions, and plays a key role in accommodating third world countries by taking their specific needs into account so as to encourage their commitment.

Conclusion: Implications of our Studies for Regime Theory

To answer my question about whether or not SPD norms are being translated into local practices and values in Turkey, one needs to come back to the distinction between participation and sustainability. With regards to the questions relative to regime theory, we can conclude that one dimension of the global SPD regime is deeply rooted in local and individual values and practices: democratic ideas about the weight of the local in the national framework and about participatory democracy generally are more legitimate than the environmental dimension. Indeed, environmental protection appears to be both a way (if not a justification) by which greater democracy can be attained and also appears as secondary in the hierarchy of values of opponents to dams. For the CSBCT the environment is mentioned as an afterthought, coming second to historical considerations, while for local businesses in Camlihemçin the important factor was the loss of revenue that would have been incurred by the loss of tourism and fishing.

While there is an impetus from the local, this part of the arguments offered by local opponents (be they intellectuals or local associations) are derived from the global arena. This suggests that strategies of persuasion and influence from the local toward the national level are based on democratic values while the NIMBY attitude regarding dams suggests, in contrast, that the environmental basis for refusal is not as strong. Thus, one can say that the global SPD regime is spreading partially since the environmental dimension remains secondary to local actors, remaining the preoccupation of a small number of the elite. Social participation thus appears to be a way to strengthen democracy while the environment is only a specific way to do this, which is the case of dams in former communist countries, such as Hungary and Slovakia (Marmorat, 2006). This conclusion also follows from a large-scale review of the literature undertaken for the French Ministry of Sustainable Development since the 1980s on the question of participation in environmental controversies, and the one undertaken for our study on dams (La Branche and Warin, 2005). It suggests that most scientific articles written on SPD presume that the refusal by local populations of large infrastructural projects is based on environmental concerns, that, in other words, the NIMBY phenomenon is necessarily ecological (La Branche, 2006). Yet, there is a problem with this: opposing a technical project, whether a dam or wind power park, does not necessarily means that environmental protection is the basis for this refusal.

Indeed, we can even raise the question as to whether a NIMBY can ever be ecological, since the earth is a closed ecosystem. How can moving an infrastructure (no matter what it is) with negative local environmental impacts, from one place to another, be ecological? While pertinent, this question is overly simplistic because other factors need to be taken into account, such as a population's capacity (economic, technical, institutional, cultural and political know-how) to handle technological innovation and risk and the links to the issue of ecological inequalities (Chaumel and La Branche, 2008); not all societies can handle pollution remediation of certain chemical contaminants or nuclear wastes and not all societies have the same capacity to mitigate and adapt to climate change, which requires the development of renewable energies, including dams.

Hence, to our questions regarding the legitimacy of an environmental regime at the local and individual level, we can add this other, highly complex issue, linked to knowledge: if one is to make reference to global norms, one needs to know that they exist and then be able to cognitively handle them and then use them in opposition strategies. In the meantime, however, one can say that among the actors studied, the two associations seemed to indeed take as legitimate, valid and even ethical those principles diffused by global dam and water management regimes, infused with SPD principles. However, it appears equally clear that among other actors, and even more among civil societies, the status of the environment in individuals' values varies greatly. Our analysis tends to show that this process of legitimation of the environmental regime is indeed under way but that it is far from being complete, as suggested by the relative depth of environmental values among actors, outside of environmental NGOs.

But even among these actors, one needs to raise a few questions for further research: can an environmental NGO that ceases to exist once the decision to go ahead or not with a dam, be qualified as environmental? Within a participatory procedure, should a decision-maker give all actors the same weight and importance even, for example, a naturalist NGO whose sole purpose is, for example, to watch birds? What is a decision-maker to do when faced with a local aesthetic argument in a global climate change context? This is a real question, since in many countries wind power development faces local opposition through democratic and participatory procedures. One argument used by such local associations is that these infrastructure spoil the natural local beauty. In researching whether these NGOs adopt global environmental norms as their own, one thus needs to reflect over whether aesthetic is, indeed, an ecological argument, which depends on one's definition and conception of environment and ecology. At issue here are these potentially conflicting views over the environment and their relative importance to democracy, which are increasingly expressed through participatory procedures that by no means guarantee more ecological decision-making.

Notes

1 This research was funded by the French Energy Council (*Le Conseil français de l'énergie)* and Hydro 21 (the European centre for sustainable development in hydraulic energy).
2 One could also mention the sociology of international organizations but they do not address the question of the links between levels.
3 For information on the Greater Anatolian Project (GAP) see Carkoglu and Eder, 1999; Ecole Polytechnique Fédérale de Zurich, 2001; on the Ceyhan Aslantas Project (CAP) see Agrin Co. Ltd., 2000.
4 This issue is left aside as is the question over potential water conflicts with Iraq, for lack of space. See Schulz (1995) for more details on these points.
5 The story was related in an interview with the Foundation's members in Ankara 21 February 2004 and with Ilhan Avci, engineering professor at the University of Istanbul, also involved in the opposition, on 17 February 2004.

References

Agrin Co. Ltd. (2000) 'Aslantas Dam and related aspects of the Ceyhan River Basin, Turkey', World Commission on Dams, Cape Town, www.dams.org

Carkoglu, A. and Eder, M. (1999) 'From hydropolitics to regional integrated development: transformation of the southeast Anatolia Development Project (GAP) and its implications', *Turkish Review of Middle East Studies*, vol 1, no 1, pp215–232

Chaumel, M. and La Branche, S. (2008) 'Inégalités écologiques: vers quelle définition?', *Espaces, populations et sociétés*, vol 1, pp3–17

Dorronsoro, G. (2005) *La Turquie conteste. Mobilisations sociales et régime sécuritaire*, Editions du CNRS, Paris

Ecole Polytechnique Fédérale de Zurich (2001) 'Sustainable management of international rivers – case study: southeastern Anatolia project in Turkey – GAP', www.eawag.ch/research_e/apec/Scripts/GAP07feb01.pdf

Hurd, I. (1999) 'Legitimacy and authority in international politics', *International Organisation*, vol 53, no 2, pp379–408

Hurrel, A. (1993) 'International society and the study of regimes: a reflective approach', in V. Rittberger (ed) *Regime Theory and International Relations,* Clarendon Paperbacks, Oxford

La Branche, S. (2003a) 'La transformation des normes de participation et de durabilité en valeurs? Réflexions pour la théorie des regimes', *Revue études internationales*, vol 34, no 4, pp611–629

La Branche, S. (2003b) 'Vers une évaluation du développement durable appliquée aux aménagements hydrauliques', *Revue Énergie*, vol 546, pp305–309

La Branche, S. (2006) 'L'incidence des normes de développement durable et participatif sur l'hydroélectricité. Les cas de la France, du Québec et de la Turquie'. Report for the French Energy Council, Paris

La Branche, S. and Warin, P. (2005) 'La concertation dans l'environnement, ou le besoin de recourir à la recherche en sciences sociales', report for the French Ministry of Ecology and Sustainable Development, Paris

Marmorat, M. (2006) 'La controverse autour du projet hydroélectrique Gabčíkovo-Nagymaros sur le Danube (Hongrie/Slovaquie). Catastrophe écologique ou tremplin pour le développement local?' PhD dissertation, Institut d'études Politiques de Paris

Rittberger, V. (ed) (1993) *Regime Theory and International Relations,* Clarendon Paperbacks, Oxford

Rosoux, V. B. (2001) *Les Usages de la Mémoire dans les Relations Internationales*, Bruylant, Brussels

Schulz, M. (1995) 'Turkey, Syria and Iraq: a hydropolitical security complex', in L. Ohlsson (ed) *Hydropolitics: Conflicts over Water as a Development Constraint*, Zed Books, Atlantic Highlands, NJ

World Commission on Dams (2000) *Dams and Development: A New Framework for Decision-making, the Report of the World Commission on Dams*, Earthscan, London

Young, O. (1989) 'The politics of international regime formation: managing natural resources and the environment', *International Organisation*, vol 43, no 3, pp349–75

15

Conclusions: Politicizing Social Participation

Eric Mollard and Kate A. Berry

While social participation may not be new rhetoric, it has resulted in a profusion of regulations and been associated with a wide variety of organizations in recent years. The water domain has been particularly drawn in, from user associations to international forums including river basin organizations, river contracts and rural and urban water conservation programmes. At the same time, water has long been the domain of conflicts ranging in scale from neighbourhood to international disputes, with increasing territorial, regional and river basin concerns. Given that society involves the co-existence of antagonistic interests, social participation, as shown in this book, is not only an institution-driven arena, but also includes many struggles to (re)organize society or, at least, to allow oneself to be heard. Consequently, two concepts can be distinguished: participation *stricto sensu*, meaning formal dialogue generally framed by regulation or policy, and participation *sensu lato*, meaning the embeddedness of participatory stages within political struggles.

Examining formal regulations and policies alone is insufficient to understand the political influence of powers. Participation must be politicized in its conceptual framing as well understood through the democratization of its solutions. This concluding chapter examines the political dimensions arising from earlier chapters in the book and the extent to which social participation has become politicized. Approaching this through a politicized understanding may provide insights into questions posed in the introduction about the balance of administrative controls within social participation, the inclusion/exclusion of certain parties, power differentials and how actual experience with social participation meshes with rhetorical appeals.

Initially, politicizing social participation means conceptualizing powers as key features in society. Note the use of the plural as power is not monolithic. Some powers are socially and politically organized, some are constructed through conflicts or even formal processes of consultation, and still others are small yet significant powers that arise because of the impossibility of curbing them through denunciation, norms or public authority. Powers can be mobilized as symbolic or material resources to convince, unify, build legitimacy or otherwise activate people or groups. Yet powers seldom appear as such, because speeches and narratives rely on general principles of justice, rationality or any general values, such as those related to the environment. Although participation is constructed through relationships between powers, it is seldom able to curb or even reveal them. In the case of social participation, this is largely due to its openness to the public, which often leads to new speeches and arguments built on general principles or specific expertise but is rarely directed toward the interplay of powers.

Politicizing social participation also leads to questions about the political nature of its origins, rooted in interventionist governance in Western countries and, more recently, within processes of globalization. While governments may see increasing representation as an assault on their powers, the political and administrative elite are often at the core of participatory governance (Cooke and Kothari, 2001). As proposed by Smyrl and Genieys (2008), we explore the idea that implementation of social participation may be used to preserve the established order, in particular for the elite. If the assumption turns out to be true, this seems to be exactly the opposite of what social participation is supposed to solve through curbing unilateral administrative or other powers.

Globalization of Social Participation

As Liang Chuan and Yue Chaoyun (Chapter 8) point out, China is committed to environmental protection and participatory irrigation policy. How could it be different, even under an authoritative regime? Indeed, the national elites feel the need to be legitimate and to justify the role of skilled leaders vis-à-vis popular masses, who are supposed to be poorly aware and with little training. However, political interpretations could reverse such an explanation, starting with a population informed through the daily media and international forums about the dire straits of the environment and the need for participation as the means to reduce crises (in particular for water, which holds a special meaning in the country). The population would not understand that leaders are engaged in such a modern worldwide move. Instead of exerting leadership, the elite may be more reactive to popular expectation than proactive. This first point suggests the possibility of understanding participation in China through a political reading of the emergence of participation by analysing the elite. The second interpretation, which the authors adopt, considers poor implementation of participation at the local level, highlighting the gap between national speeches and local reality. Bureaucracy remains intact and there is an absence of financial autonomy for associations, so effective participation has not occurred. The proof is shown by the continuation of civil resistance (refusal to pay, reluctance to maintain canals, questioning faulty administration) and while demands for state assistance continue, they are underscored by a general mood of formal apathy because of the seeming impossibility of making improvements to the hydrologic and social systems.

Conversely in Quebec, Canada, Milot and LePage (Chapter 7) found participation was imposed on the provincial government in response to a broad-based movement against neoliberal attempts to privatize water. Nevertheless, the resulting participation within new river basin organizations did not turn out to be particularly engaging. Many conflicts were contained, alternate institutional arrangements were limited and some key actors, such as the municipalities, were not fully engaged. Generally speaking, political struggle for participation in integrated water management has not been common when legislators and political leaders in countries, determine top-down both the implementation and the details of participatory devices.

In France and Israel, participation was enacted early through legislation, although effective implementation was delayed for several decades in Israel (Pargament et al, Chapter 11). Such a delay underscores the gap between speeches aimed at the appearance of modernity and the understandable tendency of the government elite to maintain its prerogatives, powers and interests. The lag also reveals the lack of widespread concern in the civil society at that time. In France, if participation did not result from social demands, its early implementation made the country a pioneer on the issue.

A theory on the programmatic elite (Smyrl and Genieys, 2008) casts light on the underside of powers. Indeed, the elite is said to produce programmes to preserve its legitimacy. However, elites are not monolithic and various groups compete for ideas and resources that the government arbitrates. In France's interventionist government of the 1960s and as a reaction to state centralization, a small group of technocrats promoted river basin management open to some participation (Warin and La Branche, 2003). This change convinced the national government, which was facing conflicts between local governments resulting from a sector-oriented approach within a welfare state. At the same time, those who did not get their way, such as other technocrats, received political compensation in order to accept the new organization (Lewis, 2001). Subsequent policies in France have reinforced social participation with watershed management plans and river contracts, the latter being designed to be more flexible with greater acceptance than master plans (Allain, Chapter 5).

In a number of cases, administrative governance has generated participatory approaches, aimed both at preventing conflicts and avoiding judicial actions as much as possible. In the case of the United States described by Fisk et al in Chapter 1 this seems to be an element of the federal government's strategy. Further investigations should explore these issues, which are political to the extent that the judicial option has often been associated with poor publicity for democratic governments. Consequently, limiting judicial actions may take varied forms, from administratively framed participation to weighty procedures. Moreover, uncertainty in verdicts and sometimes ineffective sanctions against government agencies have hindered judicial strategies and favoured participatory approaches, such as contracts with groups, associations, farmers, industrialists, fishermen and others (Gramaglia, 2008).

These examples highlight the interests of the elite in social participation. There is not an opportunity here to explore the reasons why the globalization of participation is occurring at this point, as this would require a detailed historical analysis. Yet these global processes hint at the political interests behind participation, which are underscored through its implementation.

Politics in Social Participation

Social participation practices show the determining influence of power relations. In this vein, three types of participatory dialogues are presented:

1 tough debates and stalemates arising from major asymmetries in powers;
2 intermediary situations that articulate formal participation and struggles;
3 situations perceived as successful and apolitical.

We also analyse forms of politicization by expanding an examination of actors beyond negotiation tables, speeches and formal institutions.

The chapters addressing Mexico, South Africa, Peru, Nepal, India and Brazil show the tendency for antagonistic postures between power holders and exclusion for those who have little power. In Mexico, power asymmetries associated with the absence of a countervailing civil society produces different types of negotiations:

- façade negotiations, where a powerful administration is dominating small organized groups (Mollard et al, Chapter 6);
- conflicting negotiations often resulting in recurring victories for organized groups, such as industrialists, farmers or city mayors whose territorial-based power prompts elected officials and the administration to join them;
- violence-driven negotiations when a group and/or the political administration resorts to violence (Maganda, Chapter 13).

In South Africa, successful negotiations exist, but are frequently restricted to more powerful groups of contractors, firms, mining companies or others who seek common agreements to protect their interests. On the other hand, the inclusion of disenfranchised people in suburbs or rural areas seems doomed to failure. The poorest do not manage to be heard in spite of government's efforts. Thus, being a housewife living in the tail end of an irrigation canal results in no reliable water. In addition to the fact she is not heard in water or community committees, she may have no time, money or inclination to participate in water governance. This failure to engage the disenfranchised in social participation may reflect a collective loss of values and more systemic failures. If social participation *sensu lato* involves mobilization against institutions, the fact that it is frequently directed against the state, such as the South African movement 'break the meters', also underscores a more broad-based loss in values (Wilson and Perret, Chapter 9).

Such inequalities anchored in culture include not only the domination of the rich over the poor, but also have a basis in gender. Promoting participation for dominated people is all the more difficult when it is connected to daily relations where social barriers are exceptionally strong. As Zwarteveen et al (Chapter 4) emphasize, in Peru and Nepal women's difficulties in engaging in community governance and challenges in expressing themselves in community venues or to engineers, have disconnected them from actively participating in many formal water governance structures. Yet, women's influence is expressed in many routine practices of water management with significant implications. Berry (Chapter 3) similarly calls upon gendered cleavages, in this case between non-governmental organizations (NGOs) in arid India, that want to enhance women's roles in participation, and villagers, who may be invested in the status quo. The paradox is materialized through the experiences of the NGO's fieldworkers. Social participation in rural water supply management is constrained because it can neither be at odds with villagers' ideas nor with the global forces behind donors and institutions supporting gender-based participatory approaches.

The political drives underlying negotiations over water materialize clearly in the second situation, where intermediary situations articulate formal participation and struggles. This type of situation may present itself with an artificial character of participation, which sometimes takes on an opportunistic strategy, such as when an official state agency realizes that its strategy (to build a dam, for example) fails (Forline and Assis, Chapter 2). Such a case demonstrates how much participation is at stake for powers, including those exerting a determining influence beyond the negotiation table.

Of particular salience in Latin America, the sequence of political and formal phases shows that participation is not neutral when one powerful actor decides the agenda (topics and timing to be discussed), participants and what information is considered useful and relevant. As shown in Mexico by Maganda (Chapter 13), deciding the rules for participation may be a significant issue. Indeed, the NGO-shaped Latin American Water Tribunal invited the government to neutral discussions with dam opponents at the same moment that the official World Water Forum was defending participation. The government, however, rejected the invitation and even tried to dissuade the international jury from entering the country. Social mobilization prior to formal participation may make it possible to build identity-based powers that establish, or in some cases seize, the capacity to organize participation (Walker et al, 2007).

Retaining native rights to establish protocols for participation was a key point in the Indigenous Peoples Kyoto Water Declaration of 2003 (Boelens et al, 2006). In Ecuador, Indians and peasants mobilized against government-led irrigation programmes that privileged the urban population; they organized themselves, decided rules and ultimately invited other actors to discuss water rights distribution between equals. Different from neoliberal equality, which may only represent potential, real equality is not issued, it must be seized (Boelens, Chapter 12). In Brazil, participation occurred not once but twice. In the first instance of participation, indigenous groups struggled against the construction of a dam, with mobilization and participation *sensu lato* bound up with the establishment of social identities. Indigenous success in halting the proposed hydropower project may have been temporary, succumbing to the fate of many proposed water projects that disappear and then reappear sometime later when political and economic conditions favour development. The second, more recent phase of social participation was orchestrated by state agency proponents of the project with the goal of winning local support and co-opting project opponents (Forline and Assis, Chapter 2).

Given that the two first types of participation are politicized by powers, we are allowed to wonder about the politicization of every type of negotiation, including the apparently apolitical ones, which is the third type of situation examined here. River basin negotiations in Canada, Israel and Sri Lanka show the appearance of friendly, open and effective negotiations where everyone seemed satisfied with successful results ascribed to goodwill. Theoretically, apolitical negotiation may be possible if everybody agreed and a lot of money and/or water can be distributed to avoid conflicts. In Canada, people even noticed the absence of conflicts that would make debates more focused and help preserve the watershed organizations in charge of participation (Milot and Lepage, Chapter 7). In Israel, most of the population seemed to subscribe to the restoration of a highly degraded urbanized basin (Pargament et al, Chapter 11). Many of the substantial decisions, however, may have been determined not by the participatory committee, but by the national legislature and executive agencies through providing the funds necessary for restoration and changing agricultural water rights so as to circumvent local conflicts. Moreover, there was little debate over whether local agreements were carried out at the expense of other basins. Participatory programmes in Sri Lanka, as well as in the Rio Bravo river basin in Mexico, had engaging dialogues but may have been spurred on with the benefit of substantial financing, which may have attracted and motivated all involved (Shanafield and Jayaweera, Chapter 10). Political clout may not be very evident without directed investigation in cases where participation seems successful, as even successful participation can generate tensions, clientelist public actions, attempts to political neutralize opponents or simply enable opportunistic activities.

In the three negotiating situations, participation is politicized as it is criss-crossed with power struggles. At the same time, however, participation requires that every stakeholder is fair in front of the public. For that reason, stakeholders may not be able to actively display their power and may reference general principles, rather than their own interests. Thus, the world of appearances may obscure the reality of powers, particularly when civil society is used to bear witness to the values of participation. It also seems that the more the state asserts the rule of law, the more the power struggles are obscured by refined standards, political promises and institutional control and regulation. The height of sophistication is undoubtedly France (Saul, 2000), where bureaucratic centralization was one of the first, paradoxically, to practise participation. In other cases legislatures may become involved and, as in Mexico, negotiations that did not meet the anticipated results have led to new bills aimed at changing discussions already under way. Questions about whose interests are served through official participatory actions may be reorganized to cover up processes embedded in the dialectic between the world of civic appearances and the reality of contradictory interests, in which decisions about social trade offs are made while preserving the appearance of the rule of law.

Democratizing Social Participation

Recognizing asymmetrical power relations at the core of participation leads one to wonder about more politically equal negotiations and how to regulate powers more effectively. We propose two tracks for the democratization of participation and realistic development, one in the short run, the other longer term.

By democratization, it is not only a question of creating arenas for discussion, but also better technical democracy where scientific knowledge would be accessible to many, training in dialogue such as adaptive management or social learning, or any formal devices of empowerment and inclusion of disenfranchised groups and individuals. As pointed out in the previous section the advantages of participation as it is generally practised seem insufficient to curb powers, modify values or guarantee the success of public action. Several chapters pointed out questions about the representativeness of negotiators and whose values were being represented. Generally speaking, participation does not draw the attention of most people, at least in regards to water. Participation may attract some elected officials to whom it offers a political platform or who may send assistants for electoral intelligence (Gourgues, 2007). Social participation may also bring in groups affected by a project but which are likely to disappear as soon as their objectives are achieved, as well as other NGOs, as in the case of participation in Turkey, where elites were able to restructure participation (La Branche, Chapter 14). Neither real decision-makers, lobbyists nor general civil society, however interested in water management, are spontaneously present. Typically they intervene only when their interests are concerned, sometimes apart from the dialogues of participation themselves (Massardier, 2009). Consequently, democratizing participation aims to regulate the powers of influential stakeholders to avoid their control over process, monopolization of debates or skewing of outcomes and to introduce more plural interests within participation in water management.

In the short term, democratization of participation aims to improve information flow and introduce realizations about the political dimensions of social participation. Rather than denying the determining influence of powers, it would be more effective to acknowledge them so that environment and exclusion might be explicitly taken into account when possible. This could, for example, involve traditional leaders taking personal interest in projects so as to reverse the usual direction of training. Decision-makers and technical experts could be educated about the political dimensions of water projects, rather than developing new educational community campaigns that are unaware of constraints and political dynamics bearing upon traditional practices, such as farming (Adams, 1992). If they were convinced of any electoral benefit, decision-makers might be much more attentive to water governance. For NGOs, it may be less a question of giving a voice *a priori* to excluded groups or creating formal forums than consolidating and allying with local authorities when possible. Building political alliances based on consensual and realistic objects of development, NGOs may find participation to be an effective political leverage.

In the long run, the democratization of participation aims to promote regulation of asymmetrical powers. For example, such regulation would attempt to avoid coalitions that reinforce asymmetry of powers or lead public authorities to circumvent law, while preserving appearances with promises and ad hoc programme evaluations. Democratization may rest on social struggles, such as in Brazil where anthropologists publicized indigenous struggles for recognition and against their exploitation by hydropower project proponents (Forline and Assis, Chapter 2) or in Turkey where international pressure was exerted upon a national government (La Branche, Chapter 14). Moreover, it is furthered through institutions and laws supporting the plurality of powers and checks and balances within governance structures.

Civil society in general is a good countervailing power to prevent coalitions of user groups, politicians and government agencies with disproportionate powers. Whereas civil society may seem omnipresent in negotiations, as the public is used as the basis to articulate each actor's own values, it is actually absent. Moreover, civil society does not have power to control or evaluate except by general elections, denunciations or judicial rulings. By institutionally guaranteeing the plurality of powers (for example, independent administration or less powerful governors) and by designing participation as checks and balances that obliges government not to influence coalitions and to enforce the law, the public could play a political role in water governance. All this remains to be seen, however. A plurality of powers must be the way to democratization, to social participation and to realistic water governance.

References

Adams, W. M. (1992) *Wasting the Rain: Rivers, People and Planning in Africa*, University of Minnesota, Minneapolis, MN

Boelens, R., Chiba, M. and Nakashima, D. (2006) *Water and Indigenous Peoples*, UNESCO, Paris

Cooke, B. and Kothari, U. (eds) (2001) *Participation: The New Tyranny?*, Zed Books, London

Gourgues G. (2007) 'What the areas are for participative democracy: elements of territorial engines of the deliberative requirement', Ecole thématique internationale Grenoble PACT, 20 December 2007, www.pacte.cnrs.fr/IMG/html_controverse_GOURGUES.html

Gramaglia C. (2008) 'Des poissons aux masses d'eau: les usages militants du droit pour faire entendre la parole d'êtres qui ne parlent pas', *Politix*, vol 3, no 83, pp133–153

Lewis N. (2001) 'La gestion intégrée de l'eau en France: critique sociologique à partir d'une étude de terrain (bassin Loire-Bretagne)', Master's thesis, University of Orléans, France

Massardier G. (2009) 'La gouvernance de l'eau: Entre procédure de concertation et régulation "adhocratique". Le cas de la gestion de la rivière Verdon en France', Colloque International La gouvernance à l'épreuve des enjeux environnementaux et des exigences démocratiques, Congrès de l'ACFAS, Université d'Ottawa, 14–15 May

Saul J. (2000) *The Bastard Ones of Voltaire. Dictatorship of Reason in the Occident*, Payot, Paris

Smyrl, M. and Genieys, W. (2008) *Elites, Ideas and the Evolution of Public Policy*, Palgrave Macmillan, New York, NY

Walker, D., Jones III, J. P., Roberts, S. M. and Frohling, O. R. (2007) 'Where participation meets empowerment: the WWF and the politics of invitation in Chimalapas, Mexico', *Annals of the Association of American Geographers*, vol 97, no 2, pp423–444

Warin, P. and La Branche, S. (2003) 'Une brève histoire de la participation dans l'environnement en France (1975–2003)', La concertation du public – Rapport intermédiaire – Programme CDE – July. Concertation, Décision, Environnement, Ministère de l'écologie et du développement durable de la France, www.unites.uqam.ca/sqsp/pdf/congresAnn/congres2004_labranche-warin.pdf

Index